Physical and Information

# MODELS IN GEOGRAPHY

Parts I, II and V of Models in Geography

Physical and Information

# MODELS IN GEOGRAPHY

Parts I, II and V of Models in Geography

*Edited by*

RICHARD J. CHORLEY
PETER HAGGETT

UNIVERSITY PAPERBACKS
METHUEN : LONDON

Models in Geography *first published in* 1967
*These sections first published separately as a*
*University Paperback in* 1969
*Reprinted* 1970

*Printed photolitho in Great Britain*
*by Ebenezer Baylis & Son Ltd*
*The Trinity Press*
*Worcester and London*
*SBN* 416 29850 8



*Distributed in the USA*
*by Barnes and Noble Inc. New York*

University Paperbacks are published by
METHUEN & CO LTD
11 New Fetter Lane London EC 4

# Contents

# Acknowledgments

The editors and contributors would like to thank the following learned societies, editors, publishers, universities, organizations and individuals for permission to reproduce figures and tables:

*Learned Societies*

American Geophysical Union for fig. 3.12 from the *Transactions*, fig. 5.14 from *Water Resources Research* (and M. B. Fiering): American Meteorological Society for fig. 4.8 (and D. Fultz) from the *Journal of Meteorology*: American Society of Civil Engineers for fig. 5.12A from the *Journal of the Hydraulics Division*: Association of American Geographers for figs. 3.2A, 3.2B and 16.14 from the *Annals*: Geological Society of America for figs. 3.4A, 3.4B, 3.4D, 17.2 and 17.5 from the *Bulletin*: Geologists' Association of London for fig. 17.6 (and M. J. Kenn) from the *Proceedings*: Glaciological Society for fig. 3.3C from the *Journal of Glaciology*: Institution of Civil Engineers for figs. 5.3A, 5.3B, 5.3C and 5.3D from the *Proceedings*: Institution of Water Engineers for fig. 5.2 from the *Journal*: Royal Geographical Society for fig. 16.3 from the *Geographical Journal*: Society of Economic Paleontologists and Mineralogists for figs. 3.5D, 3.5E and 17.3 from the *Journal of Sedimentary Petrology*.

*Editors*

*American Journal of Psychology* for fig. 1.1: *American Journal of Science* for figs. 3.5A, 3.5B: *Australian Geographical Studies* for fig. 5.3E: *Cartographic Journal* for fig. 16.8: *Econometrica* for fig. 17.25 (and S. Enke): *Gerlands Beitrage zur geophysik* for figs. 17.10 and 17.11: *Publications in Geography, State University of Iowa* for fig. 16.21: *The Times* for fig. 16.19B: *Tijdschrift voor Economisch en Sociale Geografie* for Fig. 16.13: *Weather* for fig. 4.16 (and F. H. Ludlam).

*Publishers*

John Bartholomew, London for figs. 16.10 and 16.18: B.B.C. Publications for fig. 17.9: Cassell Ltd., London for fig. 16.18A: Clarendon Press, Oxford for fig, 16.18C: The Controller, Her Majesty's Stationery Office (Crown Copyright Reserved) for figs. 3.3B and 4.11: Faber and Faber Ltd., London for fig. 16.18E: Harvard University Press for fig. 17.1 (and L. W. Pollack): Longmans, Green and Co. Ltd., London for figs. 17.7 and 17.8 (and J. Allen): The Macmillan Co., New York for fig. 5.4B: McGraw-Hill Book Co., New York for fig. 4.7 (and J. Namias); for figs. 5.4A, 5.4D and 5.5A from *Applied Hydrology* by R. K. Linsley, M. A. Kohler and L. H. Paulhus; for fig. 5.5B from *Hydrology* by O. E. Meinzer (Ed.); and for figs. 5.10A and 5.10B from *Handbook of Applied Hydrology* by V. T. Chow (Ed,): Pergamon Press Ltd., London for fig. 4.6: George Philip and Son Ltd., London for fig. 16.18B: Prentice-Hall Inc., New Jersey for figs, 5.10E and 5.10F: Scientific American Inc., New York for fig. 5.13 by R. Revelle (All rights reserved): University of Chicago Press for figs. 3.4C (by W. E. H. Culling), 3.8 (by M. A. Melton), 3.10 (by M. A. Melton) and 3.11 (by R. L. Miller and J. M. Zeigler) reprinted from the *Journal of Geology*: John Wiley and Sons Inc., New York for fig. 5.4C.

*Universities*

Massachusetts Institute of Technology, Department of Civil Engineering, Soil Mechanics Division, for figs, 3.3D and 3.3E: Stanford University, Department of Civil Engineering, for fig. 5.11A.

*Organizations*

The Director General, Ordnance Survey for figs. 16.9 and 16.11: The Director, Trigsurvey, Pretoria, for fig. 16.5: The Director, United States Geological Survey for figs, 3.4E, 3.4F, 3.4G, 3.5C, 5.10C and 5.10D: London Transport for fig. 16.20: Ministry of Housing and Local Government (Crown Copyright Reserved) for fig. 16.19E: Provinces of Alberta and British Columbia Highways Departments for fig. 16.7: Société Nationale des Chemins de fer français for fig. 16.4: Standard Oil Co. and State of Illinois Highways Department for fig. 16.6: Thames Conservancy for fig. 5.9A: United States Department of the Army, Waterways Experiment Station, Corps of Engineers, Vicksburg, Mississippi, for figs. 3.2C, 3.2D and 3.3A: Water Research Association, Medmenham, for figs. 5.7 and 5.8.

*Individuals*

H. Flohn of the University of Bonn for fig. 4.14: B. A. Kennedy of the Department of Geography, University of British Columbia for fig. 3.7: D. R. Stoddart of the Department of Geography, Cambridge University for fig. 1.2.

Finally, the following thanks are also due:
Chapter 3. The author acknowledges his debt to his association with L. B. Leopold, M. A. Melton, S. A. Schumm, A. N. Strahler and, particularly, W. C. Krumbein: Chapter 5. The author would like to thank T. O'Donnell and F. V. Appleby for reading the manuscript and making some suggestions. Chapter 16. The author acknowledges the assistance of the Staff of the Drawing Office and Dark Room of the London School of Economics and, particularly, the opportunity of discussing some of the matters covered in the chapter with W. C. Krumbein and with his colleagues at the Department of Geography, London School of Economics, and at the Ohio State University: Index. The editors would like to thank D. R. Stoddart of the Department of Geography, Cambridge University, for his painstaking preparation of the index, and also for employing his rare bibliographical talents in assisting with checking the proofs.

CHAPTER ONE

# Models, Paradigms and the New Geography

P. HAGGETT and R. J. CHORLEY

*Models are undeniably beautiful, and a man may justly be proud to be seen in their company. But they may have their hidden vices. The question is, after all, not only whether they are good to look at, but whether we can live happily with them.* KAPLAN, 1964, p. 288.

In concluding the previous volume in this series, we attempted to review the paths taken by various workers in moving towards what they saw to be the 'frontier' in geographical research. We argued there that the quest for a model or models was a recurrent theme in their search. This volume is a direct outcome of that conclusion in that specific workers were asked to discuss the role of model-building within their own special fields of geographical research. While we would not wish to pre-judge their findings, it will be evident from the format and arrangement of the chapters that there is: (1) some measurable contrast between their approaches to geography, various as they are, and those that characterize the great part of established geographical patterns of thinking, as evidenced by existing textbooks and syllabuses; and (2) a community of common ideas that link all contributors into what Price (1963, p. 62) would characterize as an 'invisible college' of geographical practitioners. Whether this communality is sufficient to form the basis of what Manley (1966) has termed a 'New Geography' is not for us to judge. However, it is perhaps significant that the greater part of the volume is based on work produced since 1945, and much of it since 1960. In this opening chapter we discuss what we believe to be the significance of this new search for a model-based geography.

## FACTS, MODELS AND PARADIGMS

### The nature of facts

Information in geography is capable of treatment in terms of general information theory. In this context factual information only has relevance within some more general frame of reference, and such a basic operation as the definition of a relevant fact can only be made on the basis of some theoretical framework. There are also different levels of organization of relevant information. Some information can be relevantly organized only at a small scale, whereas the orderly large-scale patterns of other information are blurred or swamped altogether on the local scale. One can therefore view geographical information registration and analysis, from one point of view at any rate, as a problem in the separation of regional and local information patterns from the more randomly-organized information which, as 'noise', obscures them (Chorley and Haggett, 1965). Of course, one may choose to regard the noise

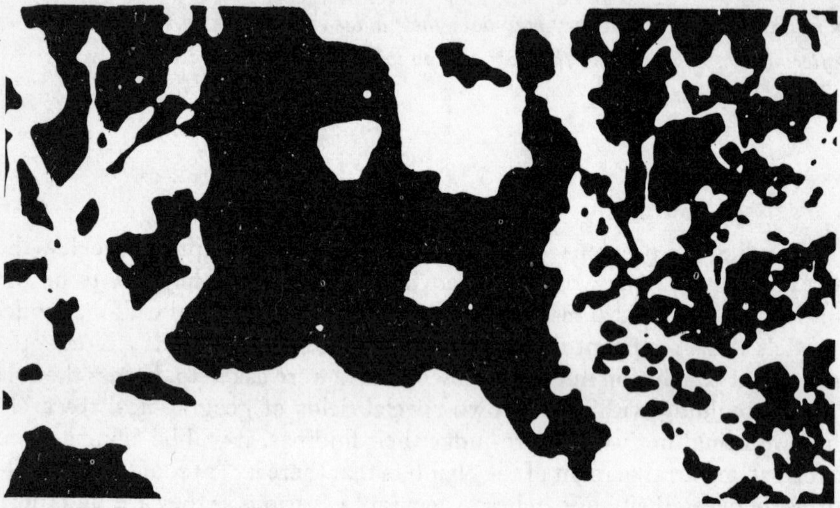

**1.1** A photograph of melting snow taken on impulse by a photographer in China just before the last war. The pattern makes no sense until it is organized as a full-face and shoulders, similar in style to a late-medieval representation of Christ; the upper margin cutting the brow and illuminated from the right (*Source: Partly from Porter, 1954*).

as the more significant element and to ask whether it is useful to try to recognize any order in reality. This results in the stress being placed on the variety of geographical information available and in attempts to subdivide information. However, it is becoming increasingly popular to ask what kinds of order are exhibited by geographical information and on what scales of space and time each operates. In short, the 'simple' registration of facts is being recognized not only as unsatisfactory but as an impossibility. Hanson (1958, pp. 8–19) has pointed out that what is observed depends not only on the context in which a particular phenomenon is set, but in the manner in which one is prepared to view it. In the words of Sigwart: 'That there is more order in the world than appears at first sight is not discovered *till the order is looked for*' (Quoted by Hanson, 1958, p. 204). Figure 1.1 gives a striking illustration both of the close apparent relationship between order and disorder, and of the subjective approach necessary to identify what it believed to be orderly.

The distinction between the idiographic and nomothetic approaches to the real world was recognized by Aristotle, although not in the terms which we currently employ, when he pointed out that poetry is more philosophical and of graver import than history because it is concerned with what is pervasive and universal, whereas history is addressed to what is special and singular (Nagel, 1961, p. 547). Today the distinction is made commonly between the 'humanities' which are primarily concerned with the unique and non-recurrent, and the 'sciences' which seek to establish general statements for repeatable events and process. Contemporary geography obviously lies athwart this apparent gulf, which must either be bridged or must lead to the dismemberment of the existing discipline. The dichotomy between the general and the particular was clearly stated by Francis Bacon in his *Maxims of the Law*; 'For there be two contrary faults and extremities in the debating and sifting of the law, which may be noted in two several manner of arguments: some argue upon general grounds, and come not near the point in question; others, without laying any foundation of a ground or difference of reason, do loosely put cases, which, though they go near the point, yet being put so scattered, prove not, but rather serve to make the law appear more doubtful than to make it more plain'. Indeed, the distinction between the idiographic and nomothetic views of geography, so strongly put by Bunge (1962), may be useful in highlighting many of the current shortcomings in the subject, but is less valuable from the more purely philosophical standpoint. Bambrough (1964, p. 100), for example, points out that all reasoning is ultimately concerned with particular cases, and that laws, rules and principles are merely devices for bringing particular cases to bear on other particular cases. 'The ideal limiting case of representation is reduplication, and a duplicate is too true to be useful. Anything that falls short of the ideal limit of reduplication is too useful to be altogether true' (Bambrough, 1964, p. 98). In short, every individual is, by definition, different, but the most significant statement which can be made about modern scholarship in general is that it has been found to be intellectually more profitable, satisfying and productive to view the phenomena of the real world in terms of their 'set characteristics', rather than to concentrate upon their individual deviations from one another.

## The nature of models

The catholic view of models taken in this volume derives largely from Skilling (1964). He argued that a model can be a theory or a law or an hypothesis or a structured idea. It can be a role, a relation or an equation. It can be a synthesis of data. Most important from the geographical viewpoint, it can also include reasoning about the real world by means of translations in space (to give spatial models) or in time (to give historical models).

*The need for idealization.* The traditional reaction of man to the apparent complexity of the world around him has been to make for himself a simplified and intelligible picture of the world. 'He then tries to substitute this cosmos of his own for the world of experience, and thus to overcome it' (Chorafas, 1965, p. 1). The mind decomposes the real world into a series of simplified systems and thus achieves in one act 'an overview of the essential characteristics of a domain' (Apostel, 1961, p. 15). This simplification requires both sensual and intellectual creativity (Keipers, 1961, p. 132). 'The mind needs to see the system in opposition and distinction to all others; therefore the separation of the system from others is made more complete than it is in reality. The system is viewed from a certain scale; details that are too microscopical or too global are of no interest to us. Therefore they are left out. The system is known or controlled within certain limits of approximation. Therefore effects that do not reach this level of approximation are neglected. The system is studied with a certain purpose in mind; everything that does not affect this purpose is eliminated. The various features of the system need to be known as aspects of one identical whole; therefore their unity is exaggerated' (Apostel, 1961, pp. 15–16). According to this view, reality exists as a patterned and bounded connexity which has been explored by the use of simplified patterns of symbols, rules and processes (Meadows, 1957, pp. 3–4). The simplified statements of this structural interdependence have been termed 'models'. A model is thus a simplified structuring of reality which presents supposedly significant features or relationships in a generalized form. Models are highly subjective approximations in that they do not include all associated observations or measurements, but as such they are valuable in obscuring incidental detail and in allowing fundamental aspects of reality to appear. This selectivity means that models have varying degrees of probability and a limited range of conditions over which they apply. The most successful models possess a high probability of application and a wide range of conditions in which they seem appropriate. Indeed, the value of a model is often directly related to its level of abstraction. However, all models are constantly in need of improvement as new information or vistas of reality appear, and the more successfully the model was originally structured the more likely it seems that such improvement must involve the construction of a different model.

*Characteristics of models.* The term 'model' is conventionally employed in a number of different ways. It is used as a noun implying a representation, as an adjective implying a degree of perfection, or as a verb implying to demonstrate or to show what something is like (Ackoff, Gupta and Minas, 1962, p. 108). In fact models possess all these properties.

The most fundamental feature of models is that their construction has involved a highly *selective* attitude to information, wherein not only noise but less important signals have been eliminated to enable one to see something of the heart of things. Models can be viewed as selective approximations which, by the elimination of incidental detail, allow some fundamental, relevant or interesting aspects of the real world to appear in some generalized form. Thus models can be thought of as selective pictures and 'a direct description of the logical characteristics of our knowledge of the external world shows that each of these pictures gives undue prominance to some features of our knowledge and obscures and distorts the other features that rival pictures emphasize. Each of them directs such a bright light on one part of the scene that it obscures other parts in a dark shadow' (Bambrough, 1964, p. 102). As Black (1962, p. 220) wrote of scale models, '. . . only by being unfaithful in *some* respect can a model represent its original'.

Another important model characteristic is that models are *structured*, in the sense that the selected significant aspects of the 'web of reality' are exploited in terms of their connections. It is interesting that what is often termed a model by logicians is called by econometricians a 'structure' (Suppes, 1961, p. 165; Kaplan, 1964, p. 267). Science has profited greatly from this *pattern seeking*, in which phenomena are viewed in terms of a kind of organic relationship.

This model feature leads immediately to the *suggestive* nature of models, in that a successful model contains suggestions for its own extension and generalization (Hesse, 1953–54, pp. 213–214). This implies, firstly, that the whole model structure has greater implications than a study of its individual parts might lead one to suppose (Deutsch, 1948–49), and, secondly, that predictions can be made about the real world from the model. Models have thus been termed 'speculative instruments', and Black (1962, pp. 232–233) has described a promising model as 'one with implications rich enough to suggest novel hypotheses and speculations in the primary field of investigation'. Similarly, Toulmin (1953, pp. 38–39) regards a good model as experimentally fertile, suggesting further questions, taking us beyond the phenomena from which we began, and tempting us to formulate hypotheses. The 'intuitive grasp' (*Gestalt knowledge*) of the capacities and implications of a model is thus the key to the exploitation of its suggestive character.

Selectivity implies that models are different from reality in that they are *approximations* of it. A model must be simple enough for manipulation and understanding by its users, representative enough in the total range of the implications it may have, yet complex enough to represent accurately the system under study (Chorafas, 1965, p. 31). In another sense, too, models represent compromises in that each has a circumscribed range of conditions within which it has relevance (Skilling, 1964, p. 389A).

Because models are different from the real world they are *analogies*. The use of hardware models is an obvious example of the general aim of the model builder to reformulate some features of the real world into a more familiar, simplified, accessible, observable, easily-formulated or controllable form, from which conclusions can be deduced, which, in turn, can be reapplied to the real world (Chorley, 1964, pp. 127–128).

*Reapplication* is a prerequisite for models in the empirical sciences. Although some mathematical model builders disclaim responsibility for the degree to which their idealizations may represent the real world, claiming that their responsibility is discharged completely and with honour if they avoid internal error (Camp, 1961, p. 22); most geographical model builders would judge the value of a model almost entirely in terms of its reapplicability to the real world.

*The functions of models.* Models are necessary, therefore, to constitute a bridge between the observational and theoretical levels; and are concerned with simplification, reduction, concretization, experimentation, action, extension, globalization, theory formation and explanation (Apostel, 1961, p. 3). One of their main functions is *psychological* in enabling some group of phenomena to be visualized and comprehended which could otherwise not be because of its magnitude or complexity. Another is *acquisitive*, in that the model provides a framework wherein information may be defined, collected and ordered. Models have not only an *organizational* function with respect to data, but also a *fertility* in allowing the maximum amount of information to be squeezed out of the data (see the 'statistical models' of Krumbein and Graybill, 1965). Models also perform a *logical* function by helping to explain how a particular phenomenon comes about. The question as to what constitutes a satisfactory explanation is a complex one, but Bridgman (1936, p. 63) put it in model terms when he wrote; 'Explanation consists of analysing our complicated systems into simpler systems in such a way that we recognize in the complicated systems the interplay of elements already so familiar to us that we accept them as not needing explanation'. Models also perform a *normative* function by comparing some phenomenon with a more familiar one (Hutton, 1953–54, pp. 285–286). The *systematic* function of model building has already been stressed in which reality is viewed in terms of interlocking systems, such that one view of the history of science is that it represents the construction of a succession of models by which systems have been explored and tested (Meadows, 1957, p. 3). This leads to the *constructional* function of models in that they form stepping stones to the building of theories and laws. Models and theories are very closely linked (Theobald, 1964, p. 260), perhaps differing only in the degree of probability with which they can predict reality. The terms 'true' or 'false' cannot usefully be applied in the evaluation of models, however, and must be replaced by ones like 'appropriate', 'stimulating' or 'significant'. Laws are statements of very high probability and, as such, all laws are models, but not all models are laws. Finally there is the *cognative* function of models, promoting the communication of scientific ideas. This communication 'is not a matter merely of the sociology of science, but is intrinsic to its logic; as in art, the idea is nothing till it has found expression' (Kaplan, 1964, p. 269).

*Types of models.* Chorley (1964) provided an initial structure for the classification of models currently used in geography and this 'model of models' has been expanded and revised with special reference to geomorphology in Chapter 3 (Fig. 3.1).

The term 'model' has been used, however, in such a wide variety of contexts that it is difficult to define even the broad types of usage without ambiguity. One division is between the *descriptive* and the *normative*; the former concerned with some stylistic description of reality and the latter with what might be expected to occur under certain stated conditions. Descriptive models can be dominantly *static*, concentrating on equilibrium structural features, or *dynamic*, concentrating on processes and functions through time. Where the time element is particularly stressed *historical* models result. Descriptive models may be concerned with the organization of empirical information, and be termed *data*, *classificatory* (*taxonomic*), or *experimental design* models (Suppes, 1962). Normative models often involve the use of a more familiar situation as a model for a less familiar one, either in a time (*historical*) or a *spatial* sense, and have a strongly *predictive* connotation.

Models can also be classed according to the stuff from which they are made, into, firstly, *hardware*, *physical* or *experimental* constructions, and, secondly, into *theoretical*, *symbolic*, *conceptual* or *mental* models. The former can either be *iconic* (Ackoff, Gupta and Minas, 1962), wherein the relevant properties of the real world are represented by the same properties with only a change in *scale*, or *analogue* (*simulation*) models, having real-world properties represented by different properties. The latter are concerned with symbolic or *formal* assertions of a *verbal* or *mathematical* kind in *logical* terms (Rosenblueth and Wiener, 1944–45, p. 317; Beament, 1960). Mathematical models can be further classed according to the degree of probability associated with their prediction into *deterministic* and *stochastic*.

Another view of models concentrates upon them as *systems* which can be defined on the basis of the relative interest of the model builder in the input/output variables, as distinct from the internal status variables. In order of decreasing interest in the status variables, many models can be viewed as *synthetic systems*, *partial systems* and *black boxes*.

The scale on which models are valuable and the standpoint from which they are constructed allow further distinctions, notably into *internalized* models which give a very parochial view of reality, and *paradigms* which are broadly significant models of value to a wide community of scholars.

*Pitfalls in model building.* The characteristics of models imply the existence of many dangers to which the model builder may fall prey. Simplification might lead to 'throwing the baby out with the bath water'; structuring to spurious correlation; suggestiveness to improper prediction; approximation to unreality; and analogy to unjustifiable leaps into different domains. Kaplan (1964, pp. 275–288) has summed up many of the dangers as problems of *overemphasis* on symbols, form, simplification, rigor and prediction. According to this view, a bad model would be heavily symbolic, present an overly-formalized view of reality, be much over-simplified, represent an attempt to erect a more exact structure than the data allows, and be used for inappropriate prediction.

Many philosophers have pointed to the dangers of craving for generality and of adopting a contemptuous attitude towards the particular case. They have often considered reality to be of too complex and multivariate a character to be susceptible to reasoning by analogy, and have asked whether the use of models introduces too great a detour into the reasoning process. In short, some hold that we should take heed of the Second Commandment: 'Thou shalt not make unto thee any graven image, or any likeness of anything that is in the heaven above, or that is in the earth beneath, or that is in the water under the earth'. In reply to this view of model building Ubbink (1961, p. 178) asks 'if this should be the case: in what sense can knowledge be true?'. Model building and reasoning are indissoluble, but 'the price of the employment of models is eternal vigilance' (Braithwaite, 1953, p. 93). Kaplan (1964, p. 276) believes that such vigilance is all the more necessary when model building is currently fashionable: 'The danger is all the greater with respect to model building because so much else in our culture conspires to make of it the glass of fashion and the mould of form. Models seem peculiarly appropriate to a brave new world of computers, automation and space technology, and to the astonishing status suddenly accorded to the scientist in government, industry and the military. It is easy to feel drawn to the wave of the future, and such tides are flowing strong today.'

## The nature of paradigms

Paradigms may be regarded as stable patterns of scientific activity. They are in a sense large-scale models, but differ from models in the sense used above in that: (1) they are rarely so specifically formulated; and (2) they refer to patterns of searching the real world rather than to the real world itself. Scientists whose research is based on shared paradigms are committed to the same problems, rules and standards, i.e. they form a continuing community devoted to a particular research tradition. In a sense then, paradigms may be regarded here as 'super models' within which the smaller-scale models are set. As such, Thomas Kuhn in his *Structure of Scientific Revolutions* (1962), has assigned to the origin, continuance, and obsolescence of paradigms a prior place in the history of the evolution of science. Progress in research requires the continual discarding of outdated models, and subsequent remodelling. The more internally consistent the original paradigm or model, the more difficult it may be to remodel an existing structure in step with changing notions and increasing data. It is usual, therefore, for the most significant intellectual steps to be marked by the emergence of completely new models (Skilling, 1964, p. 389A). As Kuhn (1962, p. 17) has argued '. . . no natural history can be interpreted in the absence of at least some implicit body of intertwined theoretical and methodological belief that permits selection, evaluation, and criticism'. Without such paradigms all the available facts may seem equally-likely candidates for inclusion. As a direct consequence there is: (1) no case for the highly-defined, fact-gathering so typical of the exact sciences; and (2) a tendency to restrict fact-gathering to the wealth of available data which comes easily to hand. The fact that much of this data is a secondary by-product of administrative systems adds further to the massive data-handling problem. Certainly most geographical accounts are strongly 'circumstantial', juxtaposing facts of theoretical interest with others so unrelated or so complex as to be outside the bounds of available explanatory models.

The importance of the paradigm lies then, in Kuhn's terms, in providing rules that: (1) tell us what both the world – and our science – are like; and (2) allow us to concentrate on the esoteric problems that these rules together with existing knowledge define. Paradigms tend to be, by nature, highly restrictive. They focus attention upon a small range of problems, often enough somewhat esoteric problems, to allow the concentration of investigation on some part of the man-environment system in a detail and depth that might otherwise prove unlikely, if not inconceivable. This concentration appears to have been a necessary part of scientific advance, allowing the solution of puzzles outside the limits of pre-paradigm thinking.

In practice such 'rules' are acquired through one's education and subsequent exposure to the literature, rather than being formally taught. Indeed a concern about them only comes to the fore when there is a deep and recurrent insecurity about the nature of the existing paradigm. Methodological debates, concern over 'legitimate' problems or appropriate methods of analysis are symptomatic of the pre-paradigm period in the evolution of a science. Once the paradigm is fully established the debate languishes through lack of interest or lack of need. Thus in contemporary economics, the most successful and sophisticated of social sciences, the early debates over the nature of economics have been replaced by rather stable – but largely invisible – rules as to what problems and methods economic science should cultivate.

## CLASSIFICATORY PARADIGMS IN GEOGRAPHY

Whatever the range of debate over the purpose and nature of geography, there is considerable communality of practice in the ways in which geographers have tackled their problems. Berry (1964) has analysed this paradigm of practice in terms of alternative approaches to a 'geographical data-matrix'. Here we look at his findings and attempt to diagnose the widespread unease generated by the continued use of this classificatory approach.

### The geographical data matrix

Although regional geography, systematic geography, and historical geography are regarded as being quite distinct types of geographical study, Berry (1964) has deftly illustrated that each may be regarded merely as a different axis of approach to the same basic geographical data-matrix.

If a matrix has only *one* column, it is commonly called a 'column vector' (Krumbein and Graybill, 1965, p. 251), in which may be stored a series of $j$ bits of information:

$$\begin{pmatrix} a_{11} \\ a_{12} \\ a_{13} \\ a_{14} \\ \cdot \\ a_{1j} \end{pmatrix}$$

Similarly we may store information about $j$ elements (i.e. temperatures, elevations, population densities, etc.) in a regional column, to give an inventory of all the available characteristics of a given location.

A matrix with only *one* row is termed a 'row vector'. Here we may store a series of $i$ bits of information:

$$(a_{11} \quad a_{21} \quad a_{31} \quad a_{41} \quad \cdot \quad a_{i1})$$

In this approach we store information about the same element but we vary the location to give the standard pattern of systematic geography, i.e. the mapping of a single feature (e.g. population densities).

By combining both the set of regions ($1 \ldots i$) and the set of elements ($1 \ldots j$) we have a rectangular array of the form:

$$\begin{pmatrix} a_{11} & a_{21} & a_{31} & a_{41} & \cdot & a_{i1} \\ a_{12} & a_{22} & a_{32} & a_{42} & \cdot & a_{i2} \\ a_{13} & a_{23} & a_{33} & a_{43} & \cdot & a_{i3} \\ a_{14} & a_{24} & a_{34} & a_{44} & \cdot & a_{i4} \\ \cdot & \cdot & \cdot & \cdot & \cdot & \cdot \\ a_{1j} & a_{2j} & a_{3j} & a_{4j} & \cdot & a_{ij} \end{pmatrix}$$

This matrix or box is termed by Berry the *geographical data-matrix* in that items containing information about the earth's surface may be stored in terms of their *regional* (or locational) characteristics and their *elemental* (or substantive) characteristics. Table 1.1A gives a formal example of this sort of matrix, and Grigg (Chap. 12, below) discusses its logical basis in terms of regional models.

TABLE 1.1

**Transformation of Vectors in Geographical Data Matrices***

| | *A* TRADITIONAL DATA-MATRIX | | *B* TRANSFORMED DATA-MATRIX |
|---|---|---|---|
| | *Major World Regions*[1] | | *Spatial Model*[2] |
| | Region 1 | | *Polar axes* |
| | Subregion 1a | | Central-place model |
| | ... | | |
| *COLUMN* | ... | | Gravity model |
| *VECTOR* | Place 1a (1) | → | ... |
| | Region 2 | | *Linear axes* |
| | ... | | Hortonian model |
| | ... | | Colonization model |
| | ... | | |
| | Region *N* | | Model *N* |
| | *Elements* | | *Geometrical Dimension* |
| | Human Elements | | Zero-order form |
| | Population | | Cities |
| | Settlement | | |
| | ... | | ... |
| | ... | | First-order form |
| *ROW* | ... | | Networks |
| *VECTOR* | Physical Elements | → | |
| | Landforms | | ... |
| | Vegetation | | Order combinations |
| | ... | | ... |
| | ... | | ... |
| | ... | | |
| | Element *N* | | Combination *N* |

By an ingenious series of row and column comparisons and by the addition of a third (time) dimension, Berry is able to reduce the great part of conventional geographical study to ten basic operations on the matrix. For example, areal differentiation is seen as column-vector comparisons and spatial covariance studies as row-vector comparisons. Comparison of a column over

* Adapted from Berry (1964, pp. 6–8).

[1] Information located with respect to *absolute* location ($X$, $Y$ co-ordinates measured from a common base, e.g. latitude and longitude).

[2] Information plotted with respect to relative *location* ($x$, $y$ co-ordinates measured from a variable base, e.g. distance and direction from a diffusion hearth).

time becomes sequent occupance, while other manipulations give the major modes of historical geography distinguished by Darby (1953). Indeed the major – and apparently fundamental – contrasts between regional, systematic and historical geography are seen by Berry (1964, p. 9) largely as a function of the relative length, breadth and depth of the study in terms of the three axes of the matrix.[1]

## Difficulties of conventional matrix operations

The need for the information in the data matrix to be structured, given coherence, generalized and made intelligible has long been recognized, and as Wrigley (1965) points out, both Von Humboldt and Ritter scorned the attitude of their predecessors which had reduced geographical studies in the eighteenth century to mere 'pigeon-holing'. There is growing evidence however that neither of the two major solutions – the study of column vectors and of row vectors – effectively meet present-day demands.

(a) *The explosion of the data matrix.* The exact rate of growth of 'information' that could conceivably be stored in a geographical data-matrix is difficult to assess exactly. We have useful figures on the speed of mapping for selected areas (Langbein and Hoyt, 1959), and are familiar with the rapid accelerations caused by the development of air-photography after World War I and of satellite scanning and remote sensing since World War II. If we add to this the growing volume of statistical material collected by international agencies, national governments, state and local administrators, then the size of the potential world 'data bank' becomes staggering. If we restrict our data bank to one fact about each square-mile unit of the earth's surface then we have a basic store of $10^9$ bits of information. If we relax this assumption to include the range of possible parameters for each unit (ranging, say, from a minimum of 1 to a maximum of $10^{10}$) and a more finely-divided world grid (say $10^{11}$ units), then we may well have a storage problem of the order of $10^{11}$ to $10^{21}$ bits on our hands.

This is of course a speculative calculation, but what gives any such figure added point is the general rate of growth of information. Price (1963, pp. 4–13), after a wide-ranging survey of the growth of scientific information, found evidence of an exponential growth of 'impressive consistency and regularity' – that is, the more information that exists the faster it grows. Depend-

[1] In more recent, and as yet unpublished work, Berry has developed a second data-matrix (an interaction matrix) in which pairs of locations (*dyads*) occupy the rows and interactions occupy the columns. By reducing the two matrices through factor analysis, the general relationships between spatial structure (Matrix I) and spatial interaction or behaviour (Matrix II) may be explored. Through this extension Berry has not only clarified the relationships between formal and functional regions, but has also laid the basis for a much more general 'field-theory' of spatial behaviour.

ing upon what we measure, it is possible to estimate that the amount of information tends to double within a period of 10 to 15 years – probably slightly shorter. If one accepts the general form of the curve then the problem facing Humboldt and Ritter was about 1/1,000 times as small as that facing the current generation of geographers. The fact that large areas of the world were still then 'blank on the map' suggests this is in fact a sizeable overestimate; that is, the rate of growth of locationally 'storable' information has been *more* rapid than that of scientific information as a whole as the ecumene itself has expanded. Stoddart (1967) has suggested that, although the intensity of geographical data generation and handling over the past 200 years has been roughly half that of science as a whole (Fig. 1.2), there are indications of a current increase of activity.

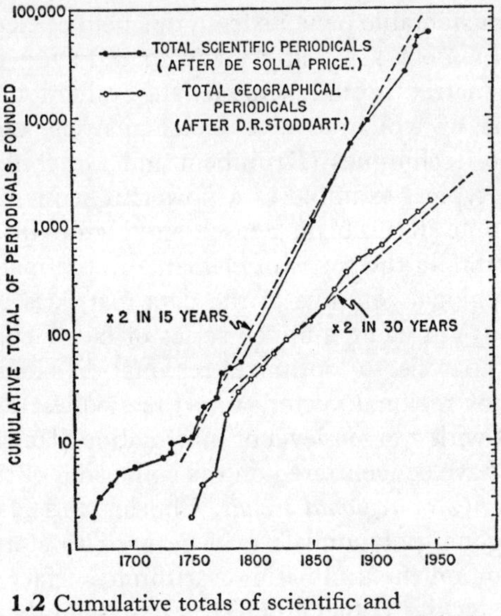


**1.2** Cumulative totals of scientific and geographical periodicals founded. The dashed lines represent doubling in 15 and 30 years, respectively. (*Source: D. R. Stoddart. See also Stoddart, 1967*).

(b) *The weakening of conventional vector analysis.* Continuing Berry's (1964) matrix algebra analogy, we may trace the problems in columnar analyses to: (1) the expanding number of locational columns as old areas were ever more finely divided and new areas added; and (2) the proven failure of local 'column' factors to explain the juxtaposition of other variables in the same column. Wrigley (1965) has shown, for example, how the development of general production techniques during the Industrial Revolution tended to destroy unique regional features and to make the geographical characteristics of these regions less explainable in purely local terms. Analysis of row vectors (i.e. systematic distribution studies) has been weakened by the tendency for individual systematic sciences to 'split off' their own rows for separate analyses. It is significant, for example, that systematic sciences are increasingly taking on their own mapping analysis – and in some cases (e.g. Perring and Walters, 1962) are themselves making substantial contributions to cartographic technique.

## Computer-orientated analysis of data matrices

One trend which runs counter to the information explosion is the increasing adaptation of digital computers to this problem. The whole field of data storage, data retrieval, data analysis, and data display is expanding rapidly and we may expect that geography, as a special type of *regional data-storage system*, will reap considerable benefits from this field of electronic technology.

Strides at the moment are being made in reducing the size and complexity of the data matrix through general data-evaluation and pattern-search programmes, some of which involve factor-analysis, cluster analysis and related multivariate techniques (Krumbein and Graybill, 1965, Chap. 15). WHIRLPOOL is a typical example of a powerful sequential multiple linear regression programme (Krumbein, Benson and Hempkins, 1964) for 'sorting out' significant factors in the series of elements in the matrix. An alternative approach to the regional columns of the data matrix is being led by Berry (1961; Steiner, 1965) in using a linked series of factor analysis, $D^2$ analysis, and discriminant analysis to 'optimize' regional divisions; thereby 'collapsing' the number of regional vectors into the smallest or most convenient number consistent with a given level of information (Haggett, 1965, p. 256). Other approaches have concentrated on the collapsing of the detailed isarithmic map into a series of *regional trends*. Chorley and Haggett (1965) have shown how orthogonal polynomials may be used to store vast amounts of regional information on the undulating isarithmic surface in terms of a very few basic equations, while Tobler (1966) has analysed other forms of numerical mapping and rapid map display.

It is clear then that technological improvements, largely in computer engineering and programming logic, are at last beginning to restore to the greatly-strained geographical data-matrix some of the order that its rapid expansion threatened to destroy. While empirical approaches to the stored material are likely to throw up a good deal of unexpected patterning, we should recall that these programmes are '. . . greatly facilitated when preceded by analysis of the problem in terms of some conceptual model – a mental picture – that defines the class of objects or events to be studied, the kinds of measurements to be made, and the properties or attributes of these measurements' (Krumbein and Graybill, 1965, p. viii). It is to these 'mental pictures' that we turn again in our last section.

## TOWARDS A MODEL-BASED PARADIGM OF GEOGRAPHY

The continuing problems of the traditional class-based paradigm of geography suggest that experimentation with alternative approaches might prove fruitful. Here we: (1) attempt to isolate some of the components in a model-based paradigm; (2) discuss the impact it might have on research; and (3) inquire whether it meets certain basic requirements that a new 'candidate' should possess.

### A model-based paradigm: a proposal

We do not propose to alter the basic Hartshorne (1959, p. 11) definition of geography's prime task, nor challenge the appropriateness of the matrix concept. We suggest, however, that it may be possible to derive a *second* matrix from the first by transforming the two basic vectors in such a way as to throw emphasis away from that of classification towards model-building. Table 1.1B gives an example of such a transformed matrix.

(a) *Vector I: locational relativity*. The search for greater accuracy along the absolute locational vector (e.g. exact latitude and longitude), and the search for significant 'fixes' in this absolute regional space (e.g. the location of the source of the Nile) is now largely centred in specialized government mapping agencies outside the university research world. However, location in an absolute sense on an isotropic surface is clearly only one way of viewing space. Bunge (1962) has urged that quite different and alternative ways are both possible and common, e.g., most ideas on *accessibility* or *isolation* refer to distance measured in a specialized way (usually in terms of *energy* translated into cost terms) and from specialized origins or axes. The selection of these *axes of symmetry* is largely determined by the spatial model being adopted: in a diffusion model the appropriate axis might be the diffusion *hearth*, and the appropriate parameter distance (accessibility) from that axis. Many of the models featured in later chapters use just such relativistic ideas of location with space measured in metric terms (see Henshall's discussion on Von Thünen's model in Chap. 11, below) or topological terms (see Haggett's review of Pitt's accessibility model in Chap. 15, below), with the measurements taken from single points, sets of points, or linear axes (e.g. Horton's model of erosion in terms of distance from the watershed). Attempts to use *relative* location rather than absolute location are a feature of spatial model-building. Although no formal attempts have yet been made to build up a 'general theory of locational relativity', the multivariate analysis of distance (measured in 'real' energy terms rather than 'neutral' *mileage* terms) and their reduction to suitable display maps pose fascinating and complex research problems. A whole new family of atlases might be envisaged showing completely different spatial patterns than the familiar absolute patterns of traditional maps. As Bunge (1964) has pointed out, on a cost-transformed surface pre-railway England reduces to coalfields grouped around an *inland* sea!

(b) *Vector II: topological-geometrical form.* The successive sub-division of the 'elements' vector has seen the constant departure of newly fledged sciences with their own geographical interpretations. Here we seem to have missed a fundamental point that these elements (e.g. vegetation, population) are not *in themselves* the objects of geography. Sten de Geer (1923, p. 2) argued that '. . . it is certain abstract qualities in the objects which are studied by Geography, and not the objects themselves'. We propose therefore that certain 'abstract properties', – the topological and geometrical form of that object or objects – be substituted for the standard 'element-class' vector. Cole (1966) has extended point-set theory to suggest that using a simple dimensional classification into zero-order (points, cities), first-order (lines, networks), second-order (areas, states), and third-order (surfaces, terrain) provides such an appropriate topological vector. Bunge (1962, p. 197) concludes his *Theoretical Geography* on a similar vein: 'the profession might, as a matter of efficiency, start dividing itself into various theoretical spatial fields, such as point problems, area problems, description of mathematical surfaces, and central place problems, rather than the current arrangement of climatology, population geography, landforms, etc.' Despite the apparent simplicity of this schema it can be rigorously extended (by set combinations) to include complex combinations of the basic topological forms: thus the four basic dimensions give fifteen combinations (Cole, 1966; see his Fig. 3.1). If we divide each of the four into only two time states (*stable* and *unstable*) then the number of possible combinations goes up to over 250. We have argued elsewhere that '. . . . many of the more successful attempts at geographical models have stemmed from this (geometrical) type of analysis' (Chorley and Haggett, 1965, p. 376) and will not repeat those arguments in detail. One such topological class – the linear network – is currently being investigated (see Haggett, Chap. 15, below; Haggett and Chorley, In preparation). In general we feel that geometrical analysis offers a logical, consistent and geographically more relevant alternative to the 'element-orientated' approach with its inevitable tendency to sub-divide geography and force it outwards towards the relevant external systematic disciplines. It not only offers a chance to weld human and physical geography into a new working partnership, but revives the central role of cartography (see Board, Chap. 16, below) in relation to the two.

Bernal (1965, p. 97) traces the change in biology through phases when it was 'primarily a descriptive science, more like geography, dealing with the structure and working of a number of peculiarly organized entities, at a particular moment of time on a particular planet', through its fission into zoology, botany and their subdivisions, and its present convergence towards common concern with chemistry and physics in molecular biology, cell biology, biophysics, biochemistry, etc. Geography has likewise shown a tendency to replace its first phase of fact gathering with a fission into physical and human geography (and their subdivisions). Whether there are signs within contemporary geography of a third converging stage are matters of debate. Certainly the general concept of systems analysis (Chorley, 1962) has been put forward by Ackerman (1963) as a fundamental integrating concept for geography, and Stoddart (1965; see also Chap. 13, below) has analysed its particular role in relation to ecological systems (ecosystems). Whatever the problems that remain to be solved, particularly problems of system identification and energy monitoring, we may expect that regional systems analysis will emerge as a major theme in geographical work over the next decade. It is not without interest that two of the most tractable systems identified and analysed by geographers – the watershed system (see Chorley, Chap. 3, below: and More, Chap. 4, below) and the nodal 'city region' (Garner, Chap. 9) – are distinctive geometrical forms with particular mathematical properties (e.g. nesting) and are organized with respect to specific axes from which relative distances are measured. That is, we may argue that distinctive *geosystems* (Stoddart, Chap. 13, below) are likely to be found at the intersections of distinctive vectors within our transformed data matrix.

It would be irrelevant here to pursue the applications of our transformation model to detailed cases. Table 1.2 (overleaf) is included to show the organization used by one of the writers in dealing with a typical nodal-system in human geography: it is based largely on topological distinctions (i.e. Vector I organization).

## Implications for research

Kuhn (1962) has shown that the period which follows the wholesale adoption of a paradigm is commonly devoted to three main classes of research problem: (1) the determination of significant facts; (2) the matching of facts with theory; and (3) the further articulation of theory.

(1) One of the curious effects of paradigm adoption is to replace a period of data abundance with one of data scarcity. The very definition of the major research problem highlights the paucity of *relevant* data and great attention is paid to the most accurate measurement of constants that would have seemed, in the pre-paradigm period, to be either too recondite or too elusive to measure. Thus a study of urban-centred fields has led to the most careful study of city sizes using stringent operational criteria, and the need for observational networks of a new rigour and accuracy makes increasing demands on research time. With further concentration we might expect that problems

TABLE 1.2

## Three-Stage Model for the Analysis of Regional Systems*

| | STAGE I<br>*System Identification* | | STAGE II<br>A. *Form Differentiation* (*Static*) | | B. *Form Differentiation* (*Dynamic*) | STAGE III<br>*System Integration* |
|---|---|---|---|---|---|---|
| *Dimensional Number* | $\left\{0, 2\right\}$ | 0 | 1 | 3 | 4 | $\int_4^0$ |
| *Geographical Form* | City (*polar* axis)<br>City region (*boundaries*) | Cities<br>Settlements<br>Urban hierarchies | Transport networks<br>Communication systems | Urban fields<br>Density gradients<br>Land-use intensity | Innovation waves<br>Frontier movements<br>Sequent occupance<br>Colonization | Regional systems<br>Internal feedbacks<br>Interregional systems<br>External feed-backs |
| *Analytical Techniques* | Numerical taxonomy<br>Local residuals<br>Regional analogues | Rank-size analysis<br>Nearest-neighbour analysis<br>Quadrat analysis | Graph-theoretic analysis<br>Connectivity<br>Network geometry | Trend-surface analysis<br>Harmonic analysis<br>Fourier analysis | Physical simulation<br>Monte Carlo models<br>Markov-chain models | Matrix analysis, factor analysis<br>Input-output analysis<br>Interregional linear pro-gramming |
| *Spatial Model* | Regional hierarchies<br>Formal, functional regions | Central-place theory<br>Gravity models<br>Weberian models<br>Basic non-basic models | Network models<br>Random graph models<br>Geodesic models | Gravity models<br>Absorption models<br>Intervening-opportunity models<br>Von Thünen models<br>Potential models | Diffusion models<br>Migration models<br>Colonization models | Regional climax models<br>Regional multipliers<br>Growth poles |
| *Major Sources for Spatial Models* | Decision theory (Psychol.)<br>Taxonomy (Biol.)<br>Discriminant analysis (Stats.) | Point set theory (Math.)<br>Organization models (Manag.)<br>Packing theory (Math.) | Graph theory (Math.)<br>Circuit design (Electr.)<br>Search theory (Math., Psychol., Zool.) | Least-effort models (Sociol.)<br>Minimum-energy models (Phys.)<br>Potential models (Phys.)<br>Game theory (Psychol.) | Epidemic theory (Medic.)<br>Diffusion theory (Fluid dyn.)<br>Rumour theory (Sociol.)<br>Colonization & succession models (Bot.) | General Systems theory<br>Ecosystems (Biol.)<br>Interregional trade theory (Econ.)<br>Multiplier models (Econ.) |

* The table is schematic only: it follows the general arrangement of topics in Haggett (1965), Chaps. 2–6 inclusive.

of interregional flow and internal energy flux would put data collection and data standardization as major consumers of research time. In the case of physical sciences such problems are commonly accompanied by the design and construction of measuring apparatus of increasing precision and complexity. The analogy in non-experimental sciences may well be the development of computer programmes for squeezing the greatest amount of information from limited data.

(2) Determination of facts that can be directly related to paradigm models form a second class of activity (Kuhn, 1962, p. 33). Although such facts cannot be said to test the model in any strict sense, they provide a focus for research areas. Thus Gunawardena (1964) directed attention at the central-place structure of southern Ceylon in an attempt to test existing central-place models. Much of the stimulus that a model provides is the drive to test (and possibly overthrow) existing models. The strength of the paradigm is that it allows the progressive evolution of models within its general terms. Kuhn argues that only when no new models (or wholly incompatible new models) can be produced is the paradigm itself likely to shift.

(3) Extension and articulation of theory therefore provides the third class of research. This may take a number of forms but will probably include: (a) the determination of constants in existing predictive equations; (b) the quantification or further mathematization of existing qualitative models; and (c) the speculative and exploratory extension of extant models into areas of unproven and even unlikely application. Olsson's (1965) study of migration rates might fall into the first category, Dacey's (1965) elegant extension of Christaller-Lösch settlement models through point-set algebra into the second, and Gould's (1965) extension of search theory to the extension of transport networks in East Africa into the third.

According to the Kuhn thesis none of these major types of research is designed to produce results entirely outside the paradigm's limits, and the range of acceptable results from the studies is small – certainly small in relation to the results that could be conceived. The internal discipline of the paradigm, its unwritten rules and traditions, guides the pattern of research and ensures by the successive cumulation of small highly-limited advances that the science as a whole will progress.

## Needed characteristics of a new paradigm

How should we recognize an efficient paradigm if we saw one? Kuhn (1962, pp. 152–158) has studied the introduction of new paradigms in fields as unlike as chemistry and electricity, and suggests three minimum ingredients for success.

(1) Firstly, the new paradigm must be able to solve at least some of the problems that have brought the old one to crisis point. We have argued above

that the most sorely-troubling feature of present geography is the explosion of the traditional data-matrix, and the forcing of geographers to study both areas and topics less and less relevant to the general shaping of the earth surface they profess to study. A new paradigm in geography must be able to rise above this flood-tide of information and push out confidently and rapidly into new data-territories. It must possess the scientific habit of seeking for relevant pattern and order in information, and the related ability to rapidly discard irrelevant information: '. . . it is the capacity for pattern-seeing and not the actual surveying of the landscape which explains this rapidity' (of scientific development). 'It explains why scientific activity remains a mystery for those devoid of theoretical insight, who see only facts' (Van Duijn, 1961, p. 67).

(2) Secondly, the new paradigm must appeal to the workers' sense of what is elegant, appropriate and simple. This somewhat aesthetic characteristic is difficult to define in specific terms and is most clearly seen in the mathematician's attraction towards elegant rather than inelegant mathematical proofs. At the smaller scale, the demonstration that Horton's law of stream numbers was a simple combinatorial system applicable to a much wider range of phenomena (Haggett: Chap. 15, below) represents an appropriate and economical simplification. Similarly Hägerstrand's (1953) overview of diffusion waves replaced many separate, clumsy, and individually articulated 'frontier' and 'sequent occupance' studies. A new paradigm for geography needs to provide a similar economical and elegant simplification for the *whole* field.

(3) Thirdly, the new paradigm must contain more 'potential for expansion' than the old. This characteristic is believed by Kuhn to be often the decisive one, albeit its adoption is based on faith in the new rather than its proven ability. Geography, coming late to the paradigm race, has the compensating advantage that it can study at leisure the 'take-off' paradigms of other sciences. There is good reason to think that those subjects which have modelled their forms on mathematics and physics – themselves 'leading sectors' in the scientific community (to continue Rostow's language) – have climbed considerably more rapidly than those which have attempted to build internal or idiographic structures. Not a little of this success stems from the great elasticity of mathematical analysis and the hierarchy of ever-more-complex equations that can be derived for observed patterns. As Kaplan (1964, p. 262) remarked: 'The use of mathematics and the construction of logical systems marks a certain coming of age'.

## EPILOGUE

In this introductory chapter we have set out to examine the nature of models and their relation to facts on the one hand and to paradigms on the other. We have looked at the traditional paradigmatic model of geography and suggested that it is largely classificatory and that it is under severe stress. We have tentatively suggested an alternative model-based approach.

In judging the success of this approach and of the models discussed in the chapters which follow, we should recall that geography must measure its progress by the number of puzzles it has effectively solved, not by the magnitude of those that remain unsolved. In welcoming Ackerman's (1963, p. 435) reminder that the philosophical goal of geography is '. . . nothing less than an understanding of the vast, interacting system comprising all humanity and its natural environment on the surface of the earth' we should recall, with Humboldt, that such a goal is utterly unattainable in any complete sense – either now or in the future. Successful application of models in geography ensures no teleological progress towards full understanding, for scientific effort does not reduce the sum total of problems to be solved – it rather increases them.

## REFERENCES

ACKERMAN, E. A., [1963], Where is a research frontier?; *Annals of the Association of American Geographers*, 53, 429–440.

ACKOFF, R. L., GUPTA, S. K. and MINAS, J. S., [1962], *Scientific Method: Optimizing Research Decisions*, (New York), 464 pp.

APOSTEL, L., [1961], Towards the formal study of models in the non formal sciences; In Freudenthal, H., (Ed.), *The Concept and the Role of the Model in Mathematics and Natural and Social Sciences*, (Dordtrecht, Holland), 1–37.

BAMBROUGH, R., [1964], Principia Metaphysica; *Philosophy*, 39, 97–109.

BEAMENT, J. W. L., (Ed.), [1960], *Models and Analogues in Biology*; Symposium No. 14 of the Society for Experimental Biology, (Cambridge), 255 pp.

BERNAL, J. D., [1965], Molecular structure, biochemical function, and evolution; In Waterman, T. H. and Morowitz, H. J., (Eds.), *Theoretical and Mathematical Biology*, (New York), 96–135.

BERRY, B. J. L., [1961], A method for deriving multifactor uniform regions; *Przeglad Geograficzny*, 33, 263–282.

BERRY, B. J. L., [1964], Approaches to regional analysis: a synthesis; *Annals of the Association of American Geography*, 54, 2–11.

BLACK, M., [1962], *Models and Metaphors*, (Ithaca, New York), 267 pp.

BRAITHWAITE, R. B. [1953] *Scientific Explanation*, (Cambridge).

BRAITHWAITE, R. B., [1962], Models in the empirical sciences; In Nagel, E., Suppes, P. and Tarski, A., (Eds.), *Logic, Methodology and Philosophy of Science*, (Stanford), 224–231.

BRIDGMAN, P. W., [1936], *The Nature of Physical Theory*, (Princeton).

BROWN, L., [1965], Models for spatial diffusion research: a review; *Office of Naval Research, Geography Branch, Contract Nonr* 1288 (33), *Technical Report*, 3.

BUNGE, W., [1962], Theoretical geography; *Lund Studies in Geography, Series C, General and Mathematical Geography*, 1.

BUNGE, W., [1964], Geographical dialectics; *Professional Geographer*, 16 (4), 28–29.

CAMP, G. D., [1961], Models as approximations; In Banbury, J. and Maitland, J., (Eds.), *Proceedings of the Second International Conference on Operational Research*, (Aix-en-Provence), 20–25.

CAWS, P., [1965], *The Philosophy of Science*, (Princeton), 354 pp.

CHORAFAS, D. N., [1965], *Systems and Stimulation*, (New York), 503 pp.

CHORLEY, R. J., [1962], Geomorphology and general systems theory; *U.S. Geological Survey, Professional Paper* 500-B, 10 pp.

CHORLEY, R. J., [1964], Geography and analogue theory; *Annals of the Association of American Geographers*, 54, 127–137.

CHORLEY, R. J. and HAGGETT, P., [1965], Trend-surface mapping in geographical research; *Transactions of the Institute of British Geographers*, No. 37, 47–67.

COLE, J. P., [1966], Set theory and geography; *Nottingham University, Department of Geography, Bulletin of Quantitative Data*, 2.

DACEY, M. F., [1965], The geometry of central place theory; *Geografiska Annaler*, 47*B*, 111–124.

DARBY, H. C., [1953], On the relations of geography and history; *Transactions of the Institute of British Geographers*, No. 19, 1–13.

DEUTSCH, K. W., [1948–49], Some notes on research on the role of models in the natural and social sciences; *Synthèse*, 7, 506–533.

GEER, S. DE, [1923], On the definition, methods and classification of geography; *Geografiska Annaler*, 5, 1–37.

GOULD, P., [1965], *A bibliography of space-searching procedures for geographers*; Pennsylvania State University, Department of Geography (Mimeographed).

GUNAWARDENA, K. A., [1964], Service centres in southern Ceylon; *University of Cambridge, Ph.D. Thesis*.

HÄGERSTRAND, T., [1953], *Innovationsförloppet ur korologisk synpunkt*, (Lund).

HAGGETT, P., [1965], *Locational Analysis in Human Geography*, (London).

HAGGETT, P. and CHORLEY, R. J., [1965], Frontier movements and the geographical tradition; in Chorley, R. J. and Haggett, P., (Eds.), *Frontiers in Geographical Teaching*, (London), 358–378.

HAGGETT, P. and CHORLEY, R. J., (In preparation), *Network models in geography*, (London).

HANSON, N. R., [1958], *Patterns of Discovery*, (Cambridge), 241 pp.

HARTSHORNE, R., [1959], *Perspective on the Nature of Geography*, (London).

HUTTON, E. H., [1953–54], The role of models in physics; *British Journal of the Philosophy of Science*, 4, 284–301.

HESSE, M., [1953–54], Models in physics; *British Journal of the Philosophy of Science*, 4, 198–214.

KAPLAN, A., [1964], *The Conduct of Inquiry*, (San Francisco), 428 pp.

KRUMBEIN, W. C. and GRAYBILL, F. A., [1965], *An Introduction to Statistical Models in Geology*, (New York).

KRUMBEIN, W. C., BENSON, B. and HEMPKINS, W. B., [1964], WHIRLPOOL: a computer programme for 'sorting out' independent variables by sequential multiple linear regression; *Office of Naval Research, Geography Branch, Technical Report* 14, Task No. 389–135.

KUHN, T. S., [1962], *The Structure of Scientific Revolutions*, (Chicago).

KUIPERS, A., [1961], Model and insight; in Freudenthal, H., (Ed.), *The Concept and the Role of the Model in Mathematics and Natural and Social Sciences*, (Dordrecht, Holland), 125–132.

LANGBEIN, W. B. and HOYT, W. G., [1959], *Water Facts for the Nation's Future*, (New York), 228 pp.

LEWONTIN, R. C., [1963], Models, mathematics and metaphors; *Synthèse*, 15, 222–244.

MANLEY, G., [1966], A new geography; *The Guardian*, March 17th, 1966.

MEADOWS, P., [1957], Models, system and science; *American Sociological Review*, 22, 3–9.

NAGEL, E., [1961], *The Structure of Science*, (London), 618 pp.

OLSSON, G., [1965], Distance and human interaction: a bibliography and review; *Regional Science Research Institute, Bibliography Series*, 2.

PERRING, F. H. and WALTERS, S. M., [1962], *Atlas of the British Flora*, (London).

PORTER, P. B., [1954], Another puzzle-picture; *American Journal of Psychology*, 67, 550–551.

PRICE, D. J. DE SOLLA, [1963], *Little Science, Big Science*, (New York).

ROSENBLUETH, A. and WIENER, N., [1944–45], The role of models in science; *Philosophy of Science*, 11–12, 316–321.

SKILLING, H., [1964], An operational view; *American Scientist*, 52, 388A–396A.

STEINER, D., [1965], A multivariate statistical approach to climatic regionalization and classification; *Tijdschrift van het Koninklijk Nederlandsch Aardrijkskundig Genootschap*, 82, 329–347.

STODDART, D. R., [1965], Geography and the ecological approach: the ecosystem as a geographic principle and method; *Geography*, 50, 242–251.

STODDART, D. R., [1967], Growth and structure of geography; *Transactions of the Institute of British Geographers*, 41.

SUPPES, P., [1961], A comparison of the meaning and uses of models in mathematical and empirical sciences; In Freudenthal, H., (Ed.), *The Concept and the Role of the Model in Mathematics and Natural and Social Sciences*, (Dordrecht, Holland), 163–177.

SUPPES, P., [1962], Models of data; In Nagel, E., Suppes, P. and Tarski, A., (Eds.), *Logic, Methodology and Philosophy of Science*, (Stanford), 252–261.

THEOBALD, D. W., [1964], Models and method; *Philosophy*, 39, 260–267.

TOBLER, W., [1966]. Numerical map generalization; *Michigan Inter-University Community of Mathematical Geographers, Discussion Papers*, 8.

TOULMIN, S., [1953], *The Philosophy of Science*, (London).

UBBINK, J. B., [1961], Model, description and knowledge; In Freudenthal, H., (Ed.), *The Concept and the Role of the Model in Mathematics and Natural and Social Sciences*, (Dordrecht, Holland), 178–194.

VAN DUIJN, P., [1961], A model for theory finding in science; *Synthèse*, 13, 61–67.

WRIGLEY, E. A., [1965], Changes in the philosophy of geography; In Chorley, R. J. and Haggett, P., (Eds.), *Frontiers in Geographical Teaching*, (London), 3–20.

CHAPTER TWO

# The Use of Models in Science

F. H. GEORGE

## INTRODUCTION

The idea of using models in science is by no means new, and Braithwaite, among many others, has analysed the use of such models and their relation to theories in science.

There is a sense in which almost anything can be used as a model for almost anything else, but as in the use of analogies or metaphors, so models, to have predictive value, must bear some measure of similarity to the structure or process being modelled.

We can classify models in various ways into at least those which model structure, those which model function and those that model both. These are *static* (for structure) and *dynamic* models (for processes or functions). We also have models which are obviously physical (e.g. a wind tunnel model) and those which are obviously symbolic (e.g. mathematical and statistical models for simulation), and so on and so forth.

Following the line of thought of Braithwaite, we might regard scientific theories as capable of being *formalized*. This process of formalization is essentially one of stripping down theories in ordinary language and showing the underlying logic, it is almost indeed a matter of making the original theory more precise, where we take some theory and rewrite in greater detail and in more 'molecular' fashion.

We can also argue that a scientific theory is made up of empirical statements which use terms capable of operational definition, whereas the model uses logical statements, which are not in themselves verifiable.

We can illustrate our general meaning by a simple example. This example shows one sort of formalization:

In Hull's theory of learning, we find a statement to the effect that:

> *the greater the similarity between the conditioned stimulus and the unconditioned stimulus, the greater the absolute value of the increment of tendency to respond to the conditioned stimulus.* (1)

To formalize this needs a lot of detailed definitions, but if $s$ is a stimulus and $r$ is a response, then the strength of association ${}_sH_r$ can be defined:

$$ {}_sH_r(t_2) = {}_sH_r(t_0) + \Sigma\delta\ {}_sH_r(t_2, s', r') \qquad (2) $$

where $(t_0, t_2)$ is a time interval, and $\Sigma\delta\ {}_sH_r$ is the change of strength of association during the interval $(t_0, t_2)$, $s'$ and $r'$ are sets of $s$'s and $r$'s that occur in the interval $(t_0, t_2)$.

We now define a function $F(x, y, z, w)$ such that:

$$ F(x, y, z, w) = \delta\ {}_sH_r(t_2, s', r') \qquad (3) $$

and where $x = \mathcal{J}(t_1, t_2)$, $t_1 = T(r')$, $y = S(s, s')$, $z = S(r, r')$ and $w = T(r') - T(s')$.

Suffice it for our purposes that the variables $x$, $y$, $z$ and $w$ can be precisely defined, but roughly speaking $x$ is concerned with drive reduction, and $y$ and $z$ are concerned with stimulus and response similarity, and finally $w$ is a function of time delay between the occurrence of the stimulus and the response.

This leads to a statement (or axiom):

$$ \frac{\partial F}{\partial y}(x, y, z, w) > 0 \qquad (4) $$

and this total argument, given here in skeleton form, is a *formalization* of Hull's original statement (1). This is formalized in (2), (3) and (4), although even this is barely adequate as an approximation to the much more precise statement still required.

Another kind of example we might mention is that of the Russell and Whitehead's formalization of number in *Principia Mathematica*.

By 'formalizing' then, we mean 'making more precise', or 'reducing statements' to their underlying logic, and clearly 'formalizing' is itself a vague word, which could well mean 'to expose the model underlying the theory'. Not only is formalizing vague, but it is also a matter of degree.

Similarly we can look at the matter the other way around, and *interpret* models in many different ways to get many different theories. Models and theories are connected by formalizing *and* by interpretation, and theories are linked directly with emperical descriptions. Let us now look at an example of interpretation e.g. the well-formed formula

$$ p \supset . p \supset q $$

or

$$ p \supset (p \supset q) $$

is from an axiomatic system $P$ and can be interpreted so that $p$, $q$ are propositions, and $\supset$ is the logical connective called 'material implication'. The dot is a bracket as illustrated by our alternative rendering of the formula. This interpretation is a part of what is called the propositional calculus.

But the identical statement $p \supset . p \supset q$ could be interpreted as a statement in a Boolean Algebra $B$. When the second interpretation is adopted, a *conventionally* different notation is used. So we have:

$$A \rightarrow (A \rightarrow B)$$

or since

$$A \rightarrow B = \text{df. } A^1 \cup B$$

where $A^1 \equiv$df. not$-A$
and $A \cup B =$df. the sum of the two classes $A$ and $B$, then

$$A \rightarrow (A \rightarrow B)$$

becomes

$$A^1 \cup (A^1 \cup B)$$

The fact is that the same model is being given two different interpretations.

Here the process could be described as taking a model or structure, which by itself has no meaning or reference, and supplying the model with the meaning. The word (name) 'Chicago' means to refer to the city Chicago, only by common agreement that 'Chicago' is the name for Chicago. The process of naming is like tying labels on physical objects.

With all this in mind, let us now turn to scientific method.

## SCIENTIFIC METHOD

The next thing that must be done is to outline the principles of scientific method. Scientific method is naturally vital to any scientific investigation of the use of models, since, as we have seen, models and scientific theories are so closely connected.

A large number of books have been written on scientific method, logic, and philosophy of science, so there will be no attempt made here to describe the whole of science and the full complications of scientific method. We shall though, try to give an abbreviated account of what scientists do, or what they should ideally do.

Let us suppose a man is walking down the street, and he sees a shop which is open. He notices the shop is selling sweets and chocolates, and from this he might infer that all sweets and chocolates shops are open for the sale of sweets and chocolates at that moment, but will this be a reasonable thing to infer? Obviously not, as he will surely be the first to admit. It could happen that the shop he has seen is open at exceptional hours or on unusual days. On the other hand, if he remembers that he has often bought sweets at that time of day before, and indeed on that particular day – a Wednesday let us say, then he will feel some added strength from this in the belief that all sweets and chocolates shops are in fact open.

In the same way, we note at what shops we can buy what sort of things, so that by and large, it is true that one buys one's vegetables at the greengrocer's, and one's groceries at the grocer's shop, and so on. We may know roughly when they open and when they shut and apply our previous experience to the present. This enables us to make predictive theories.

Science is precisely the process of using one's previous experience as a basis for prediction. The prediction is important, the whole point of having a science is to be able to make (accurate) predictions about the future state of some system, and we can only do that if we have had some previous experience of that system, or some similar systems. Not that prediction is in fact everything, we want also to understand the system, and we want furthermore to be able to control the system, usually our surroundings, to some extent. This ability to control depends at the very least on being able to make a prediction of some sort.

Our own experience sensibly used in our everyday life can be said to be identical with science and scientific method. The only differences we find are in the language used, which, for the scientist becomes more complicated, and there is the means of making the observations, which may be far more detailed than in everyday life.

Language and science then are much alike. We want to build a sort of *map* (see next section) of our surroundings and consult it when we are going somewhere to make sure we take the right route. The map of course is only one of a number of things we use – we also use time-tables, and dictionaries, and whole sets of principles that we have learned for ourselves, and we may have couched in ordinary language. Thus, 'The pub never opens until 12 o'clock on Sunday', 'The only place you can get fresh meat is the butchers'; and 'X's is the best butcher in town'. Of course we can apply deductive logic too. From the last two sentences we have written it is easy to see that we have to go to X's if we are in town and we want some fresh meat, which is also the best.

In fact we shall want to make a distinction between learning, and having learned, the language of science or any other language. Thus, for example, the learning part is to come to realize that in fact X's does have fresh meat, and then perhaps also that Y has fresh meat too, and then finally to realize that a certain type of meat shop has fresh meat and a certain type does not. This learning is what is called 'induction' or the process of making inductions or inductive generalizations.

Inductive generalizations are statements of the kind 'all . . . are –' and examples would be 'All men are featherless bipeds'. 'All birds fly', 'All cats are black' and so on, but of course often we have to be satisfied with some weaker generalization such as 'Almost all cats are black'.

In principle, a science would be ideally made up of a set of inductive generalizations which form the basis of a deductive argument. The laws of nature would be our base and then anything we wanted to know would follow from applying deductive logic to these laws. In terms of our example, it would mean being able to say that 'Charlie is black' because 'Charlie is a cat', and we know that 'All cats are black'.

Science is the business of trying to discover all the necessary inductive generalizations, and the search is likely to go on for ever. One difficulty is knowing whether a generalization is true or not. If you say 'All cats are black', then you are faced with the problem of tracking down every cat, and not only that, logicians are inclined to say 'All cats are black' means now, not only that they are black, but they have always been, and always will be, black. But how can we be sure that cats will not one day become white? One way of answering this would be to say that if they become white you should no longer call them 'cats', but if you did that, then the whole game is farcical because cats are, as logicians say, 'necessarily black', if indeed, because their being cats depends in part on their being black; so there can be no possible reason for looking around the world as they are, to put it another way, 'black by definition'.

Our learning process leads us to generalizations that are probabilistic; they cannot be known to be certain because we cannot fully confirm them. Once set up though, the inductive generalizations are precise enough for making completely certain deductions. But the certainty derived is dependent on clarity over language.

Complete certainty is here meant to mean of course that the certainty depends on our *complete understanding* of the language. Let us illustrate the point by an example: 'If there are three hundred and sixty-seven persons in a room', then deductively it follows that 'at least two persons must have the same birthday'. Of course the extra person is included in case someone mentions Leap Year. But this sort of argument is easy enough to follow, and after a little thought no one would doubt that the force of the argument quite clearly depends on our understanding of what the words involved mean. If the meaning of the word 'birthday' for example is un-clear, then the whole argument will fail to persuade. So also with a vague word, this may be the key to the whole success or failure of deductive argument.

We shall now summarize our argument; we assume that the scientist, and that means every 'rational' person, shows behaviour that is made up of learning and the application of what has been learned. This means making inductive inferences, and then arguing deductively from these inductions. Of course most people learn from other people's as well as from their own direct experience, and we cater for this by saying that some learning is by acquaintance and some is by description.

We can easily distinguish between learning by acquaintance and learning by description, if we remember that we may have learned of the existence of China by description, since we have never been there. But we may have learned of the effect of the sun on the skin by acquaintance – through actually trying to sunbathe.

The method of argument that scientists ideally use is called axiomatic, and involves the restatement of the generalizations as careful statements or axioms ('hypotheses' or postulates, are other words meaning much the same). There is of course a lot more to say on this subject but we have said enough to relate the idea of models to language and to science, and that was the main point of our current discussion of scientific method.

## STATIC MODELS

Of the many possible models that can be used to illustrate the static class of models, we shall select that of *maps*.

We can say that maps are like pictures of territory, although they are also different from pictures because they use abstractions in symbolic form of the territory itself.

It is characteristic that maps should be likened to languages and scientific theories. This is itself a reminder that languages and scientific theories are themselves related closely to each other. As a result, in view of what we said about model-theory relationships, we shall sometimes think of maps as models for languages and scientific theories. Let us look carefully at the idea of maps.

A map is any form of contour or line drawing, with or without colours, of any shape or shapes whatever. A map may be on any scale, using any projection, and may represent any detail we choose.

If we made abstract maps this would be part of a sort of universal geometry, although restricted – if we so constrict maps – to two dimensions.

Given such a set (clearly an infinite set) of maps, we could then, if we chose, interpret them as actual territories. The most obvious way of carrying out this interpretation is to choose a map with a *shape* similar to the territory. Because of projecting spherical areas onto a plane, many distortions actually occur, and in any case, we often use highly formalized maps which make no attempt to retain topographical details, only topological ones.

Maps, and we can think of them now as 'real' (i.e. ones that are used for some purpose relative to some actual territory), are to territory like scientific theories or scientific statements are to 'reality'. They either allow you to make *predictions* from territory-to-map and map-to-territory or they do not. If they do allow predictions (assuming no errors in map reading) this is because the structure of the map is in some way similar to the structure of the territory.

The map though is an abstraction – it represents roads, railways, or counties, etc. and it does so symbolically. The symbols of the map refer to abstracted bits of territory like words may refer to parts of reality.

Statements, particularly empirical statements, are highly confirmed (and are therefore probably true) if they allow accurate prediction. So the similarity between language and maps is clear enough. But words are generally non-pictorial and do not have topographical or topological significance in themselves, so in some respects maps are more like pictures. This underlines a point that is continually arising. If we use analogies (models are like analogies) then we must bear in mind *both the similarities and the differences.*

Now language is also sometimes generalized so we can refer to concepts like *reality*, *democracy*, etc. which cannot be derived from maps alone. A shape is specific or particular, a word may be a universal. So that part of language that is removed from immediate descriptions of the environment ('reality'), or sense-data statements, do not find any equivalent among maps.

A map which contained everything in the reality it mapped, would be indistinguishable from the reality itself. This is why we talk of maps as representing abstractions from the reality. If we now go back to *abstract maps* we can in some measure mimic the equivalent of universals in language.

Strictly speaking, we should not wish to choose maps as models of language because one is symbolic and not pictorial and the other is mainly pictorial with symbols attached. The correct model for language would be logic or some symbolic model which makes no specific appeal to pictorial concepts.

The usefulness of maps as models is of considerable pedagogical value, but it is also of conceptual value, and may even be of literal value. Language and maps come together in some measure in the use of graphs and algebraic geometry generally. Perhaps the most important aspect of the use of maps lies in its more direct appeal to the eye. The use of words is secondary rather than primary, and the two can clearly be used together, and when this is done, we can place the interpretation of 'scientific theory' on the result.

Maps normally do not change much, if at all, because territory does not change much. If maps were changing then they would be dynamic models, but for our dynamic models we shall look at the field of digital computer programmes.

## DYNAMIC MODELS

In using computers (or computer programmes) as dynamic models, we are using them to simulate or synthesize a certain type of problem.

The central problem we are concerned with is the programming of a computer to demonstrate various aspects of artificial intelligence. We are primarily concerned with the synthesis of artificial intelligence, but also, inescapably, we are, in doing so, bound up with the problem of human intelligence. This connection is thought to be inescapable, since any form of synthesis must, it seems, at least be suggestive of simulation, however the fact remains that synthesis as such is our primary object. Our problem is to programme a computer so that it can accept and understand questions in a natural language such as English, or possibly in a form of 'empirical' logic. It then searches through its own store to find the answer to these questions and, in general, it will state the answers if they are known to the computer, (Simmonds, 1963, Green *et al.*, 1963). This is not essential to a dynamic model as such, but essential to our use of the computer as a dynamic model of high-grade artificial intelligence.

We want also to programme a computer so that it asks questions of any 'human-like' source of information in its environment. The reason for asking these questions will, in general, result from the need for information to complete some strategy, or act as a basis of information for some decision, whether or not this decision leads to action. It should be made clear that the computer is to be thought of as acting in an environment, which contains at least one human-like source of information and also contains events which occur, some of which may be under the control of the computer, and are also capable of being described by it. The computer's problem is to learn which aspects of the environment it can control, depending on information gained by direct experience of that environment on one hand, or by a question-and-answer interplay, through language, with whatever human-like source is available on the other. It should be said that apart from the ability to collect information and store that information, the computer must have the capacity to draw inferences, of both an inductive and a deductive kind, and take whatever steps are necessary to seek information which may be needed by some other source of information.

The first such environment we are concerned with is that of a well-defined game, such as noughts and crosses (G1), or some other similar such game. The game itself is obviously trivial, and it is only used because it is one which has a decision procedure, and is also one which illustrates the way in which either decision procedure can be discovered or heuristics supplied, without obscuring the process of that derivation because of the complexity of the game.

The position is that the computer both plays the game and makes inferences about it, based on the rules it has been told, at the same time. By the phrase 'at the same time' we mean that the computer must go on-line then off-line, playing, then arguing about or reasoning about or asking questions about, the game, and then be playing the game again, and then making inferences again, and so on. This clearly implies the capacity in the computer to make inductive and deductive inferences, where the deductive process at least is one similar to that described by such writers as Newell, Shaw and Simon (1963), Gelerntner (1963) and others.

The computer must be capable of describing what it is doing and answering questions it is asked with respect to what it is doing. The second stage of this operation is to show that the computer can be given a sufficiently general vocabulary and the formal rules of some further game, G2 (e.g. checkers), so that without any more special programming technique being involved, the reasoning capacity derived from the first game, G1, plus the actual playing of the game in an environment, is sufficient to allow it to improve its performance. Initially we shall expect this to be little more than a definite improvement, without expecting it either to derive an *algorithm*, or without it necessarily being able to play a highly-skilled game.

The same technique can now be generalized and applied to any games whatever, G3, G4, . . . , Gm, whether these games be well defined such as noughts and crosses and checkers, or whether they be ill-defined, requiring optimization techniques, as well as requiring the conditional probabilities needed to play well-defined games. The computer must in the next stage of operations be able to generate new words *and* be able to understand new words which are supplied to it by the other computer or human-like sources of information in its environment. We now would like, of course, also to generalize this and say there may be many human-like sources in its environment, and to this end, we shall have to have a model not only of the environment in which the computer operates but also a model of all the features of that environment which are what we have called 'human-like' sources. This means the computer must keep a model of each human-like source and must assess the likely reliability, in the light of experience, of that model and also the likely motivation that that human-like source may have in making the particular statement it has made (Maron, *et al.*, 1964). The next stage in this programming undertaking (or dynamic modelling), surrounds the need that we do not merely describe the process in general, but supply either compiler codes or machine codes, which show precisely how this whole undertaking can actually be programmed. The next problem is to show how the computer under these circumstances of being exposed to other human-like sources, must guard itself against the possibility of being bluffed or in other ways deceived. This, in common sense terms, is easy to understand. It must depend, of course, on the nature of the ascribed motivation to the utterer of the source of information, and must depend in some respects on the reliability of that source of information, and must be a matter for processing before the computer can decide whether this source of information is a reliable one or not. The difficulty here is to show precisely what instructions you might put into a computer to show how this type of undertaking is carried out.

It is clear from what has been said already that, although the aim is to provide a synthetic system (or model) of artificial intelligence, depending on the ability to learn, to use language, and to develop and learn new languages, both from understanding external linguistic sources and from manufacturing new words internally, the overall picture is one which comes near to being a simulation of the problem solving, decision taking and planning activities of human behaviour. It should be added that a large number of programmes for a variety of different computers, including computer programmes written in ALGOL, have already been written in pursuance of the plan laid out in this section, and we will explain a little bit more of the methods in the next section of this article.

## METHODS USED IN PREPARING DYNAMIC MODELS AS PROGRAMMES

In the last section we gave an overall glimpse of the plan on which the whole of our artificial intelligence undertaking is based. We gave a few references to previous work that fitted in to this general picture, and we can remind you again at this point of some of this source work:

1 Logic Theorem Proving and Geometry Theorem Proving by Newell, Shaw and Simon (1963) on one hand and Gelerntner (1963) on the other.
2 Pattern Recognition Problems, not primarily in the field of visual pattern recognition, but as the basis for inductive inference. Here, the work of Minsky and Selfridge springs directly to mind.
3 Problems in Data Retrieval.
4 Concept Formation, which suggests especially the work of Kochen, Bannerji and his associates, among others.
5 The Confirmation of Hypotheses and degrees of factual support. This is a field which has so far impinged on artificial intelligence only very little, and one thinks in this context of the work of Carnap, Reichenbach, Popper, Hempel and Oppenheim, etc.
6 Risk analysis, where the work has been well summarized by Thrall, Coombs and Davis.

The above sample list of activities which are involved in the total process of artificial intelligence are more or less representative of the stages through which the process must go. Figure 2.2 shows this process in a very simple and generalized block diagram:

The first stage in the incoming process from the computer point of view involves the ability to recognize the nature of the input. One decision the computer must make is as to whether the input is linguistic or non-linguistic, i.e. whether it is from a human-like source in language form or whether it is from a non-human like source and represents the occurrence of events in the environment. As far as the computer is concerned, we have agreed that without the need for specific description of pattern recognition systems, a merely conventional distinction needs to be made between the linguistic and the non-linguistic inputs.

The recognition process clearly involves, among other things, comparison

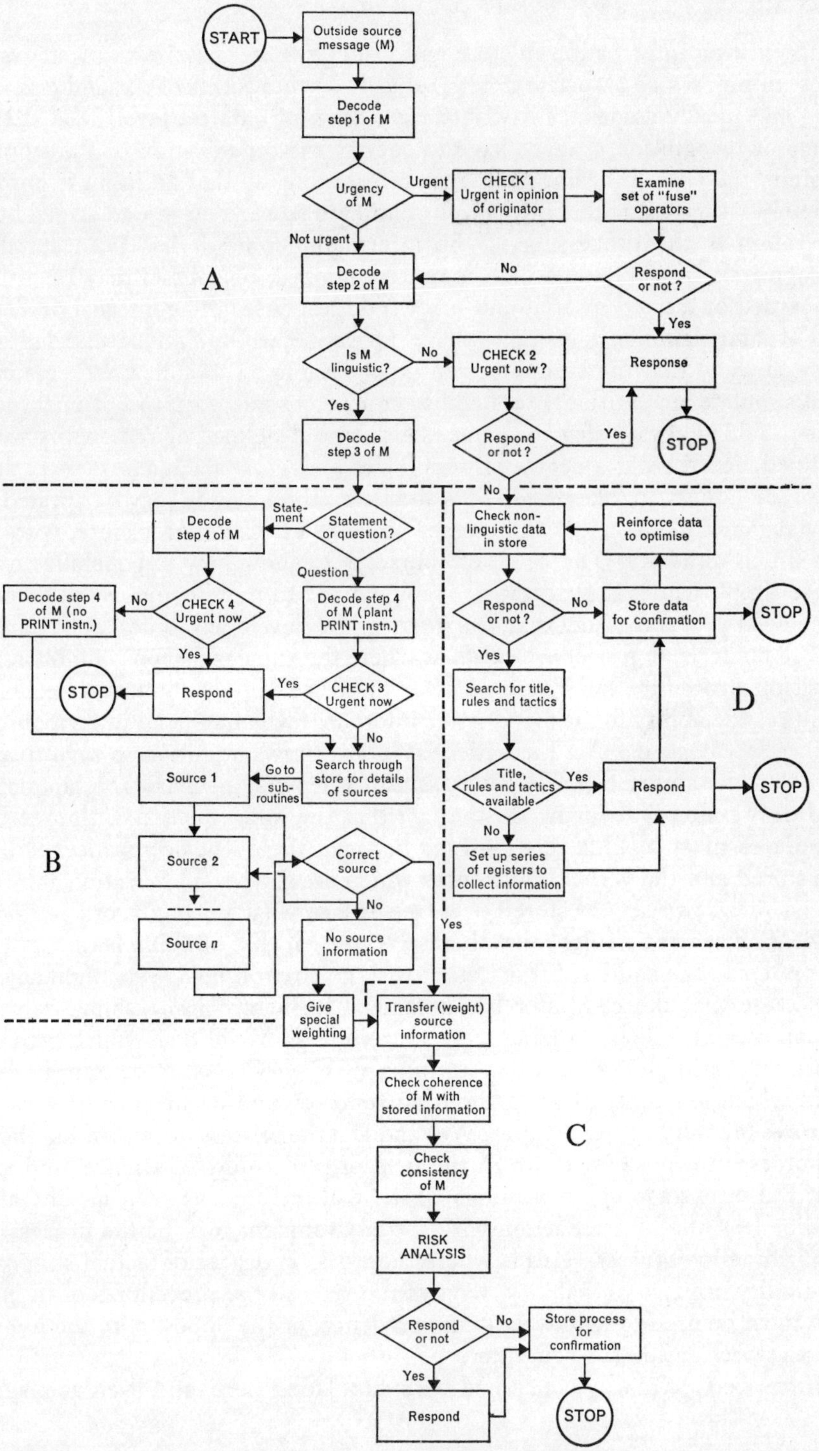
START
Outside source message (M)
Decode step 1 of M
Urgency of M
Urgent
CHECK 1 Urgent in opinion of originator
Examine set of "fuse" operators
Not urgent
Decode step 2 of M
No
Respond or not ?
Yes
A
Is M linguistic?
No
CHECK 2 Urgent now ?
Response
Yes
Decode step 3 of M
Respond or not ?
Yes
STOP
No
Decode step 4 of M
Statement
Statement or question?
Check non-linguistic data in store
Reinforce data to optimise
Question
Decode step 4 of M (no PRINT instn.)
No
CHECK 4 Urgent now
Decode step 4 of M (plant PRINT instn.)
Respond or not ?
No
Store data for confirmation
STOP
Yes
Yes
STOP
Respond
Yes
CHECK 3 Urgent now
Search for title, rules and tactics
D
No
Source 1
Go to sub-routines
Search through store for details of sources
Title, rules and tactics available
Yes
Respond
STOP
No
B
Source 2
Correct source
Set up series of registers to collect information
No
Yes
Source n
No source information
Give special weighting
Transfer (weight) source information
Check coherence of M with stored information
C
Check consistency of M
RISK ANALYSIS
Respond or not
No
Store process for confirmation
Yes
Respond
STOP

**2.1**

with existing information in store and, though there may be many stages in recognition, which associate it with recall and other forms of remembering, it also specifically associates it with the problem of data retrieval. It is at this stage of recognition that we have to specify the organization of the storage system, the cross-indexing and cross-referencing, so that an input is appropriately associated with existing information in store. The second stage of the operation is the processing of the information once it has been suitably recognized. Information which is not recognized, of course, may be either discarded or treated by trial-and-error type responses. The internal processing of information at the simplest level is event learning. Thus it is relatively easy to show that the computer can be programmed to collect information, and simulate a conditional probability computer, and learn to optimize output provided always that the game is either well-defined or reasonably well-defined, and there is a degree of periodicity of events which allows some sort of predictability to take place. This simple method, which may be applied to trivial games like noughts and crosses, leads to a decision procedure, without too much difficulty. The need, of course, is to show how to generalize over these simple games so that transfer can be made to more complicated games. The need is also to show that *heuristics* can be developed to deal with games where no decision procedure exists or where the game is far too complex for a decision procedure to be arrived at or even used in economical time. This requires the ability to make inductive inferences which depend in turn on the ability to categorize in a theoretic way. This draws attention in turn to the fact that the ability to make generalizations depends primarily on language, and draws attention further to the fact that the integral organization of the computer must include relationships between the event occurrences which are stored and the verbal occurrences which are stored. The relationship between these two sets of stored occurrences are what we might call *semantic rules*. The semantic rules must themselves be capable of being learnt by the computer in the course of transacting with its environment. We might say at this stage that the computer has organized in hierarchical fashion, sets of hypotheses or beliefs, ranging from more or less trivial ones which require more or less simple response to a stimulus, up to very much more complicated ones which are contingent on the occurrence or non-occurrence of various features in a very complex input. We can liken the process of confirming these hypotheses to those set down in the field of philosophy of science, and say that the next stage of proceedings which will interact between on-line and off-line learning by transaction with the environment, will be the process of confirming hypotheses. This is where the work on degree of factual support, credibility, inverse probability, Bayes Rule, and all the procedures of induction must be used to lend weight and credence, or the opposite, to the hypotheses stored inside the computer.

If we were primarily concerned with simulation here, and even as a synthetic undertaking, we must give some consideration to matters of priority and urgency. Thus, if our artificially intelligent system is to make decisions for us over a domain where urgency may be a prime factor, then some sort of recognition of the urgency, and this will occur right in the first stage of operations, must be brought into the picture. The fact that language is critical to this whole undertaking is a reminder in the present context that natural language will normally be transposed or translated into the logical form (functional calculi will be the general form of language used but with associated probabilities which give it the power of empirical description) and the ability to make deductive inferences on the part of the computer will result from the computer's ability to translate natural language statements not, of course, only into machine code, which is necessary for manufacturing the necessary computer orders, but also into logical language in order to make the

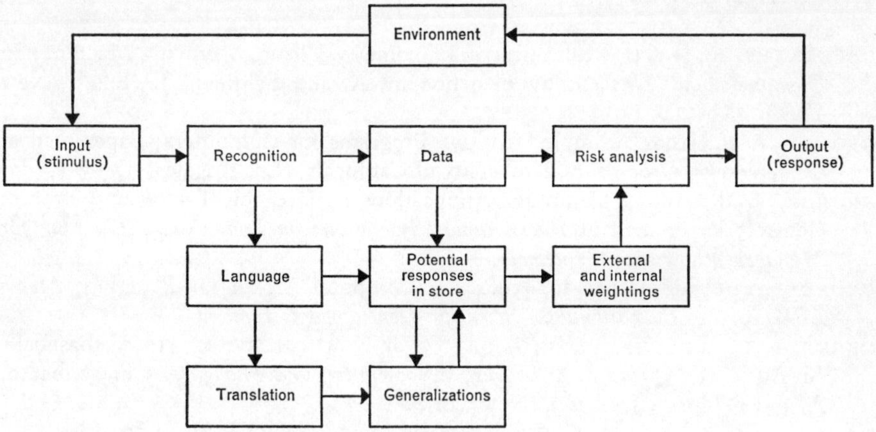


2.2

inferences with maximum convenience. At the output end it must be remembered the computer is capable of performing acts, as well as making statements or asking questions. The difference between performing an act and asking a question or making a statement is critically important to the whole success of the undertaking. Wherever an act or a statement is required as the final output on the behalf of the computer, a prior analysis of risk must be undertaken. The methods cannot be described here, but they are very closely associated with theory of games and so-called 'games against nature', and they involve the manifestly important point of assessing the risk for any particular decision. It may be that the same odds, in probability terms, in one context will be a totally different risk from the same odds in another context. It is obvious that taking a risk over a game of cards in the family context is quite different from taking a risk over the use of a nuclear deterrent, even though the probabilities of success or failure may be the same for each.

The relationship between scientific method and dynamic models should now be fairly clear, since our attempts at computer programming described above could readily be interpreted as an attempt to simulate the scientist. This all means that maps, and static models in general, are limiting special cases of the more general dynamic system.

We have lengthily described our dynamic model on the computer and we have gone well beyond the minimum requirements for a dynamic model.

It should be remembered that many programmes have already been written along the lines suggested (George, 1962, George and Gill, 1964; George, 1965), and finally we will consider a flow chart (Fig. 2.1, p. 53) of the general processes outlined in this last section.

## REFERENCES

GELERNTER, H., [1963], Realization of a Geometry-Theorem Proving Machine; In *Computers and Thought*, by Fiegenbaum, A. and Feldman, J. (Eds.), (New York).

GEORGE, F. H., [1962], Simple Adaptive Programs for Computers; Paper read at *Conference on Cybernetics*, University of California, (Los Angeles), *October*.

GEORGE, F. H., [1965], Computer Applications in Decision Taking and Process Control; Paper read at *International Symposium on Long Range Planning for Management*, (Paris), *September*.

GEORGE, F. H. and GILL, R., [1964], A Computer Model for Planning; *Memo ESD/10.3.64 Educational & Scientific Developments Limited.*

GREEN, B. F., WOLF, A. K., CHOMSKY, C. and LAUGHERY, K., [1963], Baseball: An Automatic Question Answerer; In *Computers and Thought*, by Fiegenbaum, A. and Feldman, J. (Eds.), (New York).

MARON, M. E., *Computers and Comprehension* Memorandum R.M.-40650-PR.

NEWELL, A., SHAW, J. C. and SIMON, H. A., [1963], The Theorem Proving Machine; In *Computers and Thought*, by Fiegenbaum, A. and Feldman, J. (Eds.), (New York).

SIMMONDS, R., [1962], Synthex; In *Computer Applications in Behavioural Sciences*, by Borko, H. (Ed.), Systems Development Corporation Publication.

CHAPTER THREE

# Models in Geomorphology

RICHARD J. CHORLEY

## INTRODUCTION

The changes which have taken place in geomorphology since the Second World War have been profound. This profundity does not lie so much in the flood of new factual information about the surface of the earth, its processes and its transformations, as in our basic attitudes to the subject as a whole. However, the new mass data-generation methods of remote terrain sensing (e.g. air photography, infra-red photography, airborne radar, as well as other sophisticated military techniques linked with satellites) (McCauley, 1964; Ellermeier and Simonett, 1965; Heacock *et al*, 1966; Latham, 1966; Moore, 1966; Office of Naval Research, 1966; Rouse *et al*, 1966; Rowan and McCauley, 1966; Simpson, 1966), combined with computer methods of data processing and information extraction (Krumbein and Graybill, 1965, Chaps. 13–15), are also beginning to impose the need for radical methodological and conceptual rethinking in geomorphology. Indeed, it may be asked legitimately whether the study of landforms still exists as a discrete scholarly entity, for the most important recent changes in the subject have tended to impress upon scholars the disparate character of modern researches, together with the inability of workers to identify broad common objectives of even the most general character, or even to communicate to one another their mutual objections. Although geomorphology has always been a discipline of fine diversity (Chorley, Dunn and Beckinsale, 1964), many scholars now feel that the decline in the popularity of the Davisian basis for the subject (Chorley, 1965) has produced a conceptual vacuum which has not yet been reoccupied by any comparably broad systematic approach. This deficiency has served to highlight many national preoccupations, some of long standing, with particular geomorphic objectives. Thus the development of the American style of 'dynamic-process' geomorphology, the Franco-German climatic geomorphology, the British denudation chronology/geological approach, the Polish Pleistocene-dominated geomorphology, the Russian applied geomorphology, the Swedish studies of process almost *per se*, the Eastern European morphological mapping, and the Central European tectonic bases have created a

Godot-like atmosphere of articulate introspection. Perhaps this is why many geomorphologists have become increasingly concerned with the basic structure of the subject (e.g. Strahler, 1952; Dylik, 1957; Chorley, 1962 and 1966; Schumm and Lichty, 1965; Howard, 1965), and the relationships of its parts. The present volume, with its theme of model building, seems to provide both an opportunity and a kindly environment for taking such a broad structural view of the subject.

The value of model theory in geographical studies as a whole has been stated briefly and inadequately by the present author (1964) and this paper may serve as an introduction to the more comprehensive model for geomorphological work presented in Figure 3.1. In this the thought processes of abstraction and decision-making lead to the identification of three distinct, if marginally-interlocking, systematic views of geomorphology in which all types of past and present work find conveniently-linked places. Thus the simplified conceptual model system can either be approached by being translated in time and/or space to produce a *natural analogue system*, by being dissected into some supposedly integral parts which are examined in terms of a *physical system*, or by some broad conception the phenomena being structured into a complete *general system* from the outset. Obviously these systematic approaches grade into one another and can (electrically speaking) operate either in 'parallel' or in 'series', but the characteristic members of each seem to be reasonably distinct. Such a view allows apparently very different types of geomorphic work to fall into some more general pattern, and enables us to see more clearly from this *map of geomorphic activity* the areas of previous effort and neglect, so that we may more surely chart our respective future paths through regions the opportunities and challenges of which are unfolding with awe-inspiring speed.

## NATURAL ANALOGUE SYSTEMS

One of the most common methods of illuminating a given geomorphological phenomenon is to translate its supposedly-important or characteristic features into some analogous natural system believed to be simpler, better known or in some respect more readily observable than the original. By such reasoning we see how classifications (i.e. like objects grouped together for the purpose of making some general statements about all of them) form an integral part of one kind of model building. Obviously such translations involve large intuitive leaps, for assumptions of natural analogy are often made on the most subjective bases; however, large sections of geomorphic work traditionally lie in this area. One may recognize two classes of natural analogue systems – the historical and the spatial.

Historical analogues group together geomorphic phenomena with regard

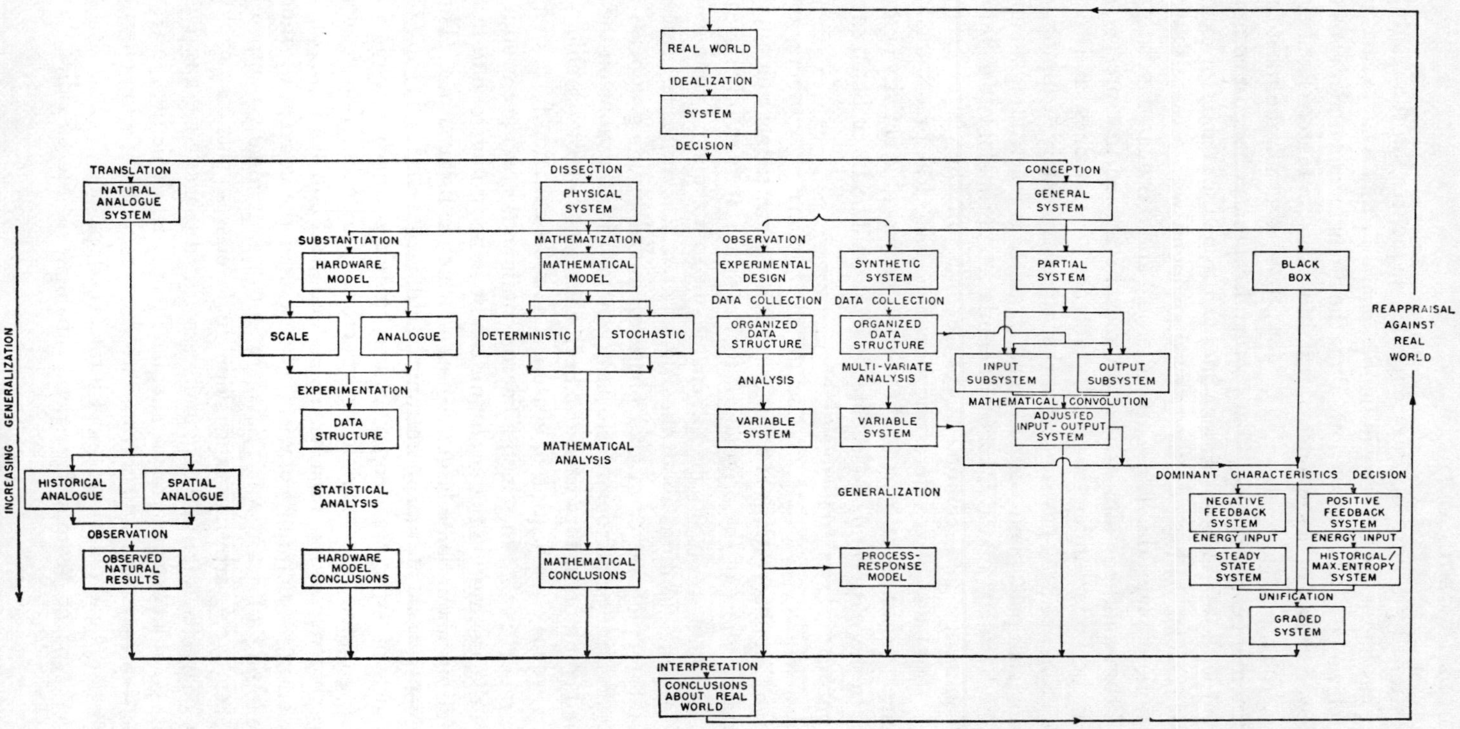


**3.1** A map of geomorphic activity.

to their assumed positions in time-controlled sequences, on the assumption that what has happened before will happen again, or that what existed in the past has relevance to what exists now. Thus the phenomenon under consideration is viewed as part of 'a sequence of real, individual but interrelated events' (Simpson, 1963, p. 25; see also Kitts, 1963). It is, after all, the most fundamental canon of geomorphology that past landscapes can be most completely understood with reference to present ones (Hutton, 1795) and, for example, Wills (1929, Chap. 7) has effectively used analogies with present deserts to describe and explain the late-Carboniferous and Permian landscapes of Britain. From another point of view, it may be argued that present landscapes can be better understood with reference to past ones, and recently Ambrose (1964, p. 850) has suggested that the sedimentary stripping of the 'sub-Ordovician paleoplain' of the Canadian shield is reactivating an ancient topography 'virtually intact'. Much of denudation chronology (e.g. Johnson, 1931; Wooldridge and Linton, 1955) depends on this latter reasoning (Chorley, 1965) and it is interesting how this historical preoccupation of the extreme 'westernmost' part of the map of geomorphic activity (Fig. 3.1) reappears in the 'easternmost' (see later Section on The Black Box), suggesting that our map 'closes' to form a globe.

Spatial analogues associate one set of phenomena with others, on the assumption that observations at another place are easier to make or simpler in character than those of the original, or that comparison with other areas believed to be in some way similar will enable one to make more confident and meaningful generalizations about the original area. It is in this last sense that the classificatory character of this phase of model building appears most clearly (Sokal, 1965), although historical (genetic) classifications on the basis of assumed common history have also been abundant in geomorphology (e.g. Johnson's (1919) classification of coasts). The most common form of spatial analogue model is that in which adjacent contiguous areas are grouped together on the assumption that each unit can be better understood in terms of generalizations about some larger region of which it forms a part. Thus on the continental scale of 'morphotectonics' (Hills, 1961 and 1963) and on the regional scale of 'structural geomorphology' (Melton, 1959) individual landforms are grouped into tectonic or structural provinces. The concept of physiographic regions, at least as understood by Fenneman (1914), has a similarly strong structural-geological basis. Under the concept of 'morphological mapping' (Tricart, 1964; Savigear, 1965), individually-mapped landscape 'facets' (e.g. arbitrary land slope angle classes, changes of slope of different character, etc.) are regionally associated (Figs. 3.2A and B). The belief that total landscapes can be adequately or usefully defined in terms of such facets springs very largely from the concepts of denudation chronology (see, for example, Wooldridge and Linton, 1955, p. 56).

Another type of geomorphic spatial analogue is the so-called 'natural model', in which what are believed to be characteristic assemblages of landform units are identified and presented as type assemblages. Beckett and Webster (1962), for example, have proposed such facet assemblages, or 'patterns', for an area around Oxford. In a somewhat similar American approach, Van Lopik and Kolb (1959) have employed the idea that one varied and accessible region (or 'gross landscape') can be divided into 'component landscapes' which are each defined in terms of a characteristic association of four measurable terrain factors (characteristic slope, relief, plan profile and occurrence of steep slopes – Figure 3.2C), to form 'terrain types' with which other regions can be compared and classified. This scheme of desert 'terrain analogues' is based on the landscape around the Yuma Test Station, part of which is shown in Figure 3.2D.

## PHYSICAL SYSTEMS

The physical systems approach is the one most obviously associated with conventional 'scientific method' and was the first to be applied to quantitative data in the earth sciences, particularly in the 1930's and 1940's. This approach is based on the view that research can best be pursued by dissecting the geomorphic problem structure into its supposedly components parts, such that the operation of each part and the interactions between the parts can be conveniently examined, leading (it is hoped) to a full synthesis of the components into a working whole. Amorocho and Hart (1964) have pointed to the dichotomy in hydrological research between what are here called the physical systems and general systems approaches, and the pivotal geomorphic work of the hydrologist Robert E. Horton (1945) may be held to belong largely to the former. However, it will be argued here that this distinction does not appear as clearly cut in geomorphology and that there is a considerable identity between experimental designs and synthetic systems, leading to the construction of 'process-response models' (Fig. 3.1). It is convenient to divide physical system investigations into three, often interrelated types: those wherein the important structural elements are substantiated into a hardware model, mathematized into a mathematical model, or subjected to field observation under some convenient experimental design.

### Hardware Models

It is a constant source of surprise that hardware models have not hitherto proved to be of greater value in geomorphic research. Indeed, with one or two notable exceptions (e.g. Friedkin, 1945), it is probably true to say that the most valuable hardware models have been those which were basically parts of unscaled reality, closely circumscribed and examined in great detail.

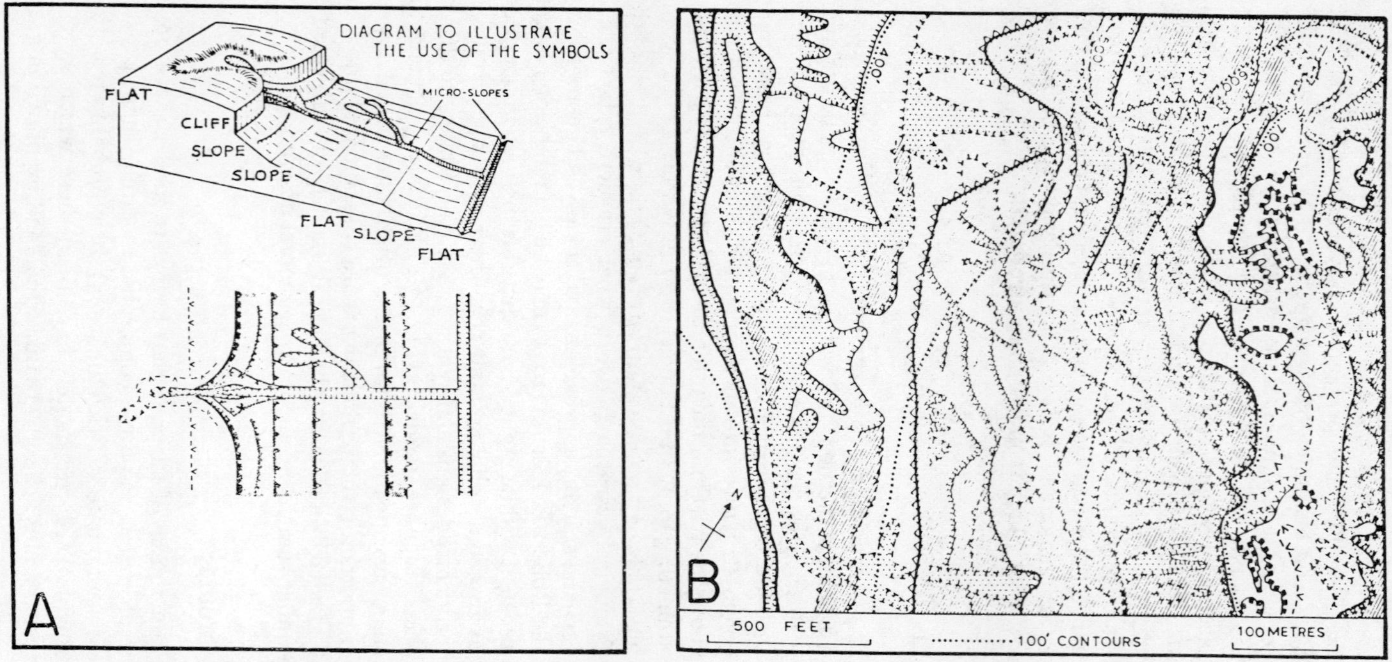


**3.2** Spatial analogues in geomorphology.

A Examples of one system of symbols used in morphological mapping (*From Savigear, 1965*).
B An area in West Cornwall, England, mapped by use of the symbols depicted in Figure 3.2A (*From Savigear, 1965*).

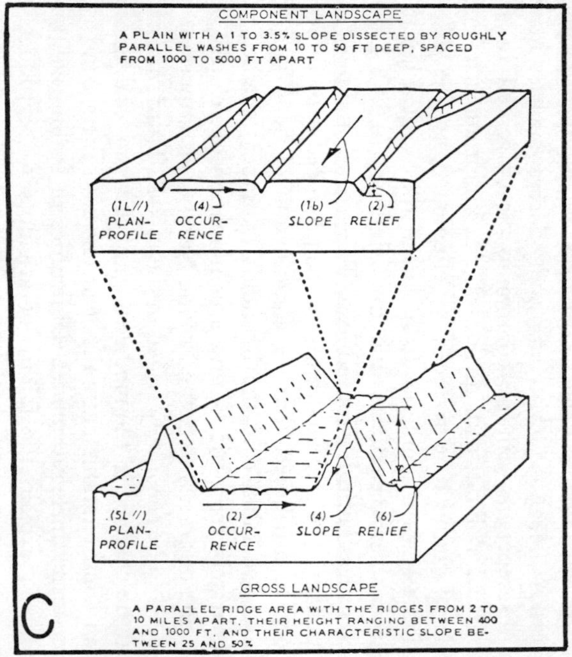


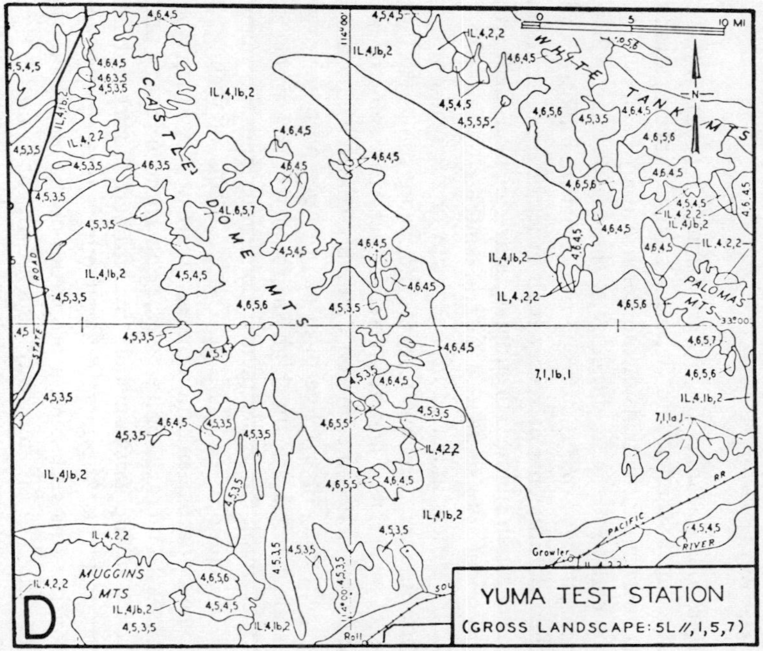


**3.2** (*continued*)

C Example of a component landscape defined in terms of four terrain factors, and the relation between a component and a gross landscape (*From Van Lopik and Kolb, 1959*).

D Part of the Yuma Test Station, Arizona, divided into characteristic component landscapes (*From Van Lopik and Kolb, 1959*).

The two most obvious instances of this are Bagnold's (1941) wind tunnel observations on sand movement and Schumm's (1956) investigation of the erosional forms and transformations of the Perth Amboy badlands. This, of course, gives the clue to the basic difficulty of the construction of hardware models in geomorphology; that the complexity of nature imposes scaling or changes of media requirements of a very high order of sophistication. It is consequently easier for an engineer to simulate a life-size man-made structure than for an earth scientist to simulate a natural complex.

Scale models are closely imitative of a segment of the real world, which they resemble in some very obvious respects (i.e. being composed mostly of the same types of materials), and the resemblance may sometimes be so close that the scale model becomes merely a suitably-controlled portion of the real world (e.g. Schumm, 1956). The most obvious geomorphic advantages of the use of scale models are the high degree of control which can be achieved over the simplified experimental conditions and the manner in which time can be compressed. The general problems of scaling and of dimensional similarity have been comprehensively treated by Murphy (1949), Langhaar (1951), Wilson (1952, pp. 317–330) and Duncan (1953), and with particular reference to the earth sciences by Hubbert (1937) and Strahler (1958). Of these problems, the most difficult one is that changes of scale affect the relationships between certain properties of the model and the real world (e.g. scale ratios) in different ways, such that, for example, the kinematic scale ratios (i.e. those involving velocities and accelerations) behave differently from linear scale ratios (i.e. those involving lengths and shapes). Similar difficulties are involved with attempts to produce meaningful dynamic scale ratios (i.e. those involving gravity forces, such as mass and inertia). Such discrepancies can be commonly circumvented in any of three interrelated ways. First, a distortion of one important attribute can (usually by rule of thumb) be reduced or eliminated by the distortion of another attribute – for example, a distortion of the vertical linear scale of river models enables the effects due to turbulence (e.g. a kinematic ratio) to be more or less faithfully reproduced. The second, and most important, way in which analogous model ratios can be produced is by dimensionless combinations of attributes. Thus a combination of density, velocity, depth and viscosity (combined in the Reynolds' Number) enables viscous effects to be accurately reproduced; and a combination of velocity, length and the acceleration of gravity (e.g. the Froude Number) is important where gravity effects need to be accurately scaled in the model. Third, one or more of the media can be changed in the model to assist the truer scaling of other effects; but such considerations naturally lead one into the second type of hardware model – the analogue model. Difficulties of scaling natural geomorphic phenomena explain the failure of attempts to reproduce whole fluvial landform associations and their transformations (e.g. Wurm, 1935),

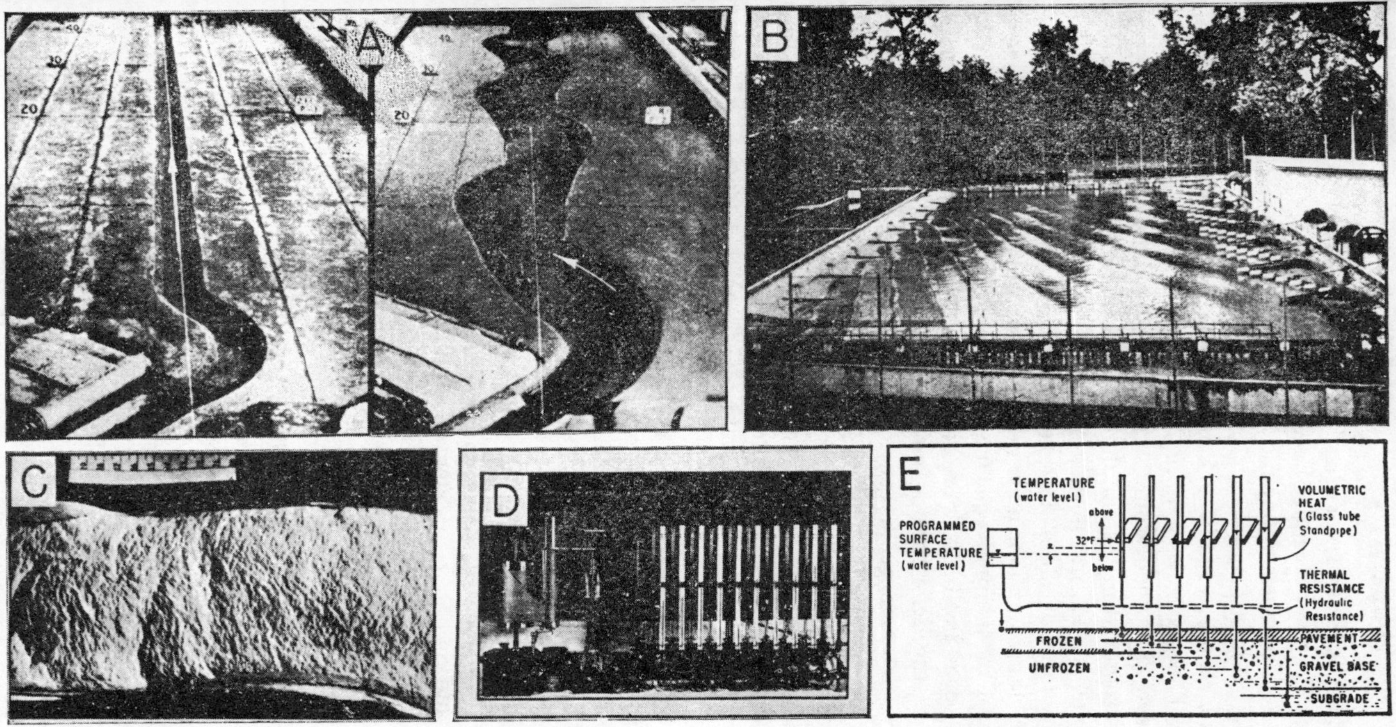


**3.3** Hardware models in geomorphology.

A The development of meanders in a laboratory tank: Left, the initial channel and, right, after 3 hours (*From Friedkin, 1945*).

B An open-air wave tank at the Hydraulics Experiment Station, Wallingford, England (*From* Hydraulics Research, *1956*) (*Crown Copyright Reserved*).

C Detail of the surface features at a bend in a kaolin glacier (scale in cms.) (*From Lewis and Miller, 1955*).

D Photograph of an hydraulic analogue computer for the study of soil freezing (*From Massachusetts Institute of Technology, 1956*).

E Diagram of the equipment shown in Figure 3.3D (*From Massachusetts Institute of Technology, 1956*).

and why the most successful work has involved attempts to reproduce more limited features like river meander bends (Friedkin, 1945) (Fig. 3.3A), nick points (Brush and Wolman, 1960) and beach segments (Saville, 1950; King, 1966A, Chap. 4) (Fig. 3.3B).

Analogue models involve radical changes in the media of which the model is constructed. They have much more limited aims than scale models in that they are intended to reproduce only some aspects of the structure or web of relationships recognized in the simplified model or idealized system of the real world segment. Such transformations are obviously rather difficult and great potential sources of 'noise' (i.e. extraneous confusion), in that great and often questionable assumptions must be made regarding the appropriateness of the changes of media involved. One example of such an analogue hardware model is Lewis and Miller's (1955) use of a kaolin mixture to simulate some features of the deformation and crevassing of a valley glacier (Fig. 3.3C). A more elaborate geomorphological analogue model is that constructed at the Massachusetts Institute of Technology (1956) to simulate the freezing and thawing of soil layers (Figs. 3.3D and E). The ground surface and successive soil layers are represented by a reservoir and glass tubes, in which the water level is programmed to correspond to the temperatures of the soil layers, the flow of heat in the soil is duplicated by the flow of water, and latent heat by suitably-placed expansion wells. Such analogue models have recently tended to be replaced by electronic computers operating in association with mathematical models or experimental designs.

Often the use of hardware models ends with the presentation of some simple visual impression of the geomorphic phenomenon (e.g. Lewis and Miller, 1955), but where quantitative experimental data results from the operation of the model (e.g. Saville, 1950) it can be built into a data structure and analysed by conventional statistical techniques (see sections on Experimental Design and Synthetic Systems) to produce hardware model conclusions.

## Mathematical Models

A mathematical model is an abstraction in that it replaces objects, forces, events, etc., by an expression containing mathematical variables, parameters and constants (Krumbein and Graybill, 1965, p. 15) involving the adoption of 'a number of idealizations of the various phenomena studied and in ascribing to the various entities involved some strictly defined properties' (Neyman and Scott, 1957, p. 109). The essential features of the phenomena are then 'analogous to the relationship between certain abstract symbols, which we can write down. The observed phenomena resemble closely something extremely simple, with very few attributes. The resemblance is so close that the equations are a kind of working model, from which we can predict features of the real thing which we have never observed' (Daniel, 1955, p. 34). The common type of geomorphic mathematical model is concerned with some simplified statement of certain important features of the real world (usually geometrical ones) which can be transformed according to assumptions regarding the basic operation of the system (usually related to changes through time), yielding, by checking the model predictions against the appropriate real world situations, some information about the basic mechanisms involved and the succession of geometric changes to which the earth's surface is subject through time. Thus mathematical analysis follows the symbolic statement of the assumed basic static and dynamic relationships, and the logical mathematical conclusions then are checked against the real world; such that the correspondence or divergence between the real world and the effects predicted by the model indicate the success which we have had in building the original model. 'The differences revealed may then lead to the discovery of further causes, and these observed facts may gradually become understood in greater completeness and detail' (Jeffreys, 1918, p. 179). Mathematical models can be commonly divided into deterministic and stochastic.

Deterministic mathematical models are based on classic mathematical notions of exactly predictable relationships between independent and dependent variables (i.e. between cause and effect), and consist of a set of exactly specified mathematical assertions (derived from experience or intuition) from which unique consequences can be derived by logical mathematical argumentation. Such models are thus intimately concerned with relationships and 'driving forces' between the factors identified in the simplified model. Jeffreys (1918), for example, developed such a model for the denudation of a land surface by runoff, and by this means deduced theoretically the form of the resulting peneplain. Strahler (1952) produced a similar model involving the time relationship governing the elevation of a given point on a graded stream (Fig. 3.4D2), assuming an exponential longitudinal stream profile (Fig. 3.4D1) and a rate of stream-bed lowering always proportional to the gradient at that point. Such assumptions are basic to the construction of mathematical models and it is these which have to be tested according to the accuracy of predictability of the model. The most common type of deterministic geomorphic model involves the transformation of slope profiles under various assumptions (Scheidegger, 1961). Here assumptions are similarly made regarding the original slope geometry and the manner of its transformation. Figure 3.4A shows diagrammatically the geometry of a subtalus rock slope as a steeply-inclined original face retreats by uniform weathering and debris removal such that the debris wedge (of greater volume than the original rock slice) accumulates at a repose angle at the slope base. Obviously a vast number of combinations of assumptions can be made, and Figure 3.4B shows a case when the depth of weathering on the original slope increases with height.

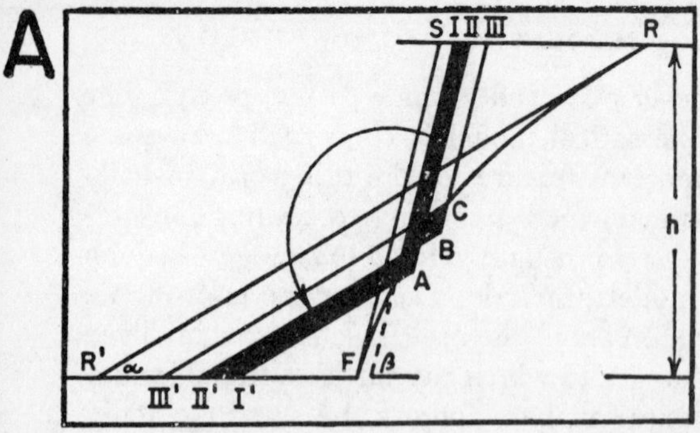


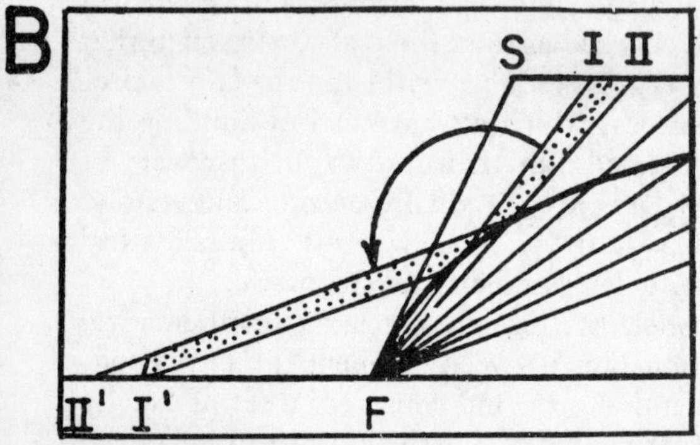


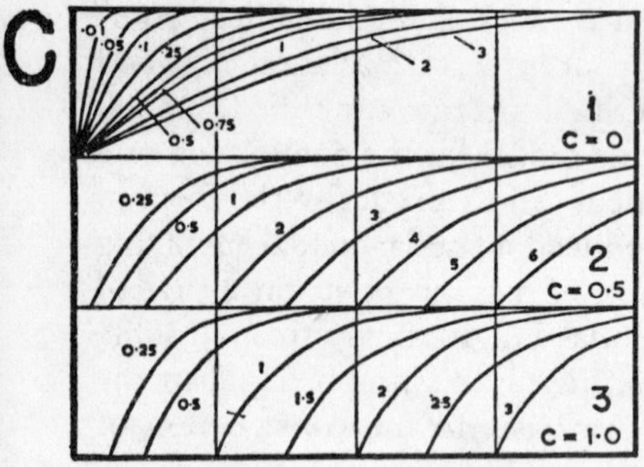


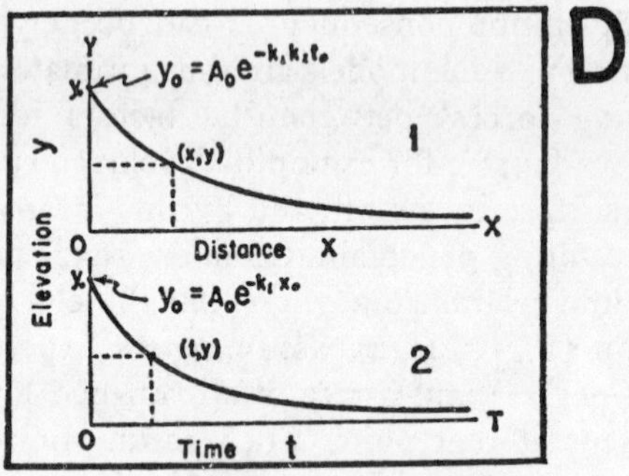


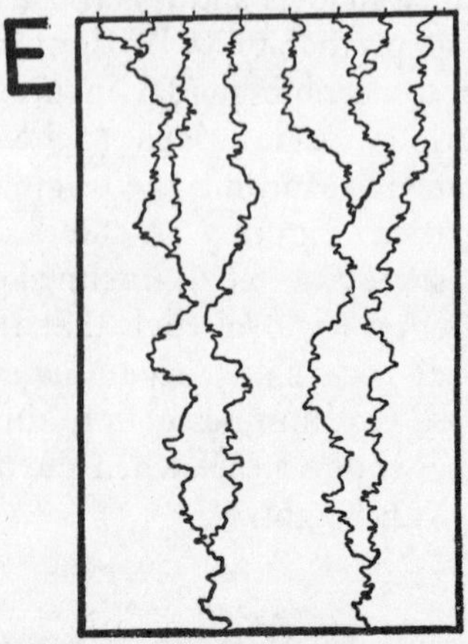


**3.4** Mathematical models in geomorpho

A Parallel weathering of a steep slope with screes collecting at the base (*After Lehmann; From Scheidegger, 1961*).

B Central rectilinear slope recession with weathering rate increasing with elevation (*After Bakker and Le Heux; From Scheidegger, 1961*).

C The progressive degradation of hill slopes by creep under three different rates (1, 2, 3) of undercutting (*c*) (*From Culling, 1963*).

D A mathematical model of a longitudinal stream profile (1), together with the degradation through time of any given point on it (2) (*From Strahler, 1952*).

E A stream-like network generated f points equally spaced at the origin by a random-walk procedure where uphill (backward) steps were exclu (*From Leopold and Langbein, 1962*)

F

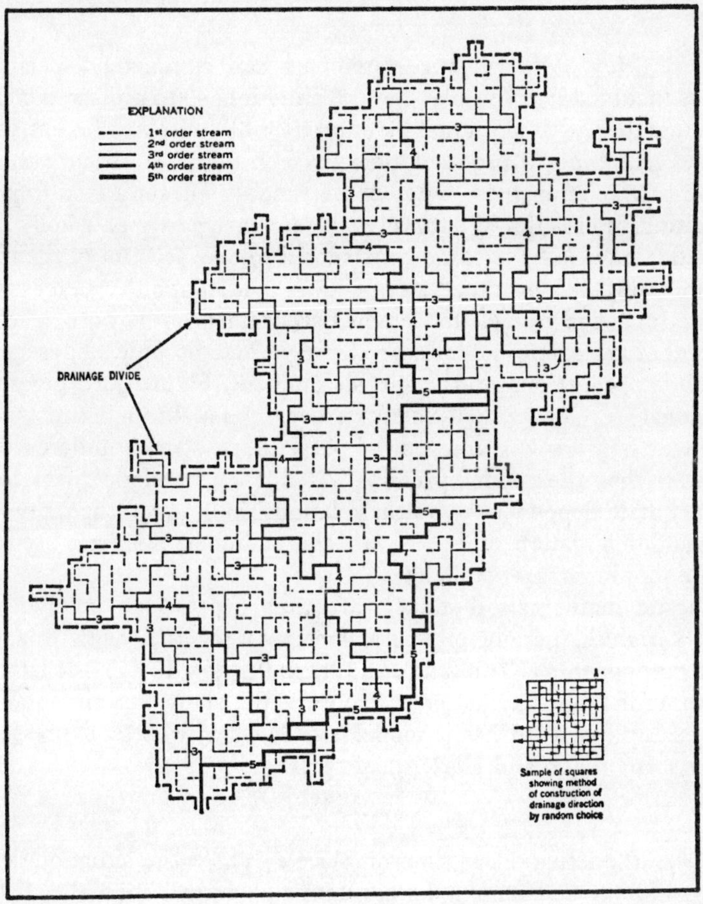


G

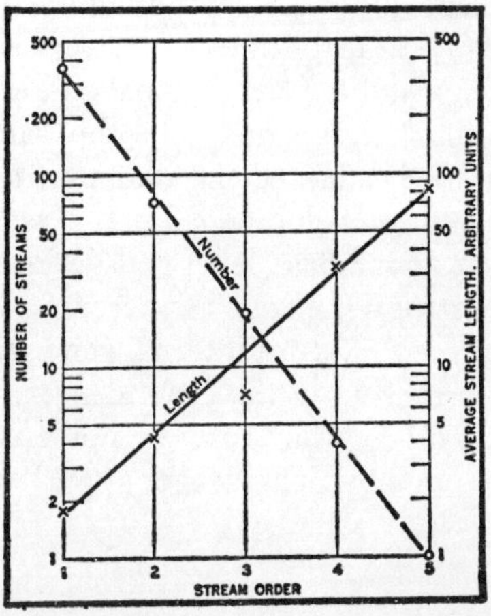


**3.4** (*continued*)

F Development of a 5th-order drainage basin network by a random-walk method from a series of squares each of which has an equal chance of draining in any of the 4 cardinal directions (*From Leopold and Langbein, 1962*).

G Plots of stream number and mean length for the network shown in Figure 3.4F, indicating their adherence to the laws of morphometry for a homogeneous region (*From Leopold and Langbein, 1962*).

However, few deterministic statements can completely specify all the variables included in a complex natural situation, so that discrepancies occur which, together with the random unpredictable effects inherent in natural processes, combine to produce 'noise' which tends to obscure the simpler deterministic relationships. Often these random effects are so important in determining the results of natural processes that partly or wholly statistical (stochastic) models have to be constructed to take account of them. Culling (1963 and 1965) has proposed a slope model based upon certain assumptions that individual soil particle movements are influenced by pore-space characteristics that are partly random, producing a macroscopic soil creep the rate of which is proportional to the surface gradient. Figure 3.4C gives the two-dimensional development of such a creep model through time with three different rates ($c$=0, 0·5 and 1·0) of basal lateral stream undercutting. It is interesting that the model predicts a tendency of slope form towards a stationary state dependent upon the relationship between the rates of undercutting and denudation, and that, once this stationary state has been reached, the slope should retreat parallel.

Stochastic mathematical models are thus expressions involving mathematical variables, parameters, and constants together with one or more random components (Krumbein and Graybill, 1965, p. 17), the latter arising from unpredictable fluctuations in observational or experimental data. A simple example of a stochastic model is a 'Markov process' expressed by the equation (Amorocho and Hart, 1964, p. 318):

$$y_{t+1}=\rho y_t+\eta \bar{E}$$

where: $y_t$=the situation at time or place $t$; $y_{t+1}$=the situation at time or place $t+1$; $\rho$=a constant; $\eta$=a constant dependent upon the distribution of $y$ in time or space; and $\bar{E}$=a random variable. It can be seen that the Markov-type phenomenon ($y_{t+1}$) is dependent partly upon a previous state of the same phenomenon, whereas in the 'Monte Carlo' model (the other main type of stochastic model) it does not depend on its previous states. The Markov-type model has found more favour with geomorphologists, Curl (1959) applying it to limestone cavern development and Melton (1962) used it to distinguish between regional and local random variations in longitudinal river profiles in Arizona. On the same subject, Leopold and Langbein (1962, pp. 4–8) adapted the statistical thermodynamic model to longitudinal stream profile development on the assumptions that absolute temperature and height above base level can be interchanged, and that there is a continuity of entropy (i.e. rate of increase of entropy in the system+rate of outflow of entropy=rate of internal generation of entropy). The first assumption has been criticized in the grounds, among others, that absolute temperature (measured on a ratio scale) is not analogous to elevation (interval scale), being defended by Scheidegger (1964), and has been subsequently elaborated by the same authors (Langbein and Leopold, 1964) to rationalize the typical longitudinal stream profile in terms of the most probable compromise between the dual tendencies towards a minimum total rate of work in the whole fluvial system and towards a uniform distribution of energy expenditure throughout the system. The same authors (Leopold and Langbein, 1962, pp. 14–17) again presented a more obvious Markov-type random-walk model to demonstrate how a combination of deterministic and stochastic terms can produce a bifurcating drainage-like network (Fig. 3.4E) which, when elaborated into a basin system (Fig. 3.4F), produces relationships akin to those found in natural stream systems (Fig. 3.4G). The possibilities of this technique have been investigated by Schenck (1963) with the aid of a computer. One of the few examples of the geomorphic application of the type of Monte Carlo model so extensively used in diffusion models of human geography has been by McConnell (1965) who examined the spacing and arrangement of topographic elements (e.g. terrain summits) in terms of Poisson and binomial distributions.

## Experimental design models

In terms of the contributions made to geomorphology, the most important aspect of the physical systems approach involves the recognition that within a given range of observational data exist certain meaningful component parts which can be identified by employing a suitable experimental design (Fig. 3.1) (Krumbein and Miller, 1953; Krumbein, 1955; Melton, 1960; Krumbein and Graybill, 1965, Chap. 9). This design, which is derived from past observation, logical deduction, intuition, or a combination of these, provides a structure within which other data are collected and then analysed by conventional statistical means (Strahler, 1954; Miller and Kahn, 1962; Krumbein and Graybill, 1965; Chorley, 1966) to produce some generalization which overcomes the inherent variance of data in the earth sciences and gives some overall statement regarding the general data properties. Such statistical generalization commonly involves the fitting to the data of simple, multiple or 3-dimensional regressions (Krumbein and Graybill, 1965, Chaps. 10, 12 and 13, respectively). In a consideration of this approach to geomorphic problems it must be recognized that we are dealing with a very border-line type of model building which, in a rather more complex and sophisticated form, also appears under the category of synthetic systems (Fig. 3.1). The latter represent more ambitious experimental designs which lead to the construction of general process-response models.

On the basis of the experimental design, constructed with reference to some conceptual model of the nature of the problem and suitable operational definitions of its component parts, the numerical data are collected, checked as regards their scalar and number-system characteristics (Krumbein and

Graybill, 1965, Chap. 3), and a 'data matrix' (*organized data structure*, or 'data model') produced. This data structure is then commonly analysed by regression-type techniques to produce a simple variable system, in which workable correlations are identified involving the direction and intensity of assumed causation. With simple experimental designs, however, the design may be so restricted, or the sources of data so constrained, that it is not possible to extend the general conclusions far beyond some comments on the original data, although indications may be given which may lead ultimately to the more general statements characteristic of synthetic-system reasoning. The following examples will suffice to show both the features of experimental design models and this gradational character with synthetic systems.

Strahler (1950) set up a simple experimental design to test the effect of basal stream corrasion on the angles of the associated valley-side slopes in the Verdugo Hills, California, and established a significant control (Fig. 3.5A). Leopold and Maddock (1953) sampled the width, depth and mean velocity of the Powder River, at Locate, Montana, and produced logarithmic relationships between discharge and these three dependent variables (Fig. 3.5C). Geomorphic periodicities have been examined for topographic cross-sections by Piexoto *et al* (1964) by means of harmonic analysis, and for meander patterns by Speight (1964) using spectral analysis. A simple 2-kilometre, rectangular sampling design was employed by Chorley, Stoddart, Haggett and Slaymaker (1966) to examine the variation of surface sand-size facies in the Breckland, Eastern England, and the third-order polynomial surface fitted to the data (Fig. 3.5D) was subsequently broadly substantiated by independent sampling. Such 'trend-surface' treatment of areal data (Krumbein and Graybill, 1965, Chap. 13; Krumbein, 1966) illustrates well the relationships between regression-type analogues and information theory, involving as it does the separation of regional and local 'signals' from 'noise' (Chorley and Haggett, 1965). A more ambitious experimental design model was adopted by Krumbein (1959) in investigating the multiple controls exercised by geometric mean diameter, phi standard deviation, moisture content and porosity over the firmness of beach sand (expressed by 'ball penetration' – Fig. 3.5E). The use of the multiple regression technique enabled reductions in the sums of squares effected by all combinations of the four independent variables (i.e. some of which are given as percentages in Figure 3.5E) to be obtained, indicating the dominance of moisture, with diameter second in individual importance. All four independent variables seemed to explain just over three-quarters of the variation in beach firmness.

With the last example it can be seen readily that we are passing into the extended range of generalization characteristic of synthetic systems, and Melton's (1957) ambitious multiple-correlation investigation of the factors controlling slope steepness and drainage density even more obviously extends

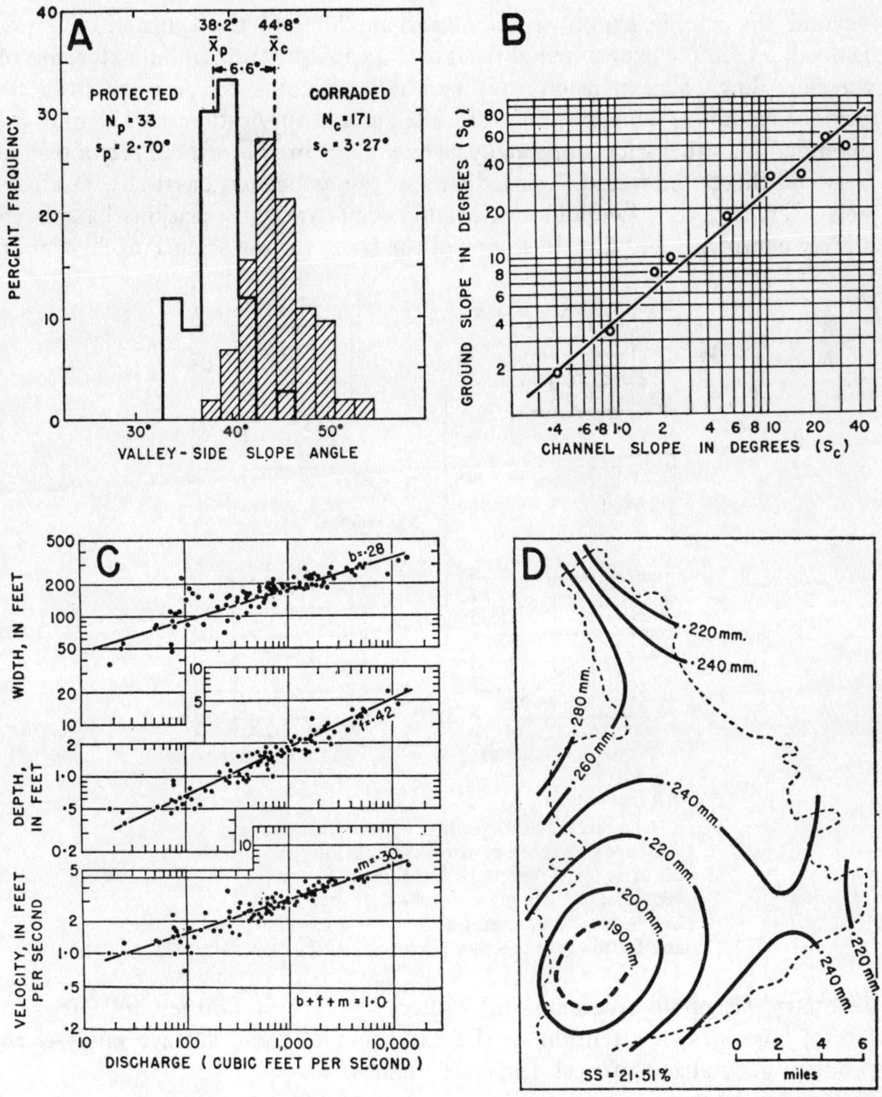


**3.5** Experimental design models in geomorphology.

A Histograms of two samples of maximum valley-side slope angles in the Verdugo Hills, California for protected and basally-corraded slopes (*From Strahler*, *1950*).

B General relationship between basal channel slope and valley-side (ground) slope for nine maturely-dissected regions (*From Strahler*, *1950*).

C Relation of stream width, depth and velocity to discharge for the Powder River at Locate, Montana (*From Leopold and Maddock*, *1953*).

D Best-fit cubic trend surface fitted to surface sand facies in the Breckland, Eastern England (*From Chorley* et al., 1966).

beyond the simple experimental design model into the synthetic system. Indeed, all simple experimental designs are capable, through an extension of the sampling, of expansion into synthetic systems to produce process-response models (Whitten, 1964) of quite general application. Strahler (1950) extended his valley-side slope analysis to suggest broad regional relationships between these slopes and the gradients of the associated basal stream channels (Figure 3.5B); Leopold and Maddock's (1953) observations have been widely confirmed and now form one of the bases for the science of 'hydraulic

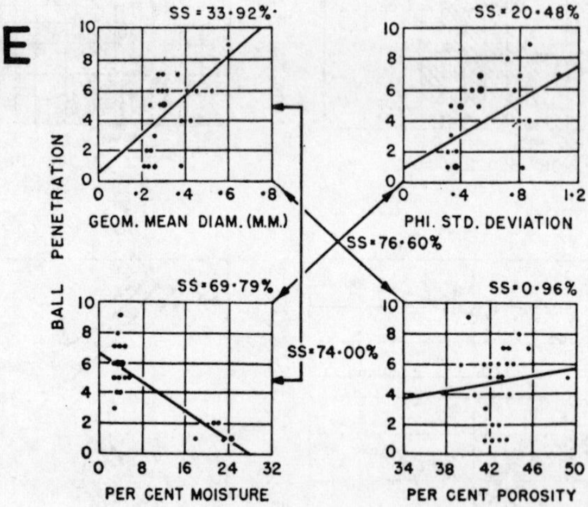


**3.5** (*continued*)

E Relations between geometric mean diameter, phi standard deviation, moisture and porosity, and beach firmness for Wilmette beach, Illinois. Per cent 'explanations' are given for the four individual independent variables, for the most important pair, and for the four together (*From Krumbein*, 1959).

geometry' (Leopold, Wolman and Miller, 1964); and Chorley and Haggett (1965) have drawn attention to the extension of trend-surface analysis to produce generalized process-response models.

## GENERAL SYSTEMS

The general systematic approach to the study of landforms is based upon some broad attitude to groups of geomorphic phenomena which is obtained as the result of experience (perhaps from some other type of analysis – e.g. experimental design models) or intuition. Here the emphasis lies in the organization and operation of the system *as a whole or as linked components*, rather than in detailed study of individual system elements (Von Bertalanffy, 1962). However, detailed knowledge of the internal operations of parts of the system (perhaps gained from experimental design models) is useful in the selection of an appropriate general system model.

A geomorphic 'system' is an integrated complex of landforms which operate together according to some discernible pattern (e.g. a drainage basin); energy and matter input into the system giving rise to a predictable system response in terms of internal organization and the resulting energy

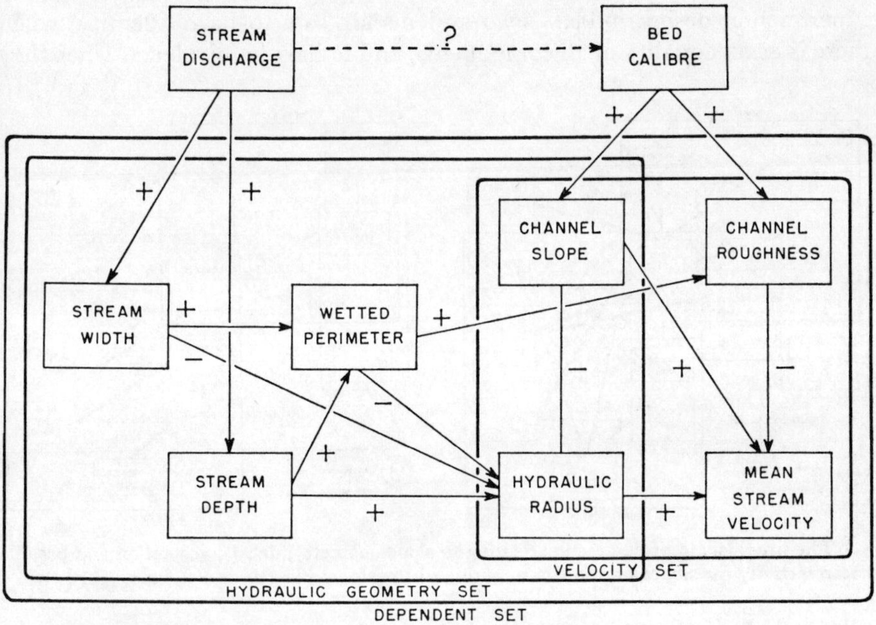


**3.6** General relationships between the major factors involved in hydraulic geometry. Arrows suggest independence and dependence; the link without an arrow yoked interaction of two variables; and the mathematical signs whether variations are direct (+) or inverse (−).

and matter output. The instantaneous condition of the system is termed its 'state', which is characterized by its composition, organization, and flow of energy and mass (Howard, 1965), and defined by system parameters. State may be steady or variable through time, and the complexity of the system is expressed by the number of dimensions in the system 'phase space' (Melton, 1958A). The systems approach emphasizes an overall operational process, the operations performed by a system being dependent on the system state, which may be externally or historically determined. A description of such a system thus involves the specification of: (1) the nature of inputs; (2) the nature of outputs; (3) the system phase space; and (4) the model relating inputs, outputs and system states in time.

Geomorphic systems can all be considered part of 'supersystems' (e.g. whole landform assemblages) and as being composed of 'subsystems' (e.g. slope or channel segments). Subsystems are thus the basic components of a system, and can be identified as distinct input-output linkages (Amorocho, 1965). In geomorphology subsystems are commonly combined by 'cascading' the output of one subsystem into another to form its input (i.e. the output of slope segments forms part of the input of the basal stream segments). When interest in the subsystem operations is very detailed, the systems approach is replaced (at least in the initial stages of the investigation) by the experimental design model. Two systems are said to have 'identity' when there is exact equality of all components, and to have 'equivalence' when they

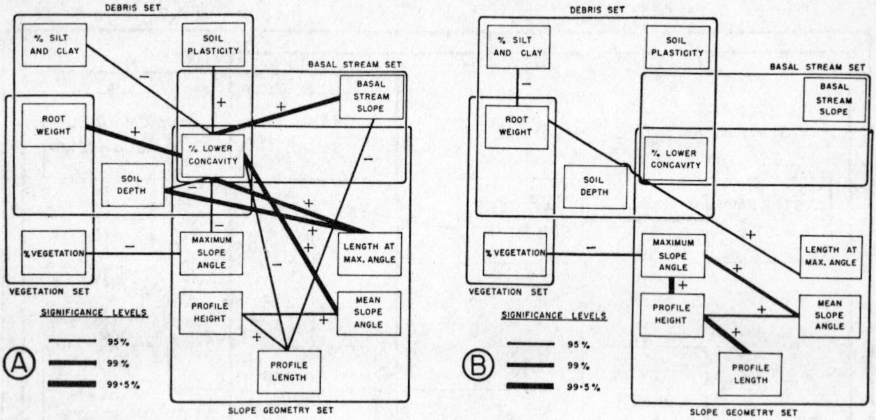


**3.7** The interlocking of parameters relating to slope geometry, debris, vegetation and basal stream activity for slopes on the Charmouthien Limestone, Plateau de Bassigny, Northern France.

A Where the basal stream is moving towards the slope base.
B Where the stream is moving away from it. (*After Kennedy*, 1965).

transform the same inputs into identical outputs. The latter is much more common in geomorphology (i.e. two basins being of similar order in a homogeneous region). The internal organization of systems can be profitably viewed in terms of 'feedback' with reference to changes in the controlling external variables (i.e. Leopold and Maddock (1953) consider discharge and sediment characteristics to be the most important external variables in stream system operations) (Fig. 3.6). 'Positive feedback' occurs whenever externally-induced changes of input produce changes in the same direction as the input change (i.e. lead to a progressively-changing, 'timebound' state). 'Negative feedback' operates when changes in the system input result in changes in other system components which regulate the effect of the changed input such as to bring about a new 'timeless' equilibrium or 'steady state'. A steady state usually manifests itself in terms of high interlocking correlations between observed subsystems (e.g. parameters) (Fig. 3.7). Langbein and Leopold (1964) have defined it in terms of balanced tendencies towards an equal areal distribution of energy expenditure and towards minimum total work. Self-regulation is thus the diagnostic feature of open systems with negative feedback. However, this regulation to changes in external variables is often complicated by: (1) 'secondary responses' which may eventually result from primary changes (e.g. a change in precipitation may change discharge which may immediately alter the hydraulic geometry; subsequently vegetational changes may further alter discharge and channel characteristics); (2) 'thresholds', the passage of the system through which involves drastic

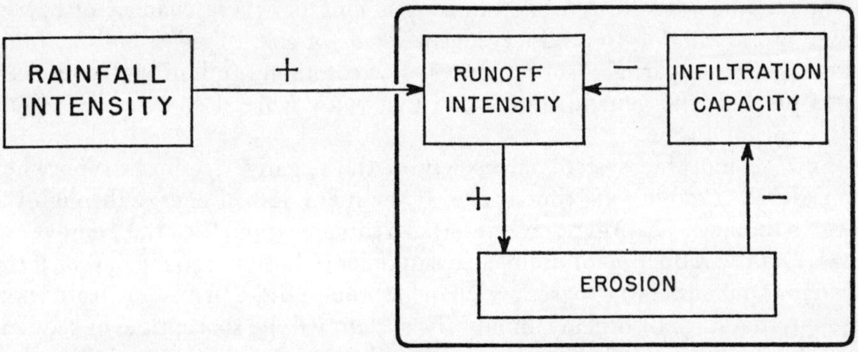


**3.8** Correlation structure of a system involving rainfall intensity, runoff intensity, erosion and infiltration capacity exhibiting dominantly positive-feedback characteristics, at least over limited timespans; in this instance as long as erosional decreases of infiltration capacity continue (*From Melton*, 1958*B*).

changes in the system state (Howard, 1965). It is profitable to view certain geomorphic systems as exhibiting dominantly positive feedback characteristics for limited periods, in particular certain drainage basin subsystems (Melton, 1958B, p. 454) (Fig. 3.8). However, most geomorphic systems operate dominantly as negative-feedback open systems of the self-regulatory type. This self-regulation involves internal reorganization of the system and is accomplished in a time period referred to as 'relaxation time' (Langbein and Leopold, 1964, p. 782; Howard, 1965). Relaxation time depends on: (1) the resistance to input changes of the individual system components; (2) the complexity of the system (i.e. the phase space or number and linkages of variables involved); and (3) the magnitude and direction of the change of input. The relaxation time of some geomorphic systems is short (i.e. the channel adjustments to changing discharge which are the subject of hydraulic geometry), but that of others is long (e.g. changes of erosion induced on granite terrains by climatic change). Relaxation time is therefore some measure of the resistance of the system to external changes. The concept is complicated in that different system parameters have different relaxation times, and that some landform elements carry a longer-term historical record than others. Just because some landscape features (e.g. those related to hydraulic geometry) are in equilibrium, this does not necessarily imply that all other features are as well (Howard, 1965, pp. 309–310). Under such a systems view, however, it is possible to associate apparently very different geomorphic phenomena by examining their tendencies towards the achievement of the steady state. Where the relaxation time is short the timeless character of the system is dominant, where it is long the timebound or historical features preoccupy one (Chorley, 1966). Many of the methodological confusions of Pleistocene and Recent geomorphology are involved with systems where the relaxation time is long or where changes of energy input are so rapid as to produce the coalescing of energy pulses and the blurring together of partial form responses. The common result of such confusion is to produce the general impression that some general changes are taking place through time.

Few geomorphic systems are solely of the negative feedback type when considered through long time spans. If the mere flow of energy through the system induces its progressive internal reorganization (i.e. the removal of mass and the reduction of drainage basin relief (Chorley, 1962, p. 3)) or if the system input suffers a progressive change through time, then some features of the internal form or organization of the system will be susceptible of sequential change through time and an historical parameter introduced (Chorley, 1962; Schumm and Lichty, 1965). Where these changes are especially important they may mask the self-regulatory characteristics of the system and it may be more convenient to view it in some other terms – i.e. as an historically-oriented sequence of operations.

It is convenient to divide general systems into *synthetic systems*, *partial systems* and '*black boxes*'.

## Synthetic systems

As pointed out earlier, synthetic systems in their initial phases bear a striking identity with experimental design models, being concerned with recognizing supposedly crucial features of the structure of geomorphic phenomena, together with their sampling and analysis. The aim of synthetic systems research, however, goes beyond this in that its goals lie in the synthesis of the analysed structure and the extension and generalization of conclusions to the point of producing process-response models.

The procedure for building up synthetic systems begins with the identification of certain key structural elements in a given geomorphic complex, together perhaps with some views as to the possible relationship between these elements (e.g. direction of causation ('impulse'); 'yoked' variables which change together; feedback from effects to assumed causes; the strength (or statistical confidence) of the bonds between elements; etc.) (Fig. 3.9). The succeeding stages are well exemplified by the work of Melton (1957 and 1958B), who used non-parametric methods (Siegel, 1956; for the parametric methods see Krumbein and Graybill, 1965, Chap. 15) to construct a 'variable system' – the mapping of highly-correlated variables into plane sets, in which the directions of causality are stated and with one or more of the

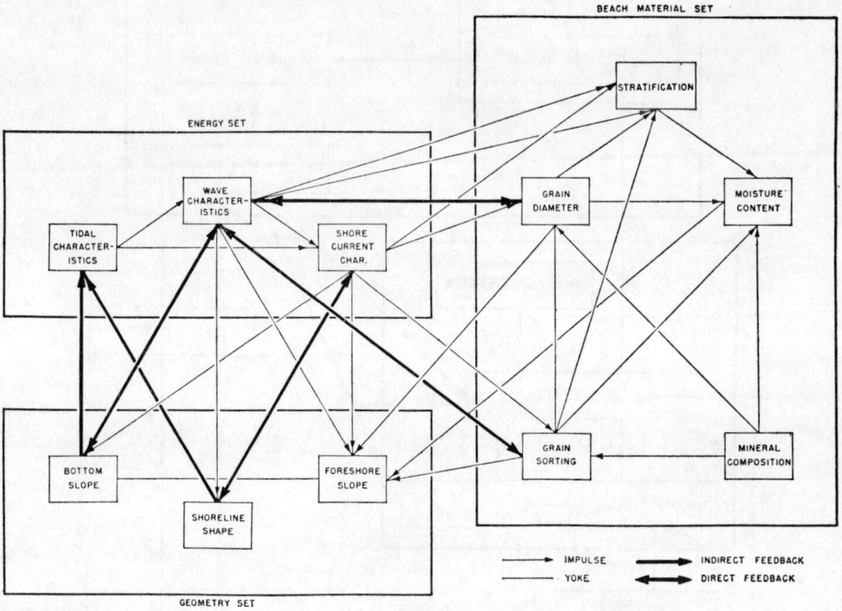


**3.9** Tentative relationships between parameters relating to wave, tide, or current characteristics, beach material and shore geometry. Feedbacks are suggested wherever parameters of beach material or shore geometry might influence the wave, tide or current energy. (*Suggested by Krumbein*, Ann. Bull. Beach Erosion Board, *17*, *1964*).

variables correlated with variables outside the sets, so that the systems are 'open' in that well-defined relationships exist between their members and the surrounding environments (Melton, 1958B, pp. 443–446) (Fig. 3.10).

Blalock (1964) has outlined certain principles for examining these correlation structures of variable systems, in which regression equations are employed as causal models. As Melton (1958B) has pointed out, the building up of correlation links between variables by techniques such as cluster analysis (Krumbein and Graybill, 1965, pp. 406–408) can give important indications as to the relationships between geomorphic variables without any preconceptions regarding the assignment of cause and effect. A difficulty, however, is that one commonly wishes to evaluate the composite effects of groups of variables, and factor analysis (Imbrie, 1963; Wong, 1963; King, 1966B) provides a means of collapsing the data matrix into a small number of

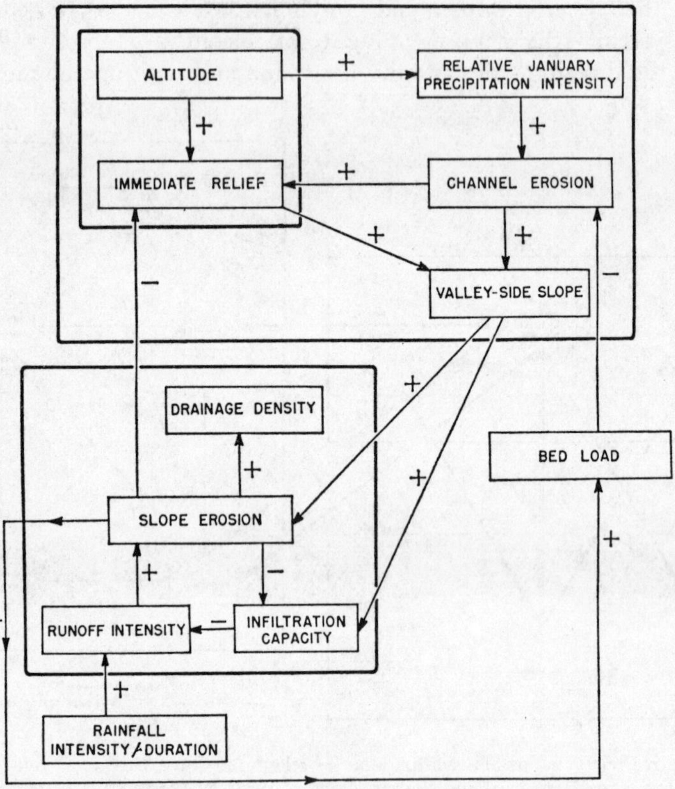


**3.10** Correlation structure of a system involving morphometric, climatic and surficial variables, emphasizing the negative-feedback loop: valley-side slope, slope erosion, immediate relief (*From Melton, 1958B*).

idealized variables which account for most of the observed variability of the data. Another means of circumventing the difficulty of handling a large number of combined effects of variables is to maintain their original identity but to eliminate all except the most predictive ones, so that their combined effects can be examined in detail. Sequential multiple linear regression, 'WHIRLPOOL' (Krumbein *et al*, 1964), is such a data-search procedure for sifting out low-correlation elements, retaining a minimum number of variables to achieve some required level of prediction, and to examine completely their interrelationships (Harrison and Krumbein, 1964).

In their most sophisticated forms such correlation structures provide process-response models (Whitten, 1964; Krumbein and Graybill, 1965, pp. 19–27; Harrison and Krumbein, 1964), which are highly-generalized frameworks of impulses and responses to which further observations can

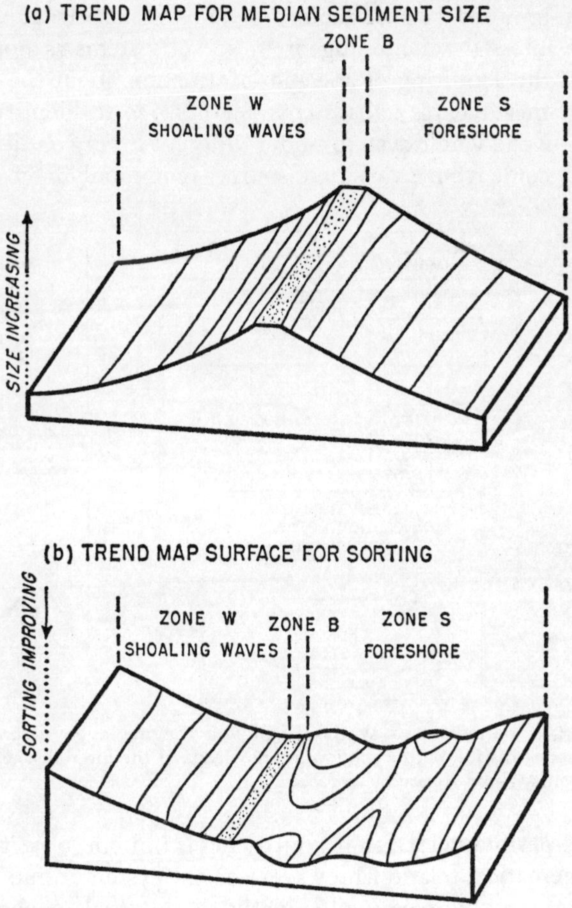


**3.11** Trend maps for nearshore breaker zone and foreshore distribution of sediment size (a) and sorting (b) characteristics representing simple process-response models (*From Miller and Zeigler*, *1958*).

be referred as an aid in identification of phenomena and as a basis for prediction. Trend surfaces, considered as 'response surfaces', are proving especially valuable in suggesting process-response models (Chorley and Haggett, 1965) and Miller and Zeigler (1958) have suggested a near-shore model for sediment characteristics (Fig. 3.11).

## Partial systems

The partial systems approach is one which has been primarily associated with the solving of essentially practical problems in the earth sciences. It has been concerned with the establishment of *workable relationships* between often coarsely-grouped sets of factors or subsystems. Detailed knowledge regarding the internal functioning of these subsystems is not regarded as necessary, but the acquiring of specific information about the interrelationships between these coarse subsystems enables one to identify and predict the behaviour of the whole system under different input conditions. This information may come from a variety of sources (commonly from the Synthetic

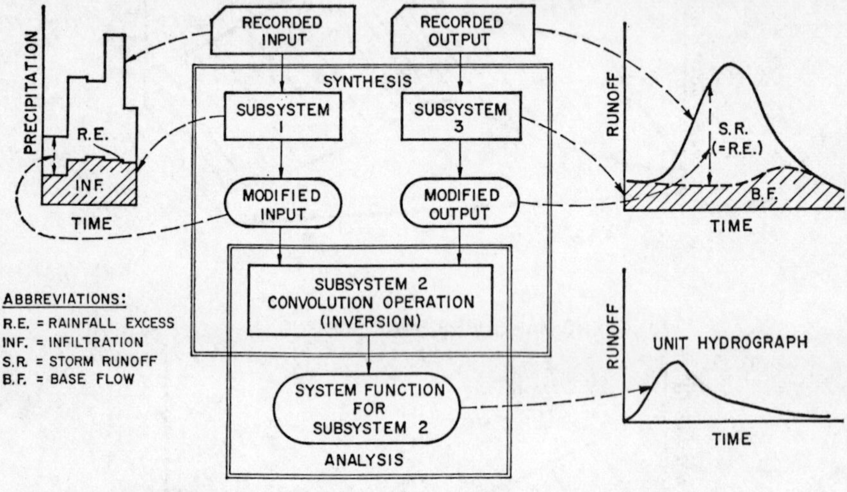


**3.12** Flow chart for partial system synthesis involving the prediction of runoff from precipitation characteristics, with a minimum knowledge of the internal operations of the system components (*From Amorocho and Hart, 1964*).

Systems or Experimental Design approaches), but the object is always to establish mathematical relationships between the system inputs and outputs, without any attempt to describe explicitly the internal mechanisms of the system. The resulting behaviour patterns are then used to predict, and to a lesser extent to understand, the behaviour patterns in other similar systems or in the same system at different times (Amorocho and Hart, 1964).

The best example of the partial systems approach is given by Amorocho and Hart (1964) (Fig. 3.12) to predict runoff patterns for different storms over a stream basin. All that is known about the complex operation of the basin system hydrological cycle is; (1) the apportionment of the precipitation between infiltration and rainfall excess, and (2) the apportionment of the resulting hydrograph between base flow and storm runoff. On the basis of past input/output relationships (i.e. storm precipitation/runoff relationships), modified inputs (rainfall excess) are related to modified outputs (storm runoff) by purely mathematical processes to produce an adjusted input-output system which, from a given pattern of storm rainfall, will predict a pattern of storm runoff from which a unit hydrograph can be obtained.

Needless to say, such a fundamentally practical partial systems approach to groups of phenomena does not find many counterparts in geomorphology. A reasonably close equivalent might be provided, however, by attempts to establish workable relationships between precipitation and drainage density based upon certain gross assumptions regarding the factors controlling drainage density and, in particular, regarding the characteristics of surface runoff (Chorley, 1957; Strahler, 1958; Chorley and Morgan, 1962; Carlston, 1963).

## The black box

The most extreme application of general systems models involves the concept of the 'black box'. This requires little or no detailed information regarding the system components or subsystems within the black box system, interest being focused upon the nature of the outputs which result from differing inputs. Thus the black box is analogous to the 'grey box' of the partial system and the 'white box' of the synthetic system.

In geomorphology the black box approach has nourished much of the important work of the last one hundred years. Such work has been characterized by broad intuitive leaps wherein decisions are taken (sometimes explicitly, but more usually implicitly) regarding the supposed dominant characteristics of broad landform assemblages. Although there has been much guesswork and speculation regarding the detailed nature of the processes at work, the characteristic patterns of landform transformations through time, and the rates of operation of geomorphic processes, we are only now making a beginning in answering these and other similar questions. Indeed, for most of the general systematic models of geomorphology the lack of such information has not been a disadvantage, its very deficiency allowing the theoretical model builders much greater scope than they would have otherwise have had. So we find that, on closer examination, such apparently disparate approaches to geomorphology as Gilbert's 'dynamic equilibrium' model (Gilbert, 1877; Chorley, 1962), climatic geomorphology, the Davis cycle, and the Penck geomorphic system are all based upon gross intuitive assumptions regarding the significant behavioural patterns of landform assemblages, all falsely give the impression that they are based upon detailed knowledge of geomorphological processes, and are all simply concerned with the landform outputs which are supposed to result from combinations of process and tectonic inputs. Not only does attention to general systems theory enable these similarities and limitations to emerge, but also makes possible the recognition that these mainstream approaches to the study of landforms can be divided into two broad groups – those wherein negative and positive feedback are dominant, respectively (Fig. 3.1).

G. K. Gilbert (1877; Chorley, Dunn and Beckinsale, 1964, pp. 546–572) has provided the most important negative-feedback black-box model for landforms for, despite his later preoccupation with fluvial processes (Gilbert, 1914), he was broadly concerned with the gross adjustments of form to process and of form with form. More modern work (especially Horton, 1945 and Strahler, 1950) has revived interest in this 'dynamic equilibrium' attitude to landforms, with the concept of 'grade' providing some clues as to the feedback mechanism. However, the work of Gilbert is mainly important in that it provided a general attitude for the study of landforms as open systems in a steady state (Chorley, 1962), through his laws of declivities, structure and divides, rather than offering a complete geomorphic framework. By stressing the timeless, equilibrium aspects of landforms Gilbert's ideas provided an important antidote to those of Davis and have also stimulated much of the modern work in dynamic geomorphology. More recently the development of climatic geomorphology is expressive of the same type of attitude, although the historical or even cyclic overtones of much of this work often tend to obscure the central input (process)/output (form) relationships. This attitude, foreshadowed by the recognition of the arid cycle by Davis (1905), has found particular fruition among Continental and associated workers (Birot, 1960; Büdel, 1963; Butzer, 1964; Tricart, 1957). Peltier (1950) and Holzner and Weaver (1965) have given statements of the underlying principle that each climatic regime possesses essentially characteristic geomorphic process assemblages which find unique expression in regional topographic distinctions which transcend structural regions. Thus gross climatic process inputs are viewed as resulting in gross geomorphic form outputs, usually irrespective of any considerations of the detailed investigation of the nature of the processes involved. Of course, considerations of climatic changes have caused many present climatic landform assemblages to be viewed as currently adjusting to recently-changed climatic inputs (Büdel, 1959) in which the assumed long relaxation times mean that 'almost everywhere, in particular in the extra-tropical zone, fossil landforms occupy a far greater area than the landforms developed by currently active processes' (Büdel, 1963). It is therefore partly through cyclical considerations and partly through those of climatic change that the timebound aspects of climatic geomorphology appear.

The other type of black-box system approach to geomorphic phenomena has proved extremely popular and, indeed, contains most of what has come to be regarded as mainstream geomorphology. This approach, too, is significantly vague as to the precise details of form and process relationships within landform assemblages, and is based upon a general attitude that the important and worthwhile aspects of geomorphology involve considerations of the geometrical transformations of landforms through time (i.e. the output) in response to changes of input (usually baselevel changes). These time-bound black-box models stress progressive changes rather than steady-state adjustments and, for this reason, have been termed positive-feedback systems (Fig. 3.1). With these dominantly historical systems we arrive at the 'east' of our *map of geomorphic activity* and find that it has acquired the features of a *globe*, with the two ends joining to form a zone of common historical endeavour, the segment of the real world forming one pole, the model conclusions about the real world the other pole, and the basic reappraisal of the model results being the central axis of the globe. The features of the Davis geographical cycle (Davis, 1909) conforming to this systems approach have been treated previously in some detail (Chorley, 1962 and 1965), with particular reference to the progressive, sequential and irreversible landform transformations attendant upon the input of one large energy pulse (i.e. uplift). Indeed, the cycle seems to operate in some respects like a closed system tending towards maximum entropy. Studies of denudation chronology (e.g. Baulig, 1928; Johnson, 1931; Wooldridge and Linton, 1955) have been dominantly concerned with interpreting geomorphic outputs (peneplain remnants, breaks of slope, etc.) in terms of a discontinuous sequence of supposed input pulses (changes of baselevel, either of diastrophic or eustatic character). The Walther Penck geomorphic system (Penck, 1924 and 1953), superficially so different from the other two, in reality simply carries this positive-feedback black-box model a step further by trying to explain the landform assemblage outputs in terms of assumed continuous inputs of varying patterns of tectonic baselevel changes. In none of the above geomorphic models was there any real need for detailed internal examination of the systems (e.g. of processes) once the initial decision had been taken regarding the supposed dominant system characteristics.

Naturally the more extreme advocates of both the negative- and positive-feedback black-box approaches have encountered criticism (see, for example Bretz's (1962) comments on Hack (1960), and Chorley's (1962 and 1965) opinions regarding the Davisian cycle), but one of the most heartening features of geomorphic work during the past three years or so has been the attempt on the part of several workers to reconcile the two approaches. Thus, Chorley (1966), Holmes (1964) and Howard (1965), among others, have been concerned with the possibility of the construction of a new model of geomophology combining the important elements of both the timeless and time-bound models, such that considerations of steady-state adjustments can find a place within a time-directed framework. This attempt has been termed, perhaps not too appropriately, the 'graded system' in Figure 3.1. It has been left to Schumm and Lichty (1965), however, to make what is perhaps the most significant contribution in this connexion. They believe that distinctions between cause and effect in the moulding of landforms depend upon the span of time under consideration. During a long period of time ('cyclic time' – Fig. 3.13A) the most important feature of the landform assemblage (or many individual morphometric features – in this case stream channel gradient) is the continual change of form attendant upon the decrease of potential energy through erosion, such that there may be no time-constant

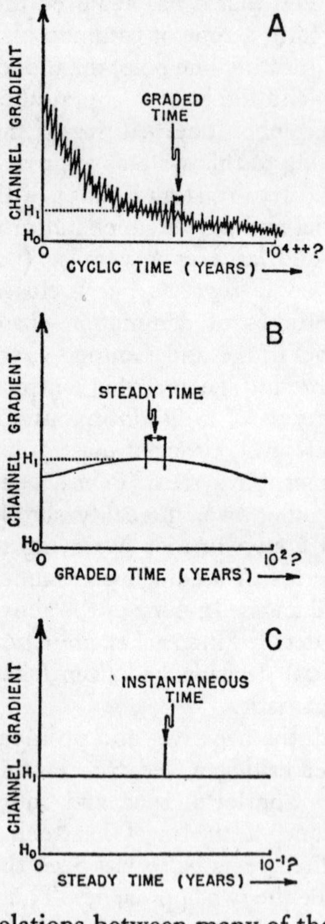


**3.13** Diagrams illustrating changes of stream channel gradient at a point through (A) cyclic, (B) graded, and (C) steady time (*Adapted from Schumm and Lichty, 1965*).

relations between many of the dependent and independent variables. Within this progressive change, however, shorter time spans ('graded time' – Fig. 3.13B) can be identified wherein a dynamic equilibrium exists between form and process, largely time-independent and dominated by negative-feedback mechanisms. This graded condition does not apply to all components of a landform system at any one time, but it might be expected that the proportion of graded features would increase with time, and that the periods of temporary graded condition for any one feature would become more frequent with the passage of time. This implies that many correlations between process and response can be fruitfully investigated at this level. During even shorter time spans ('steady time' – Fig. 3.13C) true steady states often exist with very highly-correlated relationships between form and process. Here we are well within the realm of interest of the engineer and the hydraulic geometry relationships of a stream reach having a steady and equal import and export of water and sediment provide an example of this approach. Although this fragmentation of the time-continuum does not provide a completely-integrated model, it does advance us towards a fusion of the timebound and timeless approaches to geomorphology.

## TESTING THE MODEL

The testing of the patterns, relationships and process/responses predicted by geomorphic models involves reappraisal against the real world conditions. Although such testing is undoubtedly the most important single step in successful model building, certain types of models have proved more susceptible to reappraisal than others.

The most readily testable models are found among the hardware, partial system and experimental design models. Engineering-type hardware models are constantly checked against reality during construction to allow for adequate representation of present conditions; they are often 'moulded' to known past conditions and tested as to their reproduction of known historical sequences of events; and, finally, predictions from the model usually form the basis for engineering works which themselves form costly tests for the appropriateness of the model (Price and Kendrick, 1963). Similarly, the building of partial system models involves the constant checking of real-world inputs and outputs so that the final model can usually reproduce a limited aspect of reality with reasonable faithfulness. Experimental design models can also be readily tested by collecting new data which are statistically checked against the relationships derived from the original model. The dominant feature of the simple trend-surface model of Breckland sand facies was supported by the results of subsequent sampling design of quite different character (Fig. 3.5D) (Chorley, Stoddart, Haggett and Slaymaker, 1966).

Rather more difficult to test are the process-response models resulting from the synthetic system approach, and the mathematical models. This is because of the higher level of generality presumed by these models and the difficulty of disentangling local complications (i.e. 'noise') from genuine errors in the model. An interesting example of a test of a process-response model has been given by Miller and Zeigler (1964) who tested their small-scale beach sediment model (originally developed for an area of 75 by 26 feet) (Fig. 3.11) over a larger (1,800 by 700 feet) and more complex area. They concluded that these complexities (e.g. the oblique wave attack) merely superimposed other areal sediment features which obscured the simple predictions of the original model. With regard to deterministic mathematical models few concerted attempts have been made to check their implications in the field, but recent measurements of slope transformations in the field (e.g. Schumm, 1964) are providing the kind of information to make this possible. Stochastic models (e.g. Leopold and Langbein, 1962) (Figs. 3.4F and 3.4G), however, seem to have been developed with a view to direct testing.

The most difficult types of model to test are some of the natural analogue system and the black-box general system models because their construction in the past has involved such great leaps into generality based upon decisions regarding the dominant system characteristics the origins of which may be obscure. Thus, for example, certain denudation chronology models and the cycle of erosion involve so many built-in assumptions that any testing to which they have been subjected usually develops into circular reasoning (Chorley, 1965). However, an example of a simpler type of black-box model which received striking field support after a hundred years was Darwin's theory of the development of atolls from fringing and barrier reefs by the slow subsidence of reef foundations. This model was deductively supported by Davis and, after several abortive attempts to drill deep ocean atolls, cores at Eniwetok Atoll in 1951 showed that basalt underlies shallow reef limestones dating from the Eocene at a depth of over 4,000 feet (Stoddart, Personal communication).

The reappraisal of geomorphic models is seldom entirely unequivocal, but testing, discarding and remodelling must become the centre of geomorphic interest if the subject is to develop from a subjective catalogue of phenomena into a coherent and rational discipline.

## REFERENCES

AMBROSE, J. W., [1964], Exhumed paleoplains of the pre-Cambrian shield of North America; *American Journal of Science*, 262, 817–857.

AMOROCHO, J., [1965], *Glossary on parametric hydrology* (Tentative); (Mimeo.).

AMOROCHO, J. and HART, W. E., [1964], A critique of current methods in hydrologic systems investigation; *Transactions of the American Geophysical Union*, 45, 307–321.

BAGNOLD, R. A., [1941], *The Physics of Blown Sand and Desert Dunes*, (London), 265 pp.

BAULIG, H., [1928], *Le Plateau Central de la France et sa bordure Mediterranéenne*, (Paris), 590 pp.

BECKETT, P. H. T. and WEBSTER, R., [1962], The storage and collection of information on terrain (An interim report); *Military Engineering Experimental Establishment, Christchurch, Hampshire*, 39 pp., (Mimeo.).

BIROT, P., [1960], Le cycle d'érosion sous les differents climats; *Curso de Altos Estudos Geograficos I, Centro des Pesquisas de Geografico do Brazil*, 137 pp.

BLALOCK, H. M., [1964], *Causal Inferences in Nonexperimental Research*, (Univ. of North Carolina), 200 pp.

BRETZ, J. H., [1962], Dynamic equilibrium and the Ozark land forms; *American Journal of Science*, 260, 427–438.

BRUSH, L. M., and WOLMAN, M. G., [1960], Knickpoint behavior in noncohesive material: A laboratory study; *Bulletin of the Geological Society of America*, 71, 59–74.

BÜDEL, J., [1959], The periglacial-morphologic effects of the Pleistocene climate over the entire world; *International Geology Review*, 1, 1–16.

BÜDEL, J., [1963], Klima-genetische geomorphologie; *Geographische Rundschau*, 15, 269–285.

BUTZER, K. W., [1964], *Environment and Archaeology*, (London), 524 pp.

CARLSTON, C. A., [1963], Drainage density and streamflow; *U.S. Geological Survey, Professional Paper* 422-C, 8 pp.

CHORLEY, R. J., [1957], Climate and morphometry; *Journal of Geology*, 65, 628–638.

CHORLEY, R. J., [1962], Geomorphology and general systems theory; *U.S. Geological Survey, Professional Paper* 500-B, 10 pp.

CHORLEY, R. J., [1964], Geography and analogue theory; *Annals of the Association of American Geographers*, 54, 127–137.

CHORLEY, R. J., [1965], A re-evaluation of the geomorphic system of W. M. Davis; Ch. 2 in Chorley, R. J. and Haggett, P. (Eds.), *Frontiers in Geographical Teaching*, (London).

CHORLEY, R. J., [1966], The application of statistical methods to geomorphology; In Dury, G. H. (Ed.), *Essays in Geomorphology*, (London).

CHORLEY, R. J., DUNN, A. J. and BECKINSALE, R. P., [1964], *The History of the Study of Landforms*, Vol. I, (London), 678 pp.

CHORLEY, R. J. and HAGGETT, P., [1965], Trend-surface mapping in geographical research; *Transactions of the Institute of British Geographers*, 37, 47–67.

CHORLEY, R. J. and MORGAN, M. A., [1962], Comparison of morphometric features, Unaka Mountains, Tennessee and North Carolina, and Dartmoor, England; *Bulletin of the Geological Society of America*, 73, 17–34.

CHORLEY, R. J., STODDART, D. R., HAGGETT, P. and SLAYMAKER, H. O., (1966), Regional and local components in the areal distribution of surface sand facies in, the Breckland, Eastern England; *Journal of Sedimentary Petrology*, 36, 209–220.

CULLING, W. E. H., [1963], Soil creep and the development of hillside slopes; *Journal of Geology*, 71, 127–161.

CULLING, W. E. H., [1965], Theory of erosion on soil-covered slopes; *Journal of Geology*, 73, 230–254.

CURL, R. L., [1959], Stochastic models of cavern development; *Bulletin of the Geological Society of America*, 70, 1802, (Abstract).

DANIEL, V., [1955], The uses and abuses of analogy; *Operations Research Quarterly*, 6, 32–46.

DAVIS, W. M., [1905], The geographical cycle in an arid climate; *Journal of Geology*, 13, 381–407.

DAVIS, W. M., [1909], *Geographical Essays*, (Boston), 777 pp.

DUNCAN, W. J., [1953], *Physical Similarity and Dimensional Analysis*, (London), 156 pp.

DYLIK, J., [1957], Dynamical geomorphology, its nature and methods; *Bulletin de la Société des Sciences et des Lettres de Lódz*, Classe III, Vol. VIII, 12, 1–42.

ELLERMEIER, R. D. and SIMONETT, D. S., [1965], Imaging radars on spacecraft as a tool for studying the earth; *Center for Research Inc., Engineering Science Division, University of Kansas, Lawrence, Kansas, CRES Rept.* No. 61–6, 25 pp.

FENNEMAN, N. M., [1914], Physiographic boundaries within the United States; *Annals of the Association of American Geographers*, 4, 84–134.

FRIEDKIN, J. F., [1945], A laboratory study of the meandering of alluvial rivers; *U.S. Waterways Experiment Station, Vicksburg, Mississippi*, 40 pp.

GILBERT, G. K., [1877], Land sculpture; Ch. 5 in *The Geology of the Henry Mountains*; U.S. Department of the Interior, Washington.

GILBERT, G. K., [1914], The transportation of debris by running water; *U.S. Geological Survey, Professional Paper* 86, 263 pp.

HACK, J. T., [1960], Interpretation of erosional topography in humid temperate regions; *American Journal of Science*, 258A, 80–97.

HARRISON, W. and KRUMBEIN, W. C., [1964], Interactions of the beach-ocean-atmosphere system at Virginia Beach, Va.; *U.S. Army Coastal Engineer Research Center, Technical Memo. No.* 7.

HEACOCK, R. L., KUIPER, G. P., SHOEMAKER, E. M., UREY, H. C. and WHITAKER, E. A. (Eds.), [1966], Ranger VIII and IX: Part II. Experimenters' Analyses and Interpretations; *National Aeronautics and Space Administration, Technical Rept. No.* 32–800, (Jet Propulsion Lab., Pasadena, Calif.), 382 pp.

HILLS, E. S., [1961], Morphotectonics and the geomorphological sciences, with special reference to Australia; *Quarterly Journal of the Geological Society of London*, 117, 77–89.

HILLS, E. S., [1963], Geomorphology and structure morphotectonics; Ch. 14 in *Elements of Structural Geology*, (London), 483 pp.

HOLMES, C. D., [1964], Equilibrium in humid-climate physiographic processes; *American Journal of Science*, 262, 436–445.

HOLZNER, L. and WEAVER, G. D., [1965], Geographic Evaluation of climatic and climato-genetic geomorphology; *Annals of the Association of American Geographers*, 55, 592–602.

HORTON, R. E., [1945], Erosional development of streams and their drainage basins; Hydrophysical approach to quantitative morphology; *Bulletin of the Geological Society of America*, 56, 275–370.

HOWARD, A. D., [1965], Geomorphological systems – equilibrium and dynamics; *American Journal of Science*, 263, 302–312.

HUBBERT, M. K., (1937), Theory of scale models as applied to the study of geological structures; *Bulletin of the Geological Society of America*, 48, 1459–1520.

HUTTON, J., [1795], *Theory of the Earth*, 2 Vols., (Edinburgh).

IMBRIE, J., [1963], Factor and vector analysis programs for analysing geologic data; *Office of Naval Research Project* 389–135, *Technical Report* 6, *Department of Geology, Columbia University, New York.*

JEFFREYS, H., [1918], Problems of denudation; *Philosophical Magazine*, 6th Series, 36, 179–190.

JOHNSON, D. W., [1919], *Shore Processes and Shoreline Development*, (New York), 584 pp.

JOHNSON, D. W., [1931], *Stream Sculpture on the Atlantic Slope*, (New York), 142 pp.

KENNEDY, B. A., [1965], *An Analysis of the Factors Influencing Slope Development on the Charmouthien Limestone of the Plateau de Bassigny, Haute-Marne, France*, B.A. Thesis, Department of Geography, Cambridge University.

KING, C. A. M., [1966A], *Techniques in Geomorphology*, (London), 342 pp.

KING, C. A. M., [1966B], An introduction to factor analysis with a geomorphological example from Northern England; *Bulletin of Quantitative Data for Geographers No.* 6, *Department of Geography, Nottingham University.*

KITTS, D. B., [1963], Historical explanation in geology; *Journal of Geology*, 71, 297–313.

KRUMBEIN, W. C., [1955], Experimental design in the earth sciences; *Transactions of the American Geophysical Union*, 36, 1–11.

KRUMBEIN, W. C., [1959], The 'sorting out' of geological variables illustrated by regression analysis of factors controlling beach firmness; *Journal of Sedimentary Petrology*, 29, 575–587.

KRUMBEIN, W. C., [1966], A comparison of polynomial and Fourier models in map analysis; *Office of Naval Research Project* 388–078, *Technical Report* 2, *Department of Geology, Northwestern University, Evanston.*

KRUMBEIN, W. C., BENSON, B. T. and HEMPKINS, W. B., [1964], WHIRLPOOL, a computer program for 'sorting out' independent variables by sequential multiple linear regression; *Office of Naval Research Project* 398–135, *Technical Report* 14, *Department of Geology, Northwestern University, Evanston*, 49 pp.

KRUMBEIN, W. C. and GRAYBILL, F. A., [1965], *An Introduction to Statistical Models in Geology*, (New York), 475 pp.

KRUMBEIN, W. C. and MILLER, R. L., [1953], Design of experiments for statistical analysis of geological data; *Journal of Geology*, 61, 510–532.

LANGBEIN, W. B. and LEOPOLD, L. B., [1964], Quasi-equilibrium states in channel morphology; *American Journal of Science*, 262, 782–794.

LANGHAAR, H. L., [1951], *Dimensional Analysis and Theory of Models*, (New York), 166 pp.

LATHAM, J. P., [1966], Remote sensing of the environment; *Geographical Review*, 56, 288–291.

LEOPOLD, L. B. and LANGBEIN, W. B., [1962], The concept of entropy in landscape evolution; *U.S. Geological Survey, Professional Paper* 500-A, 20 pp.

LEOPOLD, L. B. and MADDOCK, T., [1953], The hydraulic geometry of stream channels and some physiographic implications; *U.S. Geological Survey, Professional Paper* 252, 57 pp.

LEOPOLD, L. B., WOLMAN, M. G. and MILLER, J. P., [1964], *Fluvial Processes in Geomorphology*, (San Francisco), 522 pp.

LEWIS, W. V. and MILLER, M. M., [1955], Kaolin model glaciers; *Journal of Glaciology*, 2, 535–538.

MASSACHUSETTS INSTITUTE OF TECHNOLOGY, [1956], Design and operation of an hydraulic analogue computer for studies of freezing and thawing of soils; *Soil Engineering Division, Department of Civil Engineering, Technical Report* No. 62, Corps of Engineers, U.S. Army, 38 pp.

MCCAULEY, J. F., [1964], A preliminary report on the terrain analysis of the lunar equatorial belt; *U.S. Geological Survey (Preliminary), NASA Contract*, 44 pp.

MCCONNELL, H., [1965], *Randomness in spatial distributions of terrain summits*; Department of Geography, Northern Illinois University, DeKalb, Illinois, (Mimeo.).

MELTON, F. A., [1959], Aerial photographs and structural geomorphology; *Journal of Geology*, 67, 355–370.

MELTON, M. A., [1957], An analysis of the relations among elements of climate, surface properties and geomorphology; *Office of Naval Research Project NR* 389-042, *Technical Report* 11, *Department of Geology, Columbia University, New York*, 102 pp.

MELTON, M. A., [1958A], Geometric properties of mature drainage systems and their representation in an $E_4$ phase space; *Journal of Geology*, 66, 35–54.

MELTON, M. A., [1958B], Correlation structure of morphometric properties of drainage systems and their controlling agents; *Journal of Geology*, 66, 442–460.

MELTON, M. A., [1960], Intravalley variation in slope angles related to microclimate and erosional environment; *Bulletin of the Geological Society of America*, 71, 133–144.

MELTON, M. A., [1962], Methods for measuring the effect of environmental factors on channel properties; *Journal of Geophysical Research*, 67, 1485–1490.

MILLER, R. L. and KAHN, S. J., [1962], *Statistical Analysis in the Geological Sciences*, (New York), 357 pp.

MILLER, R. L. and ZEIGLER, J. M., [1958], A model relating dynamics and sediment pattern in equilibrium in the region of shoaling waves, breaker zone and foreshore; *Journal of Geology*, 66, 417–441.

MILLER, R. L. and ZEIGLER, J. M., [1964], A study of sediment distribution in the zone of shoaling waves over complicated bottom topography; Ch. 8 in Miller, R. L. (Ed.), *Papers in Marine Geology: Shepard Commemoration Volume*, (New York).

MOORE, R. K., [1966], Radar as a remote sensor; *Center for Research Inc., Engineering Science Division, University of Kansas, Lawrence, Kansas, CRES Rept. No.* 61–67, 55 pp.

MURPHY, N. F., [1949], Dimensional analysis; *Bulletin of the Virginia Polytechnic Institute, Engineering Experiment Station Series* No. 73, Vol. XLII (6), 41 pp.

NEYMAN, J. and SCOTT, E. L., [1957], On a mathematical theory of populations conceived as a conglomeration of clusters; *Cold Spring Harbor Symposia on Quantitative Biology*, 22, 109–120.

OFFICE OF NAVAL RESEARCH, [1966], Proceedings of the Fourth Symposium on Remote Sensing of Environment (12–14 April, 1966); *Infrared Physics Laboratory, Willow Run Laboratories, Institute of Science and Technology, University of Michigan, Ann Arbor, Michigan*, 871 pp.

PELTIER, L. C., [1950], The geomorphic cycle in periglacial regions as it is related to climatic geomorphology; *Annals of the Association of American Geographers*, 40, 214–236.

PENCK, W., [1924], *Die morphologische Analyse*, (Stuttgart), 283 pp.

PENCK, W., [1953], *Morphological Analysis of Landforms*, Translated by H. Czech and K. C. Boswell, (London), 429 pp.

PIEXOTO, J. P., SALTZMAN, B. and TEWELES, S., [1964], Harmonic analysis of the topography along parallels of the earth; *Journal of Geophysical Research*, 69, 1501–1505.

PRICE, W. A. and KENDRICK, M. P., [1963], Field and model investigation into the reasons for siltation in the Mersey Estuary; *Proceedings of the Institution of Civil Engineers*, 24, 473–517.

ROUSE, J. W., WAITE, W. P. and WALTERS, R. L., [1966], Use of orbital radars for geoscience investigations; *Center for Research Inc., Engineering Science Division, University of Kansas, Lawrence, Kansas, CRES Rept. No.* 61–8, 31 pp.

ROWAN, L. C. and MCCAULEY, J. F., [1966], Lunar terrain analysis; In *Lunar Orbiter-Image Analysis Studies Report, U.S. Geological Survey, NASA Contract* W-12, 123, 89–129.

SAVIGEAR, R. A. G., [1965], A technique of morphological mapping; *Annals of the Association of American Geographers*, 55, 514–538.

SAVILLE, T., [1950], Model study of sand transport along an infinitely long straight beach; *Transactions of the American Geophysical Union*, 31 (4), 555–565.

SCHEIDEGGER, A. E., [1961], Mathematical models of slope development; *Bulletin of the Geological Society of America*, 72, 37–49.

SCHEIDEGGER, A. E., [1964], Some implications of statistical mechanics in geomorphology; *Bulletin of the International Association of Scientific Hydrology*, Year 9, 12–16.

SCHENCK, H., [1963], Simulation of the evolution of drainage-basin networks with a digital computer; *Journal of Geophysical Research*, 68, 5739–5745.

SCHUMM, S. A., [1956], The evolution of drainage systems and slopes in badlands at Perth Amboy, New Jersey; *Bulletin of the Geological Society of America*, 67, 597–646.

SCHUMM, S. A., [1964], Seasonal variations of erosion rates and processes on hillslopes in western Colorado; *Zeitschrift für Geomorphologie*, Supplement 5, 215–238.

SCHUMM, S. A. and LICHTY, R. W., [1965], Time, space, and causality in geomorphology; *American Journal of Science*, 263, 110–119.

SIEGEL, S., [1956], *Nonparametric Statistics for the Behavioral Sciences*, (New York), 312 pp.

SIMPSON, G. G., [1963], Historical science; In Albritton, C. C. (Ed.), *The Fabric of Geology*, (Reading, Mass.), 24–48.

SIMPSON, R. B., [1966], Radar, Geographical tool; *Annals of the Association of American Geographers*, 56, 80–89.

SOKAL, R. R., [1965], Statistical methods in systematics; *Biological Reviews of the Cambridge Philosophical Society*, 40, 337–391.

SPEIGHT, J. G., [1965], Meander spectra of the Angabunga River, Papua; *Journal of Hydrology*, 3, 1–15.

STODDART, D. R., (Personal Communication), *Darwin's coral reef theory and deep drilling through reefs.*

STRAHLER, A. N., [1950], Equilibrium theory of erosional slopes approached by frequency distribution analysis; *American Journal of Science*, 248, 673–696 and 800–814.

STRAHLER, A. N., [1952], Dynamic basis of geomorphology; *Bulletin of the Geological Society of America*, 63, 923–938.

STRAHLER, A. N., [1954], Statistical analysis in geomorphic research; *Journal of Geology*, 62, 1–25.

STRAHLER, A. N., [1958], Dimensional analysis applied to fluvially eroded landforms; *Bulletin of the Geological Society of America*, 69, 279–300.

TRICART, J., [1957], Application du concept de zonalité à la géomorphologie; *Tijdschrift van Het Koninklijk Nederlandsch Aardrijkskundig Genootschap*, 74, 423–430.

TRICART, J., [1964], Geomorphological mapping; *Arid Zone*, No. 24, 12–14.

VAN LOPIK, J. R. and KOLB, C. R., [1959], A technique for preparing desert terrain analogs; *U.S. Army Engineer Waterways Experiment Station, Vicksburg, Mississippi, Technical Report* 3-506, 70 pp.

VON BERTALANFFY, L., [1962], General system theory – A critical review; *General Systems*, 7, 1–20.

WHITTEN, E. H. T., [1964], Process-response models in geology; *Bulletin of the Geological Society of America*, 75, 455–464.

WILLS, L. J., [1929], *The Physiographical Evolution of Britain*, (London), 376 pp.

WILSON, E. B., [1952], *An Introduction to Scientific Research*, (New York).

WONG, S. T., [1963], A multivariate statistical model for predicting mean annual flood in New England; *Annals of the Association of American Geographers*, 53, 298–311.

WOOLDRIDGE, S. W. and LINTON, D. L., [1955], *Structure, Surface and Drainage in South East England*, 2nd Ed., (London), 176 pp.

WURM, A., [1935], Morphologische Analyse und Experiment Hangentwicklung, Einebnung Piedmontreppen; *Zeitschrift für Geomorphologie*, 9, 57–87.

CHAPTER FOUR

# Models in Meteorology and Climatology

R. G. BARRY

## INTRODUCTION

Meteorological theory rests upon the basic laws of physics and hydrodynamics, and climatology, treating the seasonal characteristics of weather phenomena over specific areas in terms of general relationships between the atmosphere and the earth's surface, is dependent on the fundamentals of meteorology. In consequence climatological models are primarily statistical while those of meteorology are physical-mathematical, although the two approaches are by no means mutually exclusive. Since much research effort during the past twenty years has been devoted to problems of the global air circulation the principal themes of this chapter are 'dynamic climatology' (Bergeron, 1930; Court, 1957; Huschke, 1959) which deals with large-scale atmospheric circulations and synoptic-scale models which are a necessary adjunct for the understanding of global patterns. Models equally have their place, however, in bioclimatology and physical climatology and some important aspects of the latter are discussed in Chapter 5.

It is intended first to summarize some of the important laws and relationships of thermodynamics and hydrodynamics indicating sources which provide a fuller treatment of them. It should be noted that several of these basic principles apply to idealized (model) conditions.

(1) The first law of thermodynamics. This refers to the conservation of energy within a thermodynamic system. The law states that the heat supplied to a gas is equal to the increase in its internal energy plus the work carried out in expansion against its surroundings. When for example differential heating sets up a thermally direct circulation, such as a sea-breeze, kinetic energy is produced at the expense of internal and potential energy by the rising of warmer air and sinking of cooler air. (Willett and Sanders, 1959, p. 144; Hare, 1965.)

(2) The second law of thermodynamics. The entropy of a closed system either increases or remains constant during any process operating within the system. In meteorology, it is convenient to use potential temperature (the temperature of unsaturated air brought adiabatically to 1,000 mb pressure)

which is a direct function of entropy (Hess, 1959, p. 20; Petterssen, 1956, p. 11). Entropy and potential temperature are constant for dry adiabatic processes.

(3) The geostrophic wind equation. This is a model for wind flow based on Newton's laws of motion. It expresses the wind velocity as a balance between the horizontal pressure force and the horizontal Coriolis deflection due to the earth's rotation when there is no friction or curvature of the isobars. The velocity is proportional to the pressure gradient. The Coriolis parameter ($2\Omega \sin \varphi$) is twice the angular velocity of the earth's rotation times the sine of the latitude angle (Hess, 1959, p. 175; Willett and Sanders, 1959, p. 118; Crowe, 1965). The gradient wind equation which includes the centripetal (or cyclostrophic) term for curved isobars is usually an unnecessary refinement. Atmospheric models commonly use the geostrophic approximation which relates the wind and pressure fields and acts as a 'noise filter' removing small-scale ageostrophic components.

(4) The thermal wind equation. This relates the geostrophic wind at an upper level to the low-level geostrophic wind and the mean temperature in the intervening tropospheric layer. The thickness of this layer is proportional to its mean temperature (Sutcliffe and Forsdyke, 1951; Hess, 1959, p. 195). In analogy to Buys Ballot's wind law the thermal wind in the northern hemisphere is parallel to the thickness lines with low thickness (cold air) to the left.

(5) The continuity equation (the principle of conservation of mass). This states that for an incompressible atmosphere, where density is unchanged by motion, the horizontal divergence of the wind $\left(\frac{\partial u}{\partial x}+\frac{\partial v}{\partial y}\right)$ is balanced by downward vertical motion $\left(-\frac{\partial w}{\partial z}\right)$. $u$, $v$ and $w$ are the velocity components along the axes $x$ (eastward), $y$ (northward) and $z$ (vertical), respectively. It may be noted that geostrophic winds are non-divergent and therefore cannot be weather-producing. The complete equation of continuity shows that the local rate of change of density is determined by the net rate of mass inflow (Hess, 1959, p. 212; Petterssen, 1956, p. 7; Sutton, 1962, p. 83).

(6) Pressure tendency. Combining the continuity equation with the hydrostatic equation $\left(\frac{dp}{dz}=-g\rho\right.$, where $\frac{dp}{dz}$ is the pressure change with height, $g$ gravity, $\rho$ density), the rate of pressure change at sea-level which is termed the pressure tendency is shown to depend on the vertically integrated net mass convergence (Hess, 1959, p. 219).

(7) The vorticity equation. Vorticity is a measure of the rotation of an infinitely small fluid element. It may arise through curvature of the streamlines

or wind shear across a current. Absolute vorticity about a vertical axis ($Q$) is the sum of the vorticity due to the earth's rotation ($f=2\Omega \sin \varphi$) plus the vorticity of motion about a vertical axis relative to the earth's surface

$$\left(\zeta=\frac{\partial v}{\partial x}-\frac{\partial u}{\partial y}, \text{ positive when cyclonic}\right)$$

$$\text{i.e. } Q=f+\zeta$$

The rate of change of absolute vorticity $\left(\frac{dQ}{dt}\right)$ is proportional to the convergence (negative divergence) if the motion is frictionless and there is no lateral gradient of vertical velocity.

$$\frac{dQ}{dt}=-\left(\frac{\partial u}{\partial x}+\frac{\partial v}{\partial y}\right) Q, \text{ or } \frac{dQ}{dt}=-DQ \text{ (The vorticity equation)}$$

For large-scale horizontal motion the vertical component of absolute vorticity is approximately conserved

$$\frac{d(f+\zeta)}{dt}=0$$

(Hess, 1959, p. 247; Petterssen, 1956, p. 125; Scorer, 1958, p. 49; Sutton, 1962, p. 80; Willett and Sanders, 1959, p. 148).

These fundamental principles provide the basis for a great range of models characterizing various aspects of the complex processes in the atmosphere. The construction of conceptual, mathematical and experimental models is essential throughout the whole field of atmospheric science and the treatment in this chapter is necessarily selective. The approach is to examine first the large-scale models for the general circulation and the westerlies, then synoptic models in mid-latitudes and the quite different ones developed for the tropics, specialized meso-scale models, followed by a section on the application of models in synoptic climatology and palaeoclimatology and finally operational forecasting models.

## THE GENERAL CIRCULATION

The aim of dynamic climatology is ultimately to provide a comprehensive explanation of the general circulation; that is, the large-scale motion of the atmosphere in time and space. The behaviour of the atmosphere follows the classical laws of thermodynamics and hydrodynamics yet the complexities introduced by surface effects, transformations of energy and interacting scales of behaviour, together with observational limitations, have so far precluded the development of an adequate theoretical model. The first real progress in this direction has only recently been achieved by Smagorinsky (1963, 1964) and his associates in the United States Weather Bureau (Manabe *et al*, 1965) with a programme of successively more sophisticated models.

The time-scale appropriate to the problem of the general circulation balance of heat and momentum is often regarded as being of the order of months or years and major wind and pressure systems are treated statistically. Indeed, Mörth (1964) suggests that the mean fields of temperature, pressure and wind along a meridional cross-section can be explained in terms of radiative control by ozone in the lower stratosphere – upper troposphere and water vapour in the middle and lower troposphere. However, many circulation models represent attempts to explain those elements of individual synoptic flow patterns which occur regularly. The principal features of the circulation may be summarized following Fleagle (1957):

(1) large amplitude unstable disturbances in the westerlies in middle and high latitudes. In winter there are about eight to ten at 45° latitude and two to three near the pole.
(2) mainly slow, steady easterly flow in low latitudes, but including intense vortices.
(3) the frequent presence of jet streams in the upper troposphere over middle latitudes.

Jeffreys (1926) recognized that the mid-latitude westerlies would cease due to friction in about ten days if their momentum source were cut off and similarly Sutcliffe (1949) noted that the general dynamic and thermodynamic balance of the atmosphere appears to be maintained over periods of a few days rather than over a month or year. In view of this it seems unlikely that a satisfactory model of the general circulation can be based solely on a simple climatological view of its characteristics. In fact there are no *a priori* reasons for supposing that space – time averages of the circulation are easier to comprehend than synoptic patterns since 'filtering' by taking averages may create its own special problems of interpretation.

## Conceptual models of the general circulation

The history of attempts to characterize the essential structure of the general circulation provides illustrations of all the major categories of model building although until recently the majority have been conceptual. As early as 1686 Edmund Halley outlined a thermal circulation model with maximum heating in low latitudes and a thermally direct cell accounting for the equatorward flow of the trade winds. On a stationary earth a direct cell would produce a meridional-plane circulation as shown in Figure 4.1. This scheme was improved by Hadley (1735) who incorporated the effects of the earth's rotation into the model to explain the north-easterly and south-easterly trades (cf. Crowe,

1965). Hadley envisaged a compensatory south-westerly counter-current above the trades (Fig. 4.2) and this meridional-plane toroidal circulation is still referred to as the 'Hadley cell'.

These models overlooked the westerly wind belt and the first reasonably complete picture of the major wind systems was not produced until 1856 by Ferrel. He postulated three meridional cells in a scheme not unlike that of Rossby in 1941 (Fig. 4.3). Ferrel outlined the concept of the conservation of angular momentum and attempted to explain the location of the sub-tropical anticyclones in these terms. The concept implies that zonal rings of

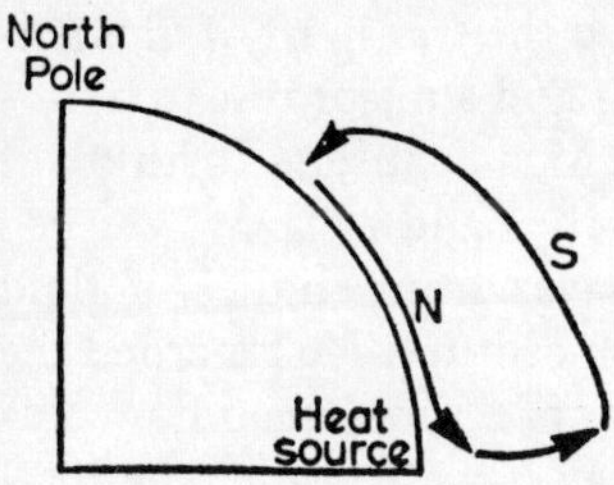


**4.1** Thermally-direct cell on a stationary earth (*Halley model, 1686*).

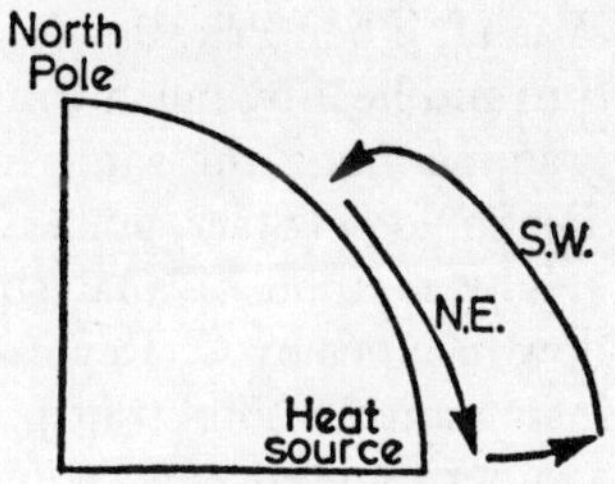


**4.2** Thermally-direct cell on a rotating earth (*Hadley model,* (*1735*).

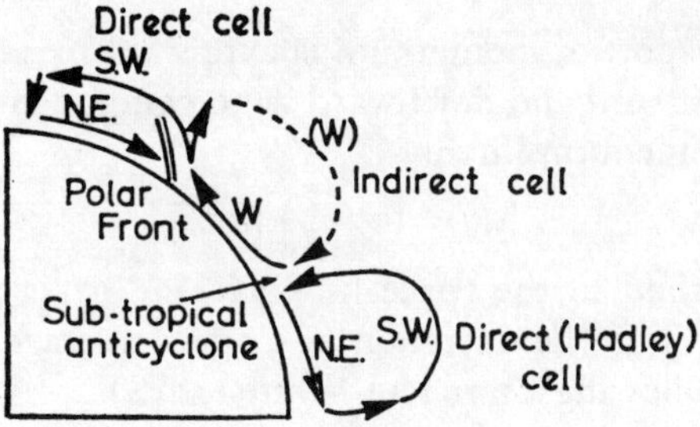


**4.3** Three-cell model of the mean circulation (*after Rossby, 1941*).

air (between two latitude circles) develop westerly winds (westerly relative momentum) if displaced polewards and easterly winds if displaced equatorwards. Ferrel's ideas did not affect the mainstream of meteorological study, though in recognition of his work the mid-latitudes westerlies are sometimes referred to as the Ferrel westerlies.

The three cell model with mean meridional-plane circulations was again put forward by Bergeron (1928) and later by Rossby (1941) in modified form. Basically, Rossby assumed that zonal circulation rings conserve their absolute angular momentum. Nevertheless, he recognized the existence of strong upper westerlies in mid-latitudes and suggested that these could be accounted for if westerly momentum from the upper branches of the tropical and polar cells (Fig. 4.3) were spread as a result of lateral mixing (in the horizontal plane) by tropospheric wave patterns. Rossby considered that the frictionally-driven mid-latitude cell was in this way obscured by the upper westerlies.

Rossby's introduction of longitudinal variations into the model had in fact been anticipated by Jeffreys who in 1926 demonstrated that the poleward transport of angular momentum required to maintain the westerlies could be carried out solely by eddy circulations in the horizontal plane. However, his ideas lay dormant until upper data became readily available after World War II. Then Priestley (1949; and 1951) calculated the transport of momentum and heat by mean winds and eddy circulations and a major research programme along the same lines was undertaken at Massachusetts Institute of Technology (Starr, 1948; Starr and White, 1951, 1952 and 1954).

The method of analysing these transports derives from studies of fluid flow in pipes by O. Reynolds in the late nineteenth century. He regarded the instantaneous total transport (or flux) as consisting of a time-mean flow plus a superimposed eddy component. In the same way if we consider the local meridional flow across a latitude circle at a given pressure level then this flow $v$ can be separated into the space-mean flow across the whole latitude circle $[v]$ and the deviation from the mean $v'$

$$v=[v]+v'$$

The poleward transport of momentum is expressed by $uv$ where $u$ and $v$ are respectively the eastward and northward wind components. Thus, the space-mean poleward momentum flux is

$$[uv]=[u][v]+[u'v']$$

Extending the method to the space-time average, where the bar denotes a time average and $[\ ]'$ is the deviation of a space average from a space-time mean, we obtain following Starr and White (1952)

$$\overline{[uv]}=\overline{[u][v]}+\overline{[u'v']}$$
$$\text{and } \overline{[uv]}=[\bar{u}][\bar{v}]+\overline{[u]'[v]'}+\overline{[u'v']}$$

$[\bar{u}][\bar{v}]$ is the transport due to mean meridional circulations

$\overline{[u]'[v]'}=\overline{[u][v]}-[\bar{u}][\bar{v}]$ represents instantaneous net meridional circulations due to time correlations between space means.

$\overline{[u'v']}=\overline{[uv]}-\overline{[u][v]}$ is the transport by large-scale horizontal eddies arising from spatial correlations between the instantaneous wind components $u$ and $v$.

Starr and White (1954) used this technique to examine the meridional transports of momentum, heat and water vapour across selected latitude circles in the northern hemisphere during the year 1950. The results of their calculations showed that the horizontal eddy component (surface cyclones and anticyclones and tropospheric troughs and ridges) accounted for a large proportion of the balance requirements of energy and momentum determined independently by other workers. In spite of these results, however, it would be incorrect to formulate a circulation model with only horizontal eddy transports. Palmén's model (1951B) of the mean winter circulation in the northern hemisphere (Fig. 4.4) has been verified by Tucker (1959) who finds evidence of a weak, indirect mid-latitude cell and an important low latitude Hadley cell for momentum transport. The role of the Hadley cell has also been emphasized in terms of the low-level equatorward transport of vapour in the tropics by Palmén and Vuorela (1963).

Finally, some of the complexities of the low latitude circulation problem may be mentioned. The gross over-simplification of the atmosphere's heat sources and sinks has been a common feature of circulation models until quite recently. In particular, the fact that the earth does not possess a simple

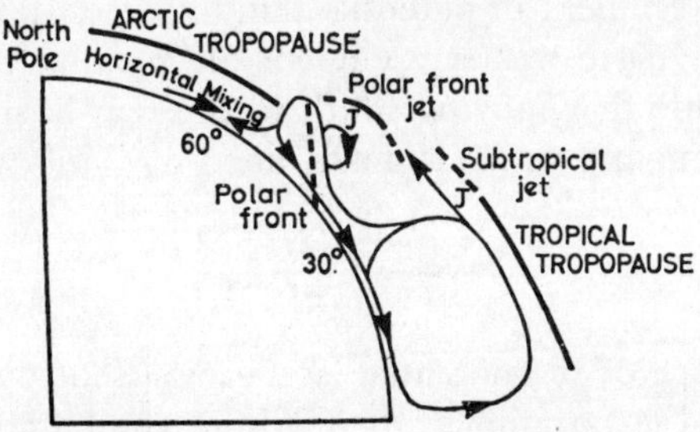


**4.4** The mean meridional circulation in winter (*after Palmén, 1951*). J indicates the positions of the jet streams. The latitude of the polar front is very variable.

heat source at the equator is usually insufficiently taken into account. Riehl (1926B), for example, shows that upward heat transfer in the equatorial zone is localized within large cumuli rather than occurring generally. The concept of a general anti-trade is also erroneous as the poleward heat flux is confined to limited longitudinal sectors. In addition to these points there is increasing evidence (Tucker, 1965) that inter-hemispheric exchanges cannot be neglected in hemispheric circulation models.

Sufficient illustration has been given of the complexity of the general circulation to show the difficulties facing those workers who seek to develop adequate models of its behaviour. The need to isolate the really fundamental processes at work in the atmosphere has encouraged the application of mathematical and experimental methods which will now be discussed.

## Mathematical models

Much current research into the operation of the general circulation uses mathematical models based on simplified hydrodynamical equations. Certain atmospheric properties are specified, many of them parametrically (by reference to empirical results or semi-intuitive ideas), appropriate boundary conditions are assumed and calculations are made of the resultant circulations. There are two basic atmospheric models. An atmosphere where isobaric surfaces and constant density (isosteric) or constant temperature surfaces are parallel is referred to as *barotropic*, while one in which these surfaces intersect is termed *baroclinic*. A frontal zone with sharp intersection of temperature and pressure surfaces is hyper-baroclinic (Fig. 4.9). If we visualize sets of isobaric-isosteric surfaces intersecting in space they form tightly packed quadrilateral tubes termed solenoids.

Some of the most significant contributions to general circulation theory have come from attempts to explain the development of unstable waves in a baroclinic current (Eliassen, 1956; Fleagle, 1957). The model assumes that potential temperature increases with height (a stable stratification) and that there is a poleward gradient of potential temperature. In a wave disturbance superimposed on a basic westerly current air moving north ahead of the trough is rising while that moving south in the rear is sinking. The north-south slope of the moving air ($\delta$) is determined by

$$\delta = \frac{\mu - \beta f / 2HS\alpha^2}{1 + f^2/2H^2S\alpha^2} \qquad \text{(Fleagle, 1960)}$$

where $\mu$ = the slope of the potential isotherms (isentropes)
$\beta = \partial f/\partial y$, the variation of the Coriolis parameter ($f$) with latitude
$H$ = the depth of atmosphere (10 km.)
$S$ = the static stability
$\alpha$ = the wave number $= 2\pi/L$ ($L$ = wavelength).

For short-waves, where $\alpha$ is large, the slope $\delta$ is much greater than $\mu$ as shown in Figure 4.5A and since the rising air has a lower potential temperature

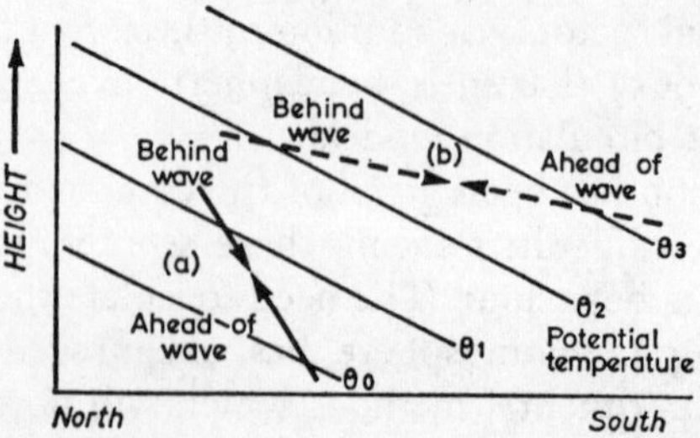


**4.5** Baroclinic instability model.
(a) short wavelength case
(b) medium wavelength case
The potential isotherms refer to arbitrary values.

than the subsiding air the potential energy is increased. This means that no energy is available for growth of the disturbance. With medium wavelengths, still with moderate vertical motion, the potential energy is being decreased by the southward movement of sinking potentially cold air and northward movement of rising potentially warm air shown in Figure 4.5B. The maximum conversion of potential energy to kinetic energy occurs for $\delta=\mu/2$ when $L=2\pi H\sqrt{(2S/f)}$. For very long waves the vertical motion is too slight to produce significant energy exchanges.

The maximum slope of potential temperature surfaces is between 30–60° latitude and since $\beta/f$ increased southwards, causing $\delta$ to approach $\mu$, the maximum baroclinic instability is found in higher mid-latitudes. Theoretically, there is a critical latitude at 35° with an essentially stable atmosphere equatorward of this. These deductions, go some way towards accounting for the observed circulation features.

One of the most interesting mathematical experiments is due to Phillips (1956) who used a two-layer baroclinic model of the atmosphere (see p. 135). It was assumed that non-adiabatic heating occurred at low latitudes with a high-latitude sink so that no net heating resulted. Starting from an atmosphere at rest with uniform temperature gradient, zonal flow developed after 130 'days' with a slow meridional cell and equatorward flow at low levels. However, after the introduction of random disturbances and longitudinal variations into the model three meridional cells developed and a marked westerly jet stream appeared. Energy calculations showed that the eddies supplied kinetic energy to the mean zonal flow, a result anticipated by Rossby (1949) and by other theoretical investigations of Kuo and Fjortoft, but contrary to the 'classical' view that the perturbations are driven by the kinetic energy of the mean flow. Smagorinsky (1963) elaborated these techniques using a two-level model with the 'primitive equations of motion' which allow for ageostrophic components and thus provide a proper representation of wind flow in low latitudes. Subsequent work is based on a nine-level model incorporating the effects of surface boundary layer fluxes, computed radiative effects and the moisture cycle (Manabe *et al*, 1965). The reduced dependence on parametric representation of such factors is a major advance on previous models. The computations provide a realistic simulation of many atmospheric features in at least qualitative terms and the quantitative agreement with a number of observed characteristics is also good. Ultimately it should be possible to describe the entire climatology of atmospheric motions by means of a time-dependent, deterministic model.

## Experimental Models

Important contributions to general circulation theory have also been made by laboratory studies of rotating fluids. The crucial problem in developing useful experimental models is a knowledge of the similarity conditions between the model and physical reality. The quantities normally requiring description are lengths, flow characteristics and fluid properties such as density and viscosity. In the case of the atmosphere scale presents a problem since the depth of the model layer cannot be reduced in the same proportion as the horizontal dimension, but such difficulties are overcome if the equations of motion for the atmosphere and the model are made directly comparable through non-dimensional parameters. An important parameter is the Rossby number (*Ro*), the ratio of the velocity relative to the absolute speed of the point on the equator.

$$Ro=u/Ce$$

When *Ro* is small (approximately $<0{\cdot}2$) the motion is quasi-geostrophic. One intractable problem of working models is that gravity is always a force acting in the same direction.

There have been three major types of experimental model (see Fultz, 1960). They are (1) the Benard cell in which a non-rotating fluid is either heated from below or cooled from above. Cellular convection develops when the temperature gradient exceeds the critical value necessary for instability. A large critical temperature gradient is required if the fluid is rotated; (2) the Hide annulus (Hide, 1953; Riehl and Fultz, 1957 and 1958) in which a tall column of water is rotated between two concentric cylinders. The temperature of the outer wall is gradually raised and that of the inner wall lowered, keeping a constant temperature in the liquid. Beyond a critical temperature gradient the initial slow, symmetrical, easterly flow (Hadley regime) breaks down first into a seven-wave westerly pattern, then with systematically fewer waves down to an eccentric westerly vortex as the temperature gradient is increased (Fig. 4.6); this sequence is referred to as the Rossby regime. With still greater temperature gradients a further Hadley regime appears with a symmetrical westerly current at the upper surface. The transition from the Hadley to the Rossby regimes also depends on the rate of rotation (Fig. 4.6). When the fluid is close to a transition between different wave numbers it is found that the pattern commonly shows a vacillation similar to the index cycle (see p. 111); (3) the 'rotating dishpan' containing two immiscible fluids with mechanically induced motion. The lower layer can be induced to form a 'polar cap' with eastward-moving waves along the fluid interface, simulating the polar front (Fultz, 1960 and 1961).

More specific studies include those relating to the effect of different shaped barriers on fluid flow in a rotating dishpan (Long, 1952 and 1955) or in a rotating hemispheric shell (Frenzen, 1955).

Energy balance calculations for experiments in a steady rotating annulus (Riehl and Fultz, 1958) produced some results which disagreed with the numerical experiment of Phillips. When calculations were made in curvilinear co-ordinates related to the jet stream axis it was found that the kinetic energy production was maintained by a single direct cell and was transferred from the mean circulation to the eddies. This type of approach is valuable in that it suggests ideas which may be tested with observations from the atmosphere. Indeed Krishnamurti (1961) subsequently demonstrated that there is a Hadley circulation around the axis of the subtropical jet stream of the northern hemisphere in winter.

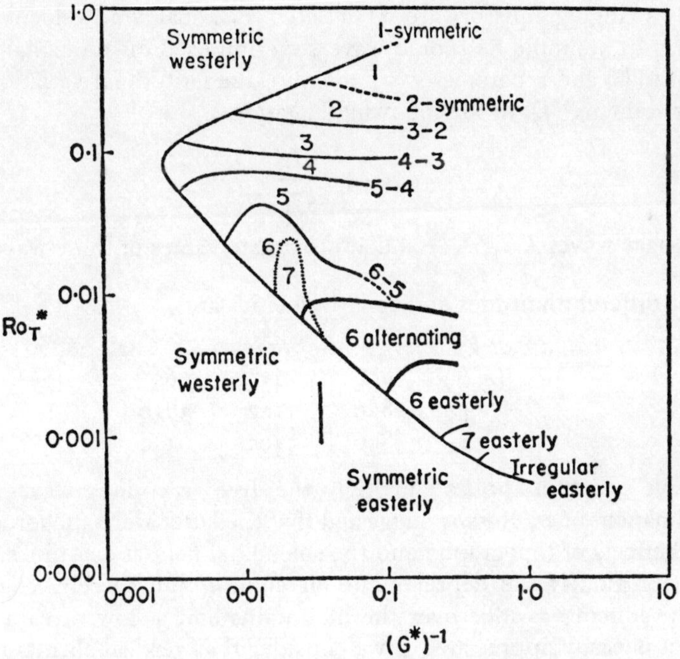


**4.6** Circulation regimes in an annulus experiment (*Fultz, 1960*). $Ro_T{}^*$ is proportional to the radial temperature gradient, $(G^*)^{-1}$ is proportional to the square of the rate of rotation of the annulus. Hadley regimes lie to the left of the main curve, Rossby regimes to the right. The numbers indicate the number of waves in the flow pattern.

## THE WESTERLIES

### Theoretical models of the long waves

Investigation of the middle and upper troposphere has demonstrated the existence of quasi-permanent long waves in the westerlies. Mean contour maps for the 500 mb surface in winter (Sutcliffe, 1951) show deep troughs over eastern North America and eastern Siberia and a minor one at about

60° E., while in summer, although the westerlies are much weaker, the first two troughs remain prominent in almost the same locations and the other is replaced by a weak ridge.

Rossby (1939 and 1940) developed the first theory of wave patterns on the global scale by assuming horizontal motion in a barotropic atmosphere. The crux of his argument was that since the Coriolis parameter increases with latitude then if absolute vorticity is to be conserved a current moving poleward must reduce its relative vorticity, thus becoming more anticyclonic and eventually swinging equatorward. He related the zonal wind velocity ($u$), the velocity ($c$) of a simple harmonic wave superimposed on the zonal current, wavelength ($L$) and a parameter $\beta$ ($=\partial f/\partial y$), the latitudinal variation of the Coriolis parameter ($f$), in the following equation

$$u-c=\frac{\beta L^2}{4\pi^2}$$

For stationary waves $L=2\pi\sqrt{\frac{u}{\beta}}$ and approximate values of these wavelengths in km. for different latitudes and zonal velocities are

| *Zonal Velocity* (*m/s*) | *Latitude* 30° | 45° | 60° |
|---|---|---|---|
| 4 | 2820 | 3120 | 3710 |
| 12 | 4890 | 5400 | 6430 |

The wave equation applies equally to the 'free', travelling waves and the quasi-permanent ones. Rossby suggested that the latter were anchored by the combined effects of topography and the solenoidal field at coastlines. Mountain barriers modify the depth of the air column and thereby change the vorticity producing a ridge over the mountains and a downstream trough. This effect is easily appreciated if we consider that vertical shrinking of the column over the barrier leads to lateral expansion (divergence) and anticyclonic vorticity (see the vorticity equation, p. 99), with the reverse as the column expands vertically downstream of the barrier. Thermal effects are due to energy sources such as the Gulf Stream and Kuro Shio encouraging low-level convergence (high-level divergence) in winter months while over the continents there is surface divergence (high-level convergence). This pattern results in upper troughs in the westerlies over the eastern continents (Palmen, 1951A).

Despite Rossby's views, two schools of thought developed in the 1950's as to the origin of these 'Rossby waves'. Charney and Eliassen (1949) and Bolin (1951) proposed that mountain barriers were primarily responsible since the troughs appeared to indicate no seasonal variation in position, although it can be shown that different interpretations result in part from different series of average contour maps. The Rockies were visualized as the primary control in the northern hemisphere, giving rise to the eastern North America trough, while Bolin found a similar pattern downstream of the Andes. Theoretical considerations (Queney, 1948) and physical models (Long, 1952 and 1955) also suggested the importance of such barriers but it must be emphasized that the calculations of Queney and Charney and Eliassen were based upon the simplified assumptions of a barotropic-model atmosphere.

Sutcliffe (1951) raised qualitative objections to these ideas and argued that since the atmosphere is baroclinic the role of solenoidal fields cannot be neglected. He presented 1,000–500 mb thickness maps and 500 mb contour charts to demonstrate the close correlation between their patterns in both January and July.

Investigations by Smagorinsky (1953) and Academica Sinica (1957 and 1958) using a variety of baroclinic models suggest that both factors need to be taken into account, although their relative importance has not been fully established. Gilchrist (1954) showed with a baroclinic model that in winter there can be a surface trough over a heat source with the upper trough to the west as a result of high zonal velocities, but in summer with low zonal velocities there is a surface trough over a heat sink with the upper trough to the east. Such changes are observed over Asia and the Pacific and may depend on wavelength adjustments due to different zonal velocities or on a different pattern of heat sources. The distribution of the latter, however, is still the subject of considerable uncertainty. For example, Clapp (1961) calculated the heat sources and sinks for the northern hemisphere in winter by two independent methods and obtained heating patterns which in one case were positively correlated with temperature and in the other negatively. Other studies indicate that the relation is positive, implying that the heat sources do generate the mean motion of the atmosphere (Sheppard, 1962). The effect of heat sources is further complicated by the fact that the heating of the atmosphere is a function of the motion of the air itself and models must make allowance for this non-linearity (Sutcliffe, 1950; Döös, 1962).

Analogue studies of the wave patterns in the two hemispheres may help to resolve some of the present differences of opinion. Lamb (1959) considers that south of 45° S. the dominant control is thermal, but data for the southern westerlies are at present too incomplete for final conclusions to be drawn.

## Harmonic analysis – a mathematical model of wave patterns

Recent studies of the wave patterns in the westerlies use the technique of harmonic (Fourier) analysis which allows a complex curve to be resolved into simple component parts comprising only sine and cosine functions of the form $\sin x$, $\sin 2x \ldots \sin nx$ and $\cos x$, $\cos 2x \ldots \cos nx$. The wave number ($n$) as specified by these functions, indicates the number of troughs along a selected latitude circle. Wave number 1 represents an eccentricity of the polar vortex, number 2 refers to two troughs 180 degrees apart and so on (Hare, 1960). The technique also allows the specification of the phase difference (longitudinal distance) between each wave, the amplitude (latitudinal extent) of the wave and the percentage contribution of each wave number to the total variance.[1]

Wave numbers 1, 2 and 3 are dominant at high latitudes (Eliassen, 1958; Saltzman and Peixoto, 1957) and in middle latitudes waves of number 5, 6 and 7 are superimposed upon the longest waves and travel almost continuously eastwards. Wave numbers 11 and 12 are effectively 'noise' as a result of the coarse data network and uncertainties in the original contour analyses. It must be noted that wave number 7 at 30° N. is equivalent to number 3 at 60° N. in terms of actual wavelength. A weakness of the method is that the wave numbers are discrete (Boville and Kwizak, 1958). Thus the dominance of a wave number $n$, accompanied by high percentages of wave numbers $n-1$ and $n+1$ may indicate that the spacing of the '$n$' troughs is not consistently $2\pi/n$.

In addition to analytical studies of the eddy pattern in the zonal circulation the technique can also be used to investigate the partition of energy between the various wave spectra at different latitudes and isobaric surfaces. For example, Wiin Nielsen *et al.* (1964) show that a large proportion of the total transport of heat and momentum in winter is due to wave numbers 1 to 4 whereas in other months these long waves are less important.

## Zonal indices and models of circulation patterns

Several attempts have been made to formulate models of hemispheric circulation patterns by reference to the overall strength and average latitude of mid-latitude westerlies. The basis of these ideas, the zonal wind velocity, is only an analytical tool, but it is necessary first to outline the method before discussing the models which derive from it.

The overall strength of the westerlies can be assessed by measuring the average hemispheric pressure gradient between 35° and 55° N. and converting this into geostrophic west wind. This zonal index may be determined for MSL isobars or upper air contours. The method was first applied by Rossby *et al.* (1939) in relation to the long-wave equation and subsequently developed in detail by Allen *et al.* (1940). The MSL index is termed high when the average pressure difference between 35°–55° N. exceeds about eight mb and low when it is less than about three mb. There is a seasonal trend, however, in response to the weaker westerlies in summer.

The belt 35°–55° N. is the mean zone of strongest westerlies though often in winter their circulation pole is displaced from the geographical pole to-

[1] Other applications include the resolution of cyclical time series into their principal harmonic components. An example is the analysis of annual precipitation profiles for North America (Lyle and Bryson, 1960; Sabbagh and Bryson, 1962).

wards 170° W. Such displacements are closely related to the strength of the zonal flow (La Seur, 1954) and since the index only reflects activity within the stated latitude belt the precise interpretation of index fluctuations may be difficult. For this reason some studies also give indices for 20°–35° N. and 55°–70° N., but La Seur advocated the use of a 'moving co-ordinate system' related to the circulation pole of the westerlies. An alternative solution is to plot a time profile of the average zonal west wind component for 5° latitude belts. Such zonal profiles reveal northward and southward trends of relative maxima and minima within the mean westerlies which may persist several weeks (Riehl *et al.*, 1952).

Simple models of the synoptic patterns associated with high and low index were indicated by Rossby *et al.* (1939) and Allen *et al.* (1940). With high index there is rapid eastward movement of cyclones and little meridional exchange of air masses despite the strong meridional temperature gradient. The subtropical anticyclones and the Aleutian and Icelandic lows are intense features and the latter tend to be eastward of their normal position. With low index the pattern is strongly cellular and longitudinal temperature contrasts are marked. The Aleutian and Icelandic lows tend to be westward of their normal position. Rossby and Willett (1948) and later Namias (1950) elaborated these ideas and showed that over a period of four to six weeks 'index cycles' occur, during which the westerlies increase in strength as the Ferrel vortex expands and the circulation breaks down into cellular patterns (Fig. 4.7).

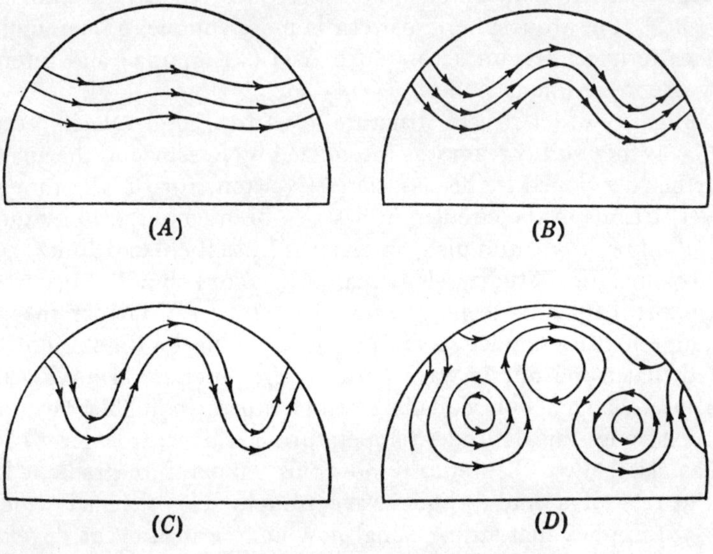

**4.7** The index cycle (*Namias, 1950*). Four stages in the change-over from zonal flow (*A*) to a meridional cellular pattern (*D*).

A more detailed model of the index cycle was put forward by Riehl *et al.* (1952) on the basis of the stages in the northward or southward drifts of relative westerly maxima and minima on the zonal profile. This more complex scheme shows that the previous model might prove a dangerous over-simplification. There are for instance four stages which broadly correspond to 'low index' but only one is closely similar to 'high index'. Each stage is characterized by a particular set of synoptic patterns at the surface and in the upper air. In spite of this clarification of the complex nature of index fluctuations and the expression of doubts about its usefulness (Forsdyke 1951), the simplified terminology of high and low index continues to be used. These extreme states are also employed in palaeoclimatology (see p. 131) and in analytical studies of the circulation. For example, Bradbury (1958) examined the behaviour patterns of cyclones and anticyclones for strong and weak circulations.

## Jet streams and general circulation models

Narrow bands of very strong winds or jet streams are regularly observed in the upper tropospheric westerlies. In winter there is a subtropical jet stream about 200 mb near the poleward limit of the trade wind cell and there may be one or more jet streams at about 300 mb associated with the major frontal zones in middle and high latitudes (Newton and Persson, 1962; Reiter, 1963; Riehl, 1954 and 1962A). Recent research has also shown the existence of a 'polar night jet' in the winter stratosphere over the arctic and antarctic (Hare, 1962). The tropospheric pattern is more complex in summer with a tropical easterly jet stream across Afro-Asia (see p. 124) and intermittent westerly jets in middle latitudes.

The similarity of jet stream structure to narrow, high velocity ocean currents such as the Gulf Stream was recognized by Rossby and the analogy has been further developed by his associates (Newton, 1959). The fundamental role of jet streams in the circulation has also been emphasized by numerical models (Phillips, 1956) and dishpan experiments (Riehl and Fultz, 1958), as well as by analytical studies (Krishnamurti, 1961), but the occurrence of jet streams still remains to be accounted for by a satisfactory theory. The most comprehensive model so far prepared is due to Rossby (1949) who suggested that tropospheric wave patterns promote lateral mixing through airmass interchange. This equalizes temperatures in higher latitudes and creates a strong mid-latitude temperature gradient. Fleagle (1957 and 1960) has also shown theoretically how this temperature gradient is built-up to the rear of a mid-latitude wave trough. The thermal wind equation (p. 98) implies that strong zonal flow and vertical shear develop with such a meridional temperature gradient, but it must be stressed that lateral mixing is the simultaneous cause of both the wind and temperature fields.

Rossby demonstrated that the mixing leads to constant absolute vorticity in higher mid-latitudes and a latitudinal profile of zonal wind similar to that observed in the northern hemisphere as shown in Fig. 4.8 (Fultz, 1960; Tucker, 1962). At the equatorial limit of this region of constant absolute vorticity (about 35° latitude) there is strong lateral shear such that on the equatorward side the profile of wind velocity approximates to that for constant absolute angular momentum (zero absolute vorticity).

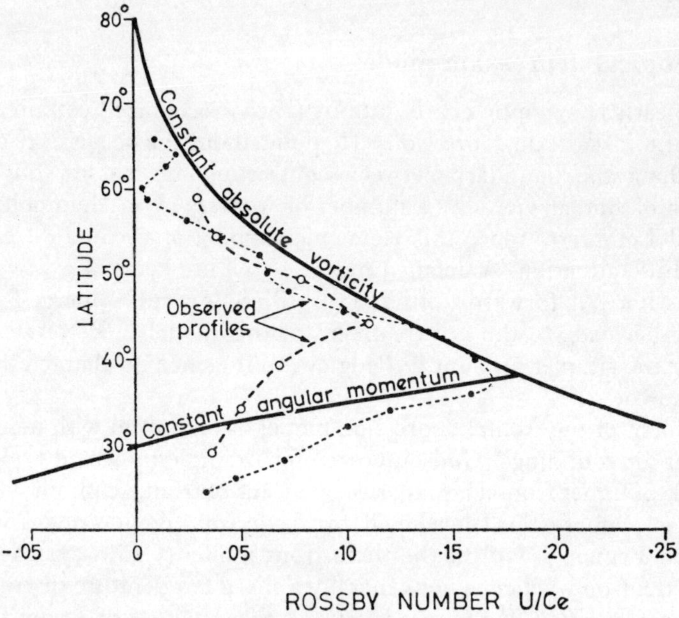


**4.8** Theoretical profiles of wind velocity according to the principles of the conservation of absolute vorticity and the conservation of absolute angular momentum compared with observed profiles near the tropopause (*Staff Members, University of Chicago, 1947*). The profile shown with open circles refers to a nine-day average over North America. Wind velocity (u), is expressed as a non-dimensional ratio by dividing by the linear equatorial velocity of the earth Ce.

This model is still far from accounting for many observed features, particularly the existence of double westerly jet streams in winter. Palmén (1951) regards the subtropical jet as related to the high-level poleward flux of momentum associated with the Hadley cell (Figure 4.4) and it may be that Rossby's scheme is more appropriate for the polar-front jet.

## SYNOPTIC-SCALE MODELS

*A synoptic model is essentially a standard configuration of the field of pressure or flow with which is associated a characteristic weather pattern.*

FORSDYKE, 1960

### Extra-tropical depression models

From the earliest synoptic charts Fitzroy (1863) recognized certain essential features of mid-latitude depressions. He pointed out the occurrence of cyclogenesis along the boundary between contrasting air masses, but despite reiteration of similar views by a number of workers (Von Helmholtz, 1888; Shaw and Lempfert, 1906) this elementary model of cyclone development received little attention. A detailed model of the life cycle of a wave disturbance was not put forward until 1922 by Bjerknes and Solberg. Reference need only be made to the full treatment of this model in Petterssen (1956, p. 217) or the simpler account in Pedgley (1961) since its characteristics are widely known.

Refinement of the frontal depression model has followed with more intensive upper air soundings. Godson (1950) for example indicated typical configurations of upper frontal boundaries by means of frontal contour charts and these ideas were further developed by Anderson, Boville and McClellan (1955) and Penner (1955) in the three-front model (Galloway, 1958). The nature of the front, which is not generally a sharp temperature discontinuity, has also been clarified. Figure 4.9 illustrates the structure of a typical frontal zone in winter. The meridional temperature gradient is most intense in the frontal zone but is not confined to it. It may be noted that the presence of sharp frontal zones in the middle and upper troposphere is commonly due to subsidence and not to air mass origin (Sawyer, 1958). Detailed models of frontal structure and air motion relative to the frontal surface are given by Sansom (1951) and Miles (1962).

Important improvements of the model of a developing frontal wave are due to the work of Holmboe and Bjerknes (1944) on the relationships between upper tropospheric waves and surface waves. They demonstrate that the continued fall of pressure at the centre of the depression is made possible, despite active low-level convergence during the deepening stage, by active divergence aloft. Upper troughs have gradient winds of sub-geostrophic value whereas in ridges they are super-geostrophic as a result of the opposite effect of the centripetal term. As a result the transport of air in the contour channels of troughs is less than in ridges causing divergence to be most commonly located ahead of upper troughs in the westerlies. The divergence due to this curvature effect generally exceeds the convergence in poleward flow which results from the increasing Coriolis parameter. The theory of cyclogenesis has been greatly extended by the work of Sutcliffe and others on divergence – vorticity relationships (Sutcliffe, 1947; Sutcliffe and Forsdyke, 1951; Petterssen, 1956, p. 320 *et seq.*). Sutcliffe analysed the vorticity equation (p. 99)

$$\frac{d\underset{\sim}{Q}}{dt} = -DQ$$

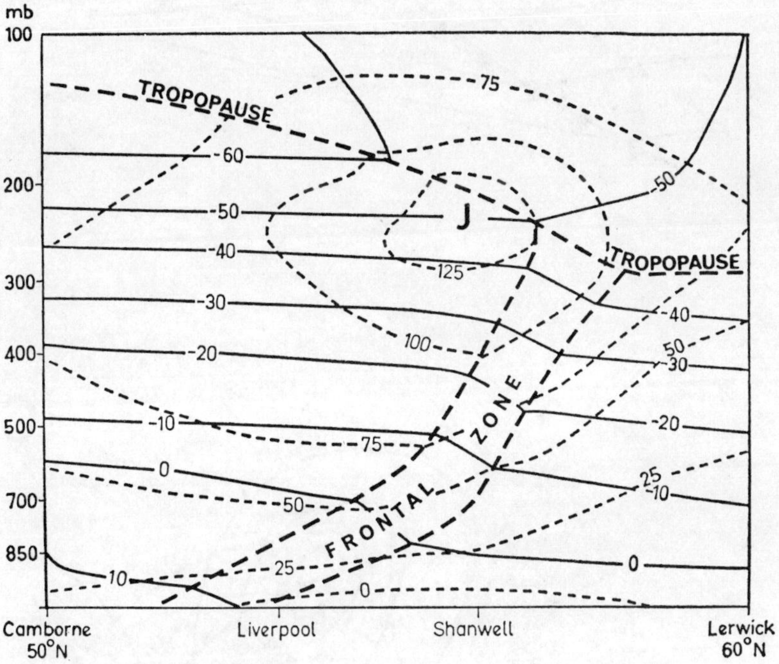


**4.9** Model of a frontal zone (based on 26 September 1963)
Full lines – isotherms (°C.).
Dashed lines – isotachs of the westerly wind component (knots)

and showed that cyclonic development ($-D$) comprises two major components – the advection of vorticity in the middle troposphere and thermal advection. Their role can be illustrated with reference to the model of a developing depression in Figure 4.10. In the initial stages (*a* and *b*) the primary control is considered to be the positive vorticity advection ahead of an advancing upper trough. Subsequent intensification at the centre of the surface wave is mainly dependent on thermal advection (shown by the schematic thickness lines) in the warm sector strengthening the production of vorticity. The original Bjerknes – Solberg model emphasized the upgliding of warm sector air at the warm front but this is now regarded as of secondary importance. Development may also be affected by non-adiabatic heating or

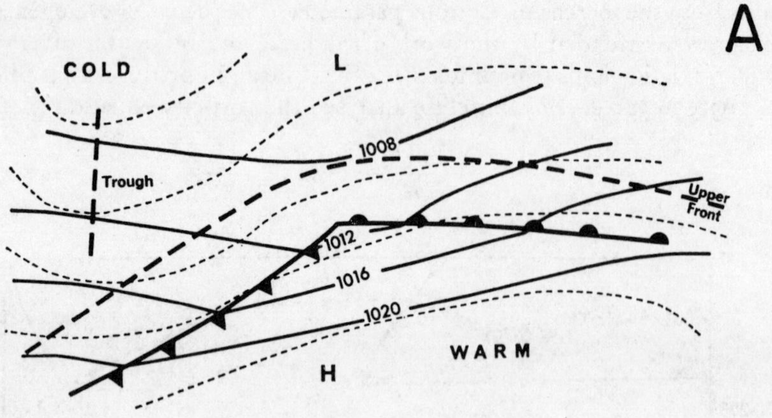

A
COLD
L
1008
Trough
Upper Front
1012
1016
1020
WARM
H


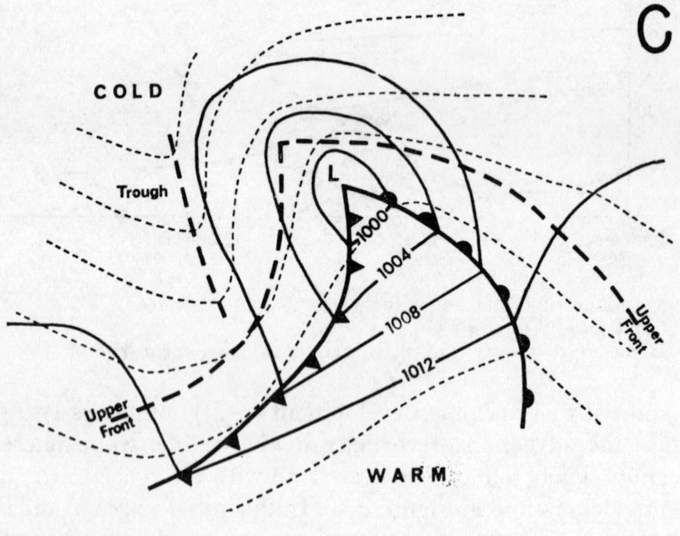

C
COLD
Trough
L
1000
1004
1008
1012
Upper Front
Upper Front
WARM

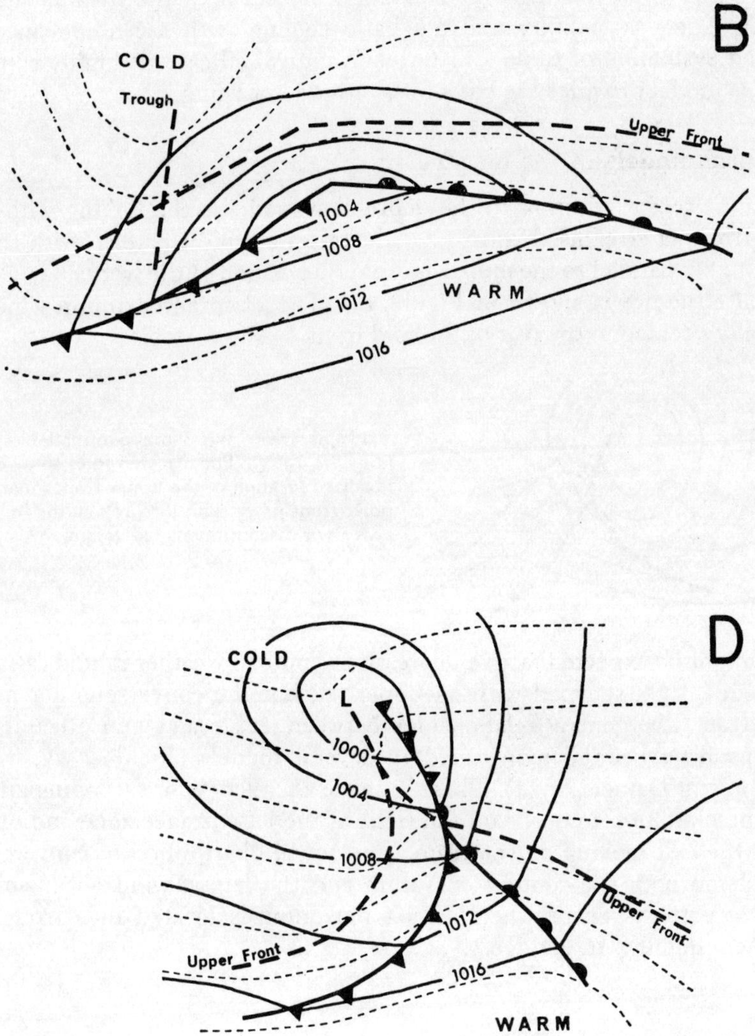


**4.10** Idealized frontal wave development. The full lines are surface isobars, the pecked lines relate to schematic 1000–500 mb thickness lines and the dashed lines show the axis of the upper cold trough and the position of the upper front at about 500 mb.

cooling due to air moving over heterogeneous surfaces and by adiabatic influences such as stability and adiabatic cooling with ascending air. The detailed evaluation of these four terms in individual cases is highly complex but this model provides the basis for routine forecasting.

### Synoptic models of the jet stream

The most recent addition to the depression model is the linking with it of polar-front jet streams. Figure 4.11 illustrates a typical pattern with the jet axis roughly parallel to the mid-tropospheric position of the frontal zone. The vertical structure is shown in Figure 4.9. The jet core (maximum winds) is generally situated to the rear of the cold front.

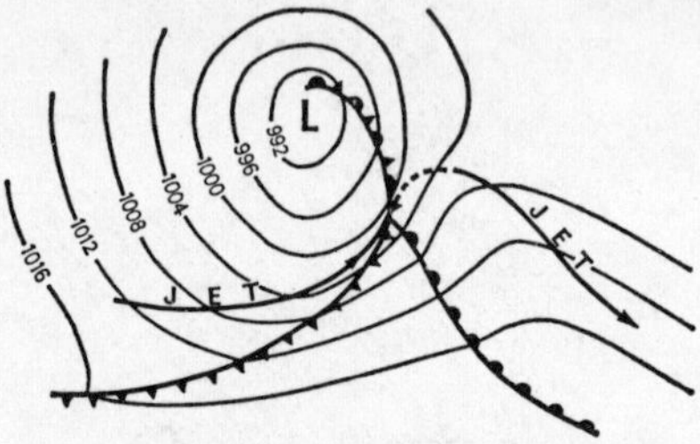


**4.11** Jet stream and depression model (*Boyden, 1963*). The arrowed lines show the idealized location of the upper tropospheric polar front jet stream, the broken line indicates a weak or discontinuous jet stream.

As might be expected, active depressions and bad weather in mid-latitudes are frequently associated with an upper jet, but the converse is not necessarily true. The general relationships between jet streams and precipitation patterns have been demonstrated in several models (Riehl *et al.*, 1952; Johnson and Daniels, 1954). These indicate a tendency for ascending air and more precipitation to occur on the right of the jet entrance zone and on the left of the exit looking downstream. The actual distribution of convergence depends on both the contour curvature and the lateral wind shear so that different patterns ensue when the jet maximum is located in a trough or (more commonly) in a ridge.

## TROPICAL MODELS

Intensive study of the wind systems and disturbances of the Tropics dates only from about 1940. The earlier literature is for the most part characterized by faulty concepts based on inadequate observations and frequently biased by ideas developed in middle latitudes. Only Riehl's unsurpassed text on tropical meteorology (1954) and Trewartha's recent book (1961) provide any comprehensive coverage of modern views on the tropical atmosphere.

Palmer (1951) refers to the early workers as mainly belonging either to the 'climatological school' or the 'air mass – frontal school'. The former considered the weather in terms of diurnal and seasonal controls plus local effects while the latter transferred the models of Bjerknes and Bergeron to low latitudes, (for example Garbell, 1947). A specific illustration may be cited. In 1936 the trade wind inversion was explained by Ficker in terms of two distinct air masses, a view unfortunately repeated by Pédelaborde (1958). However, Riehl *et al.* (1951) demonstrated that in the north-east Pacific the inversion is due to subsidence and that particle trajectories cross the inversion boundary. More serious problems appeared when attempts were made to reconcile the Bjerknes model of the frontal depression with observations of tropical disturbances and the polar front with the inter-tropical front. These problems are by no means solved at the present time, but it is possible to indicate the more realistic models which have been proposed during the last twenty-five years.

The inter-tropical discontinuity, also termed the inter-tropical convergence zone (ITCZ) or, to avoid genetic implications, the equatorial low pressure trough, is generally not a frontal discontinuity in the mid-latitude sense since air mass differences in the tropics are frequently insignificant. This is particularly so with regard to density and it is mainly in west Africa and over the Indian sub-continent that the term inter-tropical front (ITF) has some validity. Equally, convergence of the two trade wind systems is far from continuous in space or time and the synoptic occurrence of convergence along the ITCZ is to a significant extent controlled by perturbations in the easterlies (Crowe, 1949 and 1950; Palmer, 1952).

## Wave disturbances

Attention was first drawn to the occurrence of wave disturbances in the trades by Dunn (1940) working in the Caribbean. He referred to these features as isallobaric waves, but this term is not now used. Dunn showed that bad weather develops behind the low pressure troughs, the converse of extra-tropical depressions. He noted that the great majority of the waves were stable and that development appeared to depend on the breaking-down of the trade wind inversion. Riehl (1954) was the first to examine these phenomena in detail by means of streamline analysis. This is necessary in view of the (assumed) non-geostrophic balance, and hence ambiguous relationships between wind and pressure fields, in low latitudes. The classical model of the easterly wave is shown in Figure 4.12 with its associated pattern of convergence and divergence when the easterlies are travelling faster than the wave form, a frequent but not habitual case. The explanation of the configuration rests upon the equation for the conservation of potential vorticity assuming adiabatic motion due to C-G. Rossby (1940):

$$\frac{(\zeta+f)}{\Delta p}=k$$

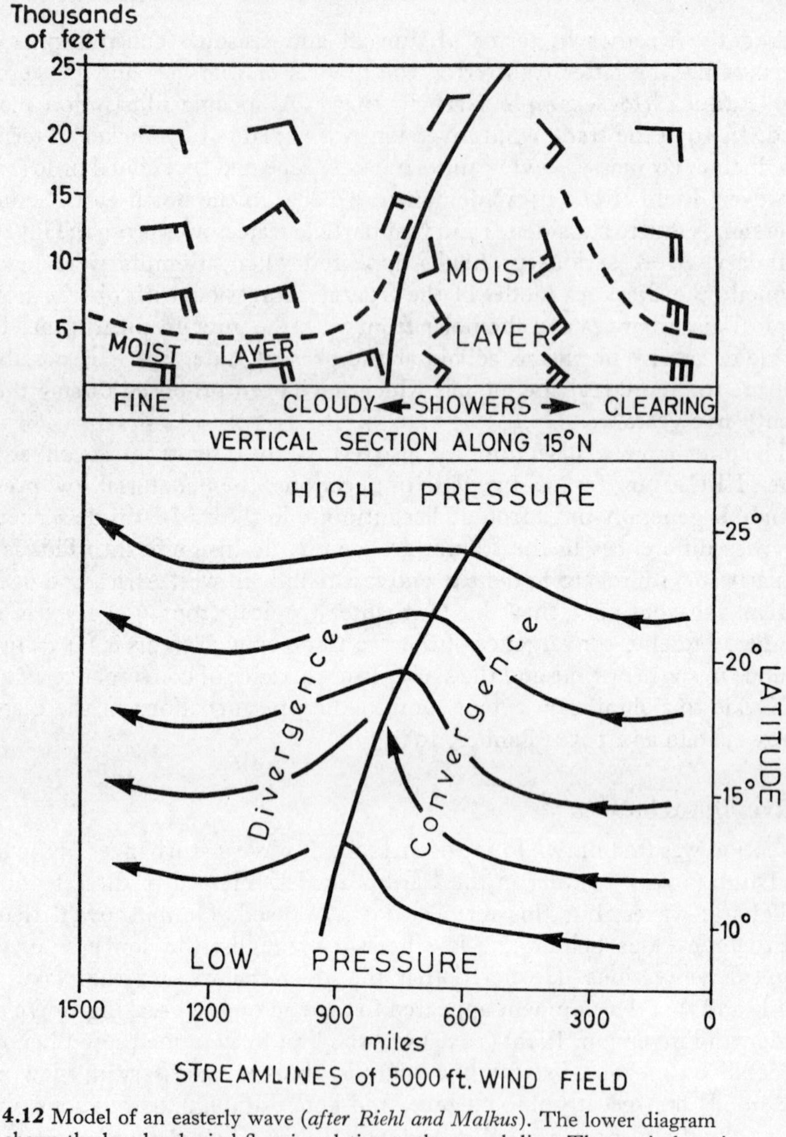


**4.12** Model of an easterly wave (*after Riehl and Malkus*). The lower diagram shows the low-level wind flow in relation to the trough line. The vertical section shows the wind direction (where the plotting convention assumes north is at the top of the diagram) and speed, the depth of the moist layer (dashed line) and the general weather sequence.

where $k$=a constant

$f$=the Coriolis parameter

$\zeta$=relative vorticity (for cyclonic vorticity $\zeta$ is positive)

$\Delta p$=the thickness of the layer between two isentropic surfaces.

The expression implies that for air moving northward from the rear of the wave towards the trough and acquiring greater absolute vorticity $(\zeta+f)$ due to increase of $f$ and $\zeta$, there must be concomitant increase in $\Delta p$. Such vertical expansion necessitates horizontal contraction i.e. convergence. Conversely there is divergence for air moving southward ahead of the wave and undergoing anticyclonic curvature. It must be noted that this argument assumes that curvature is more important for vorticity than lateral shear.

If the wave is travelling through the wind field it can be shown that the distribution of convergence is reversed. Thus on occasions when the easterlies decrease with height it is possible for divergence to overlie low-level convergence to the rear of the wave trough, creating optimum conditions for the development of bad weather as a result of vertical motion. Recent aerial surveys (Malkus and Riehl, 1964) emphasize the importance of upper tropospheric divergence for cloud build-up, especially when it is accompanied by a low-level wave disturbance.

Similar perturbations are observed in the central and western equatorial Pacific (Palmer, 1952), although there, as noted by La Seur (1960), the equatorial easterlies are convergent in the mean whereas the trades are divergent in the mean. Thus, when waves are embedded in the equatorial easterlies there are only isolated areas of divergence. A further point is that a double system of cyclonic circulation develops when a wave in the equatorial easterlies straddles the equator.

Freeman (1948) pointed out that the easterly wave theory fails at the equator, since there is no horizontal Coriolis component, yet disturbances still occur. He developed a physical-mathematical analogy between incompressible airflow beneath an inversion and the high-speed flow of a compressible fluid. Freeman suggested that jump-type waves were possible in the easterlies and gave an example from the New Guinea area. These ideas have not been further applied in this particular context, mainly perhaps as a result of the difficulty of using them for an operational model in areas of generally sparse data. However, the analogy has been developed with reference to meso-scale squall-lines and planetary blocking waves (Freeman, 1951; Fujita, 1955). The 'disturbance lines' of West Africa (Eldridge, 1957) are probably one such category of squall-line. They travel westwards at between 20–30 kt. (about the speed of the upper easterlies at 300 mb) and may be several hundred miles in length roughly from north to south.

It is probable that a wide range of models will be required for the varied perturbation types in the tropics (Forsdyke, 1960; La Seur, 1960). In this connection the use of satellite cloud pictures is proving a useful analytical tool and should eventually provide information on the more frequent patterns. On this basis for example Merritt (1964) suggests that Riehl's classical model of the easterly wave occurs less frequently than previously supposed and he illustrates a range of linear and vortical perturbations. A frequent disturbance type has a wave at low levels with a closed cyclonic circulation about the 600 mb level.

## Vortices

The development of tropical cyclones is a crucial problem for tropical meteorologists, but no satisfactory forecast model has yet been put forward. Early theoretical models based on convection cells or frontal concepts are now regarded as erroneous, and it is generally considered that the storms originate from pre-existing wave disturbances. Indeed in a few cases the development of Atlantic hurricanes has been successfully traced back to west Africa. Erickson (1963) reports that Hurricane Debbie of the 1961 season originated in this area and deepened after moving over the Cape Verde Islands.

The main problem is the mechanism responsible for triggering full hurricane development since many tropical vortices fail to attain such intensity (Malkus, 1958). The essential characteristic of a hurricane model in this respect appears to be the presence of a warm core with upper tropospheric temperatures 5°–8° C. warmer in the core than the surroundings. Without this feature the storm cannot develop very low surface pressure and the accompanying high wind speeds nor high level outflow. Two factors are thought to be necessary for the growth of the warm core. One is the release of latent heat by rising air, undiluted by entrainment of drier air of external origin. The 'hot towers' of cumulonimbus clouds may occupy less than one per cent of the disturbance and Scorer (1966) considers that a persistent shower area some 100 to 200 miles across is required to maintain an adequate latent heat supply. The second is that the rising air must have a higher heat content than normal tropical air in order to generate the requisite surface pressure decrease. The additional heat energy (approximately 2·5 cal/g) is acquired in the form of sensible and latent heat as the air spirals inward at high speed. Hence the whole process is dependent upon a feedback mechanism (Malkus, 1962, pp. 231–253; and Yanai, 1964). In the opinion of some authorities, moist adiabatic cooling is by itself insufficient to explain the observed warmth of the core and the effect of descending air in the eye from the ring of cloud forming the 'eye wall' probably accounts for the additional warmth. It may be noted that the warming process appears to operate from the upper troposphere downwards (Estoque, 1964). The inception of the eye is clearly a vital feature of any hurricane model but details of this process are still largely unknown.

The initial development of the cyclone prior to eye formation is dependent on interaction between divergence in the upper troposphere and a pre-existing low-level wave (Riehl, 1951 and 1954) but no complete model of this stage is available. One hypothesis put forward by Sawyer (1947) which linked high level outflow with negative absolute vorticity ('dynamic instability') as a starting mechanism is not now considered to be applicable. Baroclinic instability (see p. 104) is a more probable source of the initial growth of cyclones (Yanai, 1964). General requirements for development include sea-surface temperatures exceeding about 27° C. and a location several degrees from the equator to ensure cyclonic rotation. Once developed the system is self-maintaining through release of latent heat as long as there is a moisture supply and the air diverging aloft descends well away from the centre so as not to destroy the temperature difference between the core and its surroundings. In these respects the mature hurricane is a prime example of an open system maintaining dynamic equilibrium. When the storm moves over land or cool seas it decays rapidly.

## Monsoons

The classical model of the monsoon depicts the regime as a large-scale land-sea breeze phenomenon. Text-book statements about the Asian monsoon refer to it typically as follows:

> In January, decidedly high pressure develops over central Siberia, and the doldrums are somewhat south of the equator in the Indian Ocean. Steady north-east winds therefore blow out of Asia in an anticyclonic circulation across the Indian Ocean to the equator (Blair, 1942, p. 65),

although later in the book (p. 327) Blair noted the effect of the mountains separating the cold interior of Asia from the plains of India. However, as recent studies demonstrate, even a modified thermal model of summer heating over northern India leading to monsoonal inflow, and local outflow from the subcontinent in winter, differs considerably from reality.

The construction of three-dimensional models using upper air data is now beginning to clarify the process of seasonal change-over. According to Yin (1949) the winter flow pattern in the middle troposphere characteristically features a westerly jet stream both north and south of the Tibetan Plateau owing to the mechanical effect of the barrier. Downstream over China these currents re-unite. Analogues of this pattern have been obtained with laboratory models by Long (1952 and 1955). Subsidence tends to occur beneath the southern side of the branch of the westerlies over northern India, creating persistent dryness in the low-level north-easterlies. In early summer the southerly branch weakens and disintegrates leaving the main westerlies now over central Asia. This breakdown of the middle and upper tropospheric flow is linked in time with the onset of the south-west monsoon over India and a close causal relationship between these events was at first assumed.

Subsequent investigations cast doubt upon the barrier effect of the Tibetan Plateau. Chang Chia-Ch'eng (1960) shows that in fact there is a low frequency of wind maxima just north of the Tibetan Plateau and that two jet streams are also commonly observed to the west of the Plateau. Nevertheless, the branch over northern India is remarkably stable and Chia-Ch'eng demonstrates the existence of a strong latitudinal gradient of solar radiation between 20°–40° N. over 35°–80° E. from November to April which he holds responsible for the presence of the westerly maximum.

The breakdown in May–June is undoubtedly linked with the seasonal readjustment of the global circulation but Flohn (1960A) postulates a new role for the Tibetan Plateau. Much of the area is close to the 600 mb level with mountain peaks well above this. Absorption of solar radiation by this elevated source raises the air temperature considerably above that of the surrounding free air (Academica Sinica 1957, 1958 and Ramakrishnam *et al.* 1960). Since a warm air column has a less rapid decrease of pressure with height than a cold one high pressure should form over the area and Flohn (1960A) and Pisharoty and Asnani (1960) amply demonstrate the existence of an anticyclone at the 500 mb level. Paradoxically, the solenoidal field produces at the same time descent away from the plateau and rising air above the Tibetan anticyclone (Lockwood, 1965). The inception of this summer pattern undoubtedly contributes to the disruption of the upper westerly flow over northern India, since the anticyclonic circulation assists the development of an easterly jet, with a core about 150 mb, over southern Asia between 10°–15° N. as shown in Figure 4.13 (Koteswaram, 1958). The high level (Pacific) easterlies (above 8 km.) appear first over Indo-China in May, as a col develops about 95° E. in the upper troposphere between the western Pacific anticylonic cell and that over Africa-western Asia, and they push north-westward towards northern India in June and July (Ramanthan, 1960) aided by the easterly flow round the southern side of the Tibetan high at 500 mb. This model has recently been criticized, however, by Frost and Stephenson (1965) who find no evidence of a trough in the upper westerlies.

Breaks in the summer monsoon may occur even when it is apparently fully established. Ramaswamy (1962) shows that they are associated with the temporary redevelopment of a westerly jet stream over northern India and the removal of the Tibetan high by deep troughs during low index circulation. His results lend considerable weight to the foregoing models stressing the role of the upper tropospheric circulation.

A more general model of the circulation reversal over southern Asia in summer can now be examined. The reversal, which usually takes place in early June, is primarily a manifestation of the seasonal shift of the equatorial low pressure trough. Its movement permits the rapid northward expansion of the equatorial westerlies to produce the south-west monsoon. These westerlies are commonly regarded as south-east trades of the southern hemisphere deflected by the changed sign of the Coriolis parameter at the equator, but as Flohn (1960B) points out Halley in 1686 recognized that in the Indian Ocean the change of direction occurs at 2–3° South as a discontinuity not as a gradual transition. Equally, over Indonesia in the southern hemisphere summer the change is at 2–3° North. In Flohn's view (1960B, C and D) the

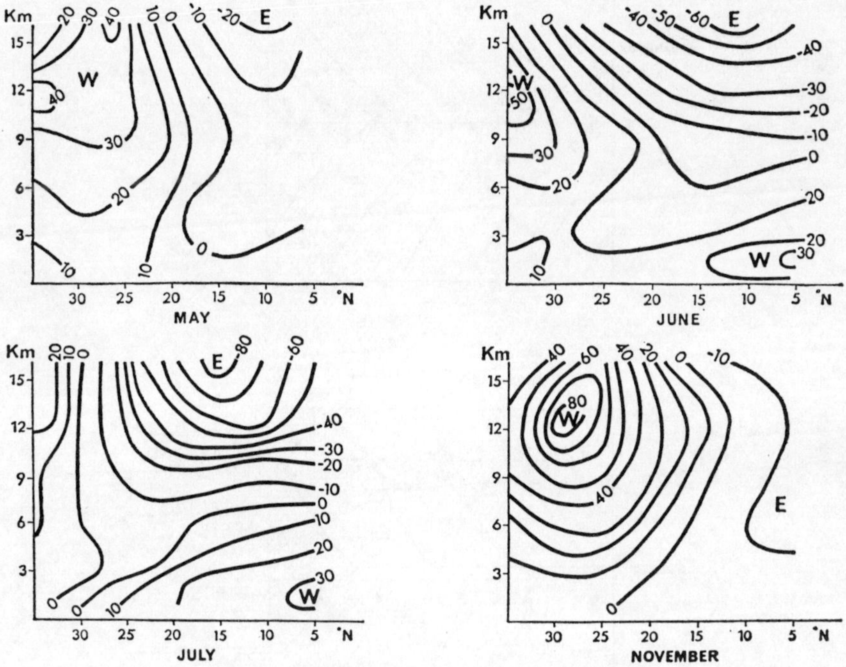


**4.13** Average zonal wind components (knots) for longitudes 75°–80° E. (*after Ramakrishnan, Sreenivasaiah and Venkiteshwaran, 1960*). June and July relate to the south-west monsoon season, November represents the north-east monsoon and May the transition to the south-west monsoon.

equatorial westerlies represent a planetary wind belt of the summer hemisphere, which would be continuous on a continental globe. This wind system results from the poleward displacement of the equatorial trough in summer, a movement which is demonstrably large over the land masses of the northern hemisphere. The strong pressure gradient between the equator and the trough induces deflection of the westerlies before they cross the equator due to the inertia of the predominant meridional current near the equator. On an oceanic globe the trough would move only a few degrees of latitude and no westerlies would intervene between the two Trade systems, due to the small

Coriolis component near the Equator. The circulation models for hypothetical extremes are illustrated in Figure 4.14. In a sense therefore, the thermal low still enters into the explanation but the new models differ sharply from the old in that they link the surface and upper tropospheric circulations over southern Asia and take account of global as well as regional factors.

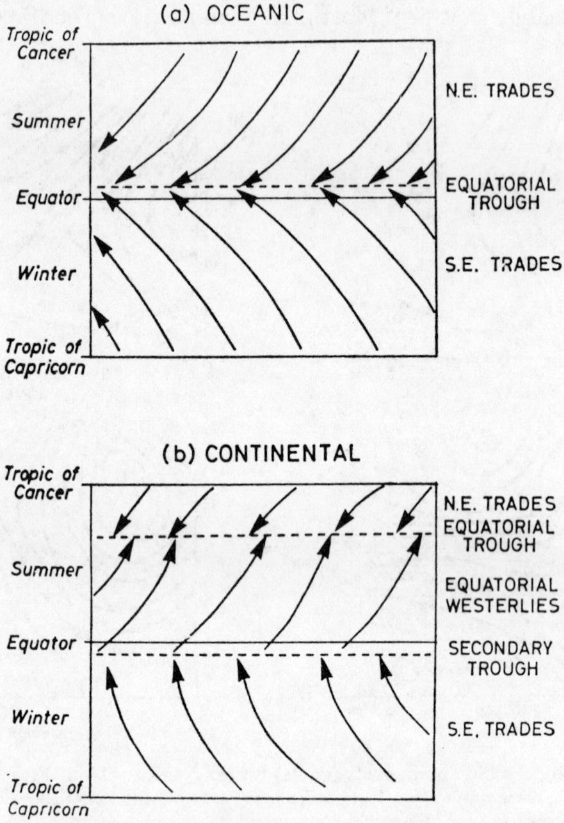


**4.14** Schematic circulation models (a) for an oceanic globe, (b) for a continental globe (*Flohn, 1960*).

## Equatorial circulation models

Attempts to apply the existing models of wave disturbances in African forecasting have generally failed and this has led to quite different ideas from those developed in other sectors of the tropics. A study of Johnson and Mörth (1960) provides a valuable illustration of model building apart from yielding new insight into equatorial flow patterns. The occurrence of daily rainfall over east Africa was shown not to be closely related to advective changes of moisture or stability conditions, but study of contour charts during periods of steady pressure fields suggested certain relationships between pressure, wind and weather. Hypothetical flow models and deduced patterns of convergence and divergence were found to show reasonable agreement with actual situations and the large-scale distribution of precipitation.

Johnson and Mörth term the two basic types of flow pattern the 'duct', where ridges in each hemisphere lead to easterly flow, and the 'drift' where there is trans-equatorial flow from high to low pressure (Fig. 4.15). The fundamental question is how does the air north of the equator 'know' whether pressure is high or low to the south of the equator, since a particle responds to instantaneous forces and in each case the contours parallel the equator. In view of the negligible curvature term it is concluded that the pressure force must be balanced by the Coriolis deflection even near the equator. Acceleration of air in the confluence zone of the duct entrance leads to convergence near the equator and this is most effective if the anticyclones are intensifying. Ageostrophic components may also encourage convergence through friction or the inability of the Coriolis term to maintain geostrophic balance within about three degrees of the equator (Johnson, 1963). In the drift pattern subgeostrophic velocity causes equatorward airflow and recurvature takes place at several degrees south as geostrophic equilibrium is recovered. This pattern regularly occurs over east Africa between January–March with fine weather over Kenya but convergence and rain in the westerlies over southern Tanzania.

Other patterns discussed by Johnson and Mörth are the 'bridge', with stable quasi-geostrophic westerly flow in both hemispheres and convergence near the equator, 'zonal downgradient flow' with meridional orientation of the pressure cells (Fig. 4.15) and the 'step' with geostrophic westerly flow in one hemisphere and geostrophic easterly flow in the other. The latter is more common in the upper troposphere.

## MESO-SCALE MODELS

The development of models for systems with dimensions of the order of 10–100 miles requires the assistance of radar meteorology and a close network of stations. Consequently this field of study dates mainly from post-World War II. The idea that localized surface heating controls all convective storms is still a common misconception although it has long been recognized that many thunderstorms follow clearly defined paths usually in association with cold front zones.

Investigations by Byers and Braham (Byers, 1959, Chap. 19) and especially Fujita (1955) have produced much information about the structure and life

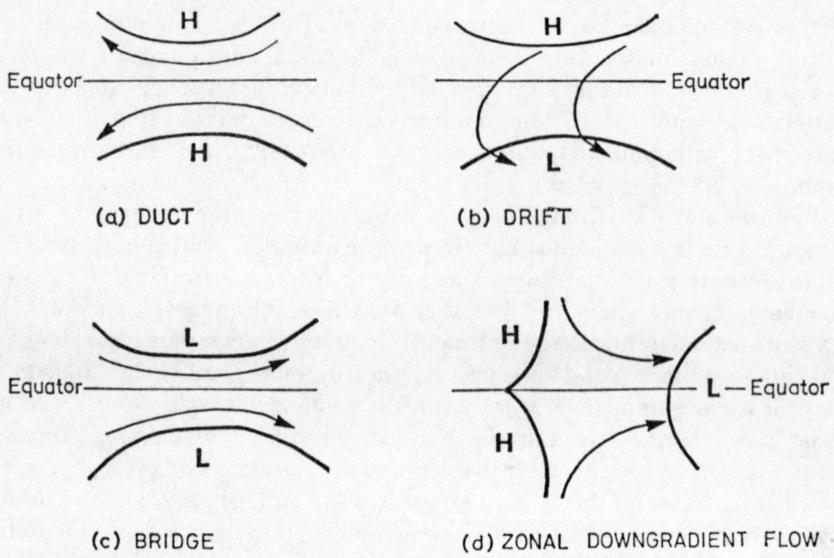


**4.15** Equatorial circulation models (*after Johnson and Mörth, 1960*).

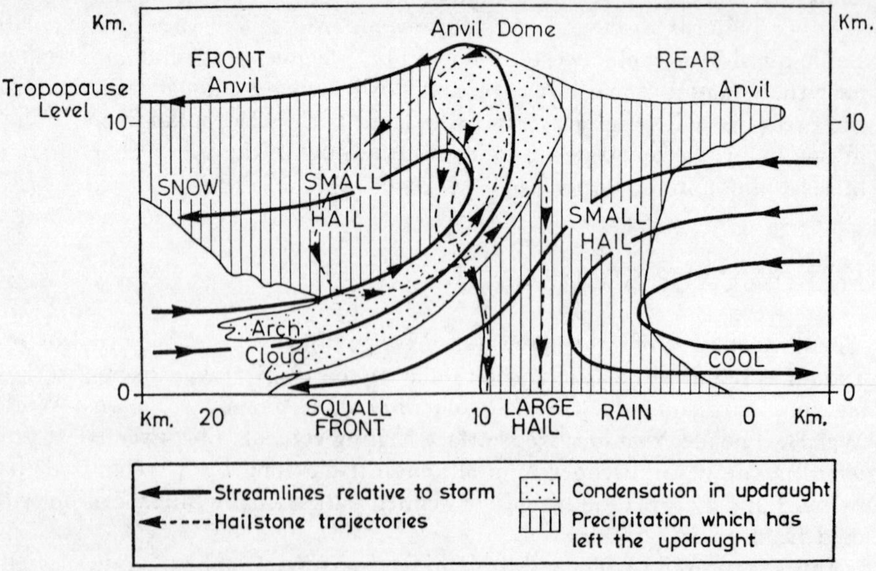


**4.16** An idealized hailstorm (*Ludlam, 1961*). The storm is travelling from right to left.

cycle of storms. Models of cold front storms show that in front of the storm there is a line of pressure surge behind which is a pressure excess of about +3 mb with a 'wake depression' (a deficit of about −2 mb) in the rear. The thunderstorm high is associated with a cold downdraught initiated by the drag of precipitation drops and the low is induced in the wake of the moving high, unless the general low-level winds are travelling faster than the pressure surge line. More detailed models including a three-dimensional one of airflow within a severe hailstorm have been presented by Ludlam (1961) and Browning and Ludlam (1962) on the basis of radar studies. The essential feature of their model (Fig. 4.16) is an updraught which is tilted backwards with low-level air ahead of the squall-line being shovelled upwards by undercutting (potentially) cold air moving from higher levels at the rear of the storm. At low levels flow relative to the storm is towards the squall-line. Given potentially cold air overlying potentially warm air and strong vertical wind shear the system, once organized, is self-maintaining.

The storm model readily accounts for the growth of large hail (an inch or more in diameter). The concentric structure of clear and opaque layers of ice was previously explained in terms of a fluctuating updraught but the idea of very unsteady currents with velocities of 30–40 m/sec able to carry large hailstones was never satisfactory. The new model, on the other hand, shows that hailstone embryos are blown ahead of the storm by strong upper tropospheric winds after their initial passage through the updraught and again fall back on to the scoop. Some small hailstones may now grow sufficiently quickly for their increasing fall-speed to slow down the rate of lift while small fluctuations in cloud water content and the transverse path of stones through the upcurrent are considered to be the means of growth of clear and opaque ice (Ludlam 1961). Large stones eventually fall from the updraught in a narrow belt to the rear of the storm.

## CLIMATOLOGICAL IMPLICATIONS

### Air masses and synoptic climatology

The classical studies of Bergeron (1928 and 1930) on air masses and their dynamic relationships with the major frontal systems were a logical development of the synoptic concepts of J. Bjerknes. The ideas provided a new basis for regional climatology removed from the earlier statistical abstractions and consequently air mass study has been applied in many parts of the world (Willett, 1933; Belasco, 1952). Nevertheless, the occurrence of fronts makes a complete climatological analysis using air masses impossible. Frisby and Green (1949) for example omitted consideration of 60 per cent of days with frontal systems in a study of air mass properties over Britain.

The air mass approach overlooks the fact that the air in any vertical column comprises particles which at different levels may have followed quite different trajectories. Furthermore, vertical stability and vapour content are far from being conservative properties. The concept remains of value for descriptive purposes if applied to the lowest 10,000 feet or so of the atmosphere but a more flexible approach is to analyse airflow pattern types. These may refer only to the pressure pattern and general airflow (Abercromby, 1883; Lamb, 1950) or may also take some account of fronts (Barry, 1960). Large-scale tropospheric steering patterns are referred to as *Grosswetter* (Baur, 1951). The complexities of weather-map pressure patterns raise their own classification problems, but synoptic-climatological models allow climatological averages to be calculated on a realistic synoptic basis rather than for arbitrary time periods (Barry, 1963).

The classification of synoptic maps involves the recognition of analogues, but most of the subjectivity involved in this procedure can now be eliminated by one of several mathematical specification techniques. Lund (1963) illustrated the use of correlation coefficients for this purpose. The correlation of pressure values over the north-eastern United States was calculated by computer for 445 days giving a (445 × 445) matrix. The pattern of the day which had the largest number of maps correlated with it (for a coefficient of at least +0·7) was selected as type A. After abstraction of these cases the day with next highest number of related maps was designated type B and so on. Ten such types accounted for 89 per cent of the period. The degree of similarity within each type is specified by the size of the correlation coefficient; a coefficient of 0·7 implies that 49 per cent of the variance of the pattern is common to a pair of maps.

A more elegant method involves the fitting of trend surfaces to the pressure map to obtain the maximum reducation of variance with a limited number of elementary surfaces (see Chap. 3). Examples of this approach are given for polar pressure patterns by Hare (1958).

The selection of synoptic patterns is only a first step towards analysis of the synoptic climatology but a more direct approach has been indicated by Grimmer (1963). By means of factor analysis he abstracted the major components of patterns of mean monthly temperature anomalies over Europe and thereby indicated the influence of continentality and blocking patterns. The technique generates the patterns of each component, thus avoiding the inherent tendency of trend surface analysis to predetermine the answer. At the same time it must be emphasized that the derived 'factors', in this case component patterns, are mathematical abstractions and are not necessarily uniquely related to an individual physical factor. Nevertheless, used with care, this technique holds considerable promise for study of the bases underlying climatic anomaly fields. A recent study by Steiner (1965) illustrates the application of this statistical model to the climatic regionalization of the United States.

## Models in palaeoclimatology

The key problem in palaeoclimatology is the relationship between the meridional temperature gradient and the intensity of the atmospheric circulation with its consequent effects upon the evaporation-condensation cycle. There are, for example, various models of the effect of solar radiation changes on the circulation. Simpson (1957) argues for an increase of solar radiation producing a steeper temperature gradient and a stronger circulation. Increased cloud and precipitation imply lower summer temperatures in higher latitudes and this may allow the persistence of snow-cover and eventually ice-caps. However, calculations for a model atmosphere by Sawyer (1963) indicate that increased solar radiation can give greater temperature rise near the poles and weakened extratropical circulation.

At present most theories of climatic change are unverifiable. For this reason, and because these ideas are treated in several standard texts, attention is concentrated on the more immediate question of the range of behaviour of the circulation, since we assume that changes in the mean circulation represent differing frequencies of various circulation types. The present as the key to the past is probably a quite inadequate guide here as we have no knowledge of the pattern of circulation on a globe without polar ice-caps and this situation appears to have been 'normal' throughout most of geological time. Furthermore our period of instrumental records covers only very minor climatic fluctuations during the last hundred years, so that these minor changes assume an importance disproportionate to their magnitude and duration. Recent energy budget calculations for an ice-free Arctic (Donn and Shaw, 1966; Fletcher, 1965) may provide a solution to the first of these problems but the data shortage cannot be fully remedied.

The improved models of the general circulation discussed earlier provide an initial basis for postulating the probable circulation patterns during recent palaeoclimatic epochs. Interest has centred primarily on glacial-interglacial conditions and much use has been made of zonal index concepts, particularly the simple high-low index model. Many workers hold the view that strong westerly flow in higher middle latitudes (high index) is an interglacial and interpluvial pattern (Butzer 1957A) while the circulation responsible for the glacial maxima resembled the low index model with an expanded circumpolar vortex and well-developed meridional components (Leighly, 1949; Willett, 1950). Butzer (1957B) considers that the retreat stages of the glacials were also characterized by high index circulations. More recently Willett and Sanders (1959, pp. 194–5) suggested that the type of circulation most conducive to glaciation is the low-latitude phase of the zonal pattern which in the index cycle (Riehl *et al.*, 1952) commonly precedes the northward shift of the westerlies and the establishment of high index. The extreme cellular type is thought to produce only localized extremes of temperature and precipitation.

Contrary views have been expressed by Flint and Dorsey (1945). They held that high index conditions were responsible for the inception of glaciation in eastern North America and similar suggestions have been advanced by Rex (1950) for Scandinavia. Rex argued that increased storminess associated with strong zonal flow aloft over Europe is required for glacierization and showed that blocking anticyclones over Scandinavia are unfavourable for glaciation. However, the high-latitude blocks described by Rex are associated with intensified zonal flow upstream and the 500 mb zonal index for 35°–55° N. remains above normal during the blocking period. As Willett (1950) pointed out this type of Scandinavian blocking needs to be differentiated from that in lower middle latitudes.

Many of the apparently conflicting views could probably be resolved if the circulation models, particularly index phases, employed in palaeoclimatological studies were carefully defined. Models of the circulation changes during the Pleistocene also need to take account of the time-lag between primary circulation changes and the development of a major ice-sheet over a period of approximately 20,000 years and secondary circulation changes consequent upon the presence of an ice-sheet. Again, these stages are not always clearly differentiated. A further consideration is that longitudinal differences of Pleistocene climates must have been considerable (Mather, 1954). For this reason over-simplified hemispheric models may be more misleading than specialized regional ones such as Butzer (1958 and 1961) has tried to develop for the Mediterranean area.

Another type of palaeoclimatic model concerns the relative chronology of the mid-latitude glacials and the lower latitude pluvials. Simpson (1957) and Butzer (1958) envisaged quite different relationships as shown in Table 4.1 and necessarily this affects the related circulation models which they proposed.

TABLE 4.1

| *Simpson* (1957) | | *Butzer* (1958) | |
|---|---|---|---|
| | | Late Würm | Post-Pluvial |
| | | Main Würm | Minor Pluvial |
| Würm | | Early Würm | Major Pluvial |
| Warm wet Inter-Glacial | Pluvial (Würm, Warm wet Inter-Glacial, Riss) | Inter-Glacial | Inter-Pluvial |
| Riss | | Riss | Pluvial |
| Cold dry Inter-Glacial | Inter-Pluvial | Inter-Glacial | Inter-Pluvial |
| Mindel | | Mindel | Pluvial |
| Warm wet Inter-Glacial | Pluvial (Mindel, Warm wet Inter-Glacial, Günz) | | |
| Günz | | | |
| Cold dry Inter-Glacial | Inter-Pluvial | | |
| Glacial | | | |
| Warm wet Inter-Glacial | Pluvial (Glacial, Warm wet Inter-Glacial, Glacial) | | |
| Glacial | | | |

A very different approach to climatic change has been outlined by Curry (1962) who uses stochastic (probability) models based on queue theory to suggest that the atmospheric circulation could shift from its present state of energy balance to a different equilibrium without there necessarily being any significant changes of energy input. The crucial factor is the amount of energy stored at any given time in the oceans and ice-caps. Over the short-term of months or a few years 'feed-back' processes may operate to affect the circulation through such physical factors as snow-cover, sea-surface, temperatures and soil moisture (Namias, 1963). Persistent changes in the zonal or meriodional flow components over longer time scales are likely to be controlled by the degree of anchoring of the long waves by solenoidal fields and orographic barriers. Models of this type are only just beginning to be developed but they appear in the present state of knowledge to provide a more realistic approach to the question of climatic changes than the numerous theories invoking unique causative agents.

## PREDICTION MODELS

Operational models for weather prediction (Bergeron, 1959) must be considered in terms of the particular time-scale involved and it is pertinent in this connection to summarize the relative time-orders of the major weather systems. These are:

| | |
|---|---|
| Mesoscale features | 1– 6 hours |
| Synoptic systems | 1– 3 days |
| Long waves | 7–10 days |
| Index cycle | 4– 6 weeks |
| Monsoon circulation | 3– 6 months |

The general categories of forecast are short-period ones covering 12–24 hours, extended-range for 5–7 days ahead and long-range outlooks for about a month. Naturally the type of model which can be applied is greatly controlled by the essential predictability of the atmosphere's behaviour over these different periods.

Short-range forecasts were originally based on rule-of-thumb extrapolation estimates of the movement of pressure systems and synoptic models of characteristic weather patterns of the type outlined by Abercromby and Marriott (1883) for the depression. The major developments up to the 1940's were improved synoptic models such as the Bjerknes-Solberg frontal wave. Even when numerical methods are fully developed there will undoubtedly remain a need for specialized models in short-range forecasting particularly in regions of sparse data, and in this respect the models of the easterly wave, the three-front depression and the jet stream are invaluable. Such models also help to maintain continuity of map analysis.

Different approaches are necessary in extended and long-range forecasting. Considerable stimulus was provided after about 1940 by the Rossby long-wave model and the zonal index concept. During the same period also analogue methods of synoptic climatology were being developed in search of operational models (Jacobs, 1946; Baür, 1951). Their application to long-range outlooks has been assisted by the introduction of physical principles including the specification of anomalies of surface temperature and ice distribution (Willett, 1951), but such forecasts remain primarily statistical. Selection of suitable map analogues on which to base forecasts has, however, been greatly facilitated by computer methods. Thirty-day forecasts of temperature and precipitation anomalies in the United States, published since 1947, are derived from calculated changes of the 700 mb contour pattern using kinematic methods and synoptic-climatological models (Namias, 1953; Klein, 1965). Attempts have been made to detect decadal trends on the basis of time-series analysis, including cycles of solar activity (Willet, 1961), but there is at present little prospect of developing a satisfactory method for such very long-range predictions (Craddock, 1964).

The most significant advances in forecasting are due to the development of numerical models based on the fundamental hydrodynamical equations. The methods demand accurate knowledge of the initial state of the atmosphere and equations which adequately predict the atmospheric motion while remaining tractable. In 1922 Richardson tried to use this approach to calculate sea-level pressure tendencies by applying the equations of horizontal motion and the continuity equation at short time intervals over a spatial (finite difference) grid. However, the real changes were swamped by cumulative errors due to 'noise' resulting from the effect of sound and gravity waves on the equations.

Post-World War II developments in this field initially used very simplified structures. One of these was the barotropic 'one-parameter' model (Eliassen, 1956; Haltiner and Martin, 1957; Sutton, 1960). The principle of the conservation of absolute vorticity is applied to the 500 mb surface which is approximately the mean level of non-divergence such that divergence (convergence) below this level is compensated by convergence (divergence) above. It is assumed that the flow is geostrophic and that the motion at 500 mb is independent of that at other levels. Under these assumptions the vorticity equation (p. 99) becomes

$$\frac{\partial Q}{\partial t}+u\frac{\partial Q}{\partial x}+v\frac{\partial Q}{\partial y}=0$$

where $f+\zeta=Q$ (the absolute vorticity)

$$\zeta = \frac{g}{f}\left(\frac{\partial^2 h}{\partial x^2} + \frac{\partial^2 h}{\partial y^2}\right), \; u = -\frac{g}{f}\frac{\partial h}{\partial y} \text{ and } v = \frac{g}{f}\frac{\partial h}{\partial x}$$

where $h$=the height of the 500 mb surface
$g$=gravity

The forecast equation

$$\left(\frac{\partial^2}{\partial x^2} + \frac{\partial^2}{\partial y^2}\right)\frac{\partial h}{\partial t} = -u\frac{\partial Q}{\partial x} - v\frac{\partial Q}{\partial y}$$

must be solved numerically from grid values of contour heights to obtain the pattern of the height tendency over the grid (the left hand side of the equation). At each grid point finite difference approximations are used to evaluate the terms on the right (Hess, 1959; Knighting, 1958; Sutton, 1960). The height tendency for a given time interval may then be added to the initial contour field and the computations are repeated. This process is carried out until the period of the forecast has been covered.

Bolin (1955) compares barotropic forecasts and actual 500 mb contours for 24, 48 and 72 hour periods. The model is shown to be reasonably satisfactory for the short periods as long as the patterns are displaced with little development because the model specifies that the vorticity is only advected by the geostrophic wind. The results are also affected by the restrictions imposed by the size of the grid area.

Subsequent models include consideration of two or more levels and more realistic atmospheres with additional parameters. For example, height changes at the 750 mb and 250 mb levels allow the calculation of vertical motion at 500 mb and determination of thickness changes in the 750–250 mb layer introduces baroclinicity into the model (Eliassen, 1956; Haltiner and Martin, 1957). The use of three levels provides better estimates of vertical motion and allows the effects of horizontal variations of static stability to be included. The results of such studies indicate that operational numerical forecasting has a bright future and although there is a long way to go before the weather elements themselves can be predicted directly progress is being made along these lines.

## REFERENCES

ABERCROMBY, R., [1883], On certain types of British weather; *Quarterly Journal of the Royal Meteorological Society*, 9, 1–23.

ABERCROMBY, R. and MARRIOTT, W., [1883], Popular weather prognostics; *Quarterly Journal of the Royal Meteorological Society*, 9, 27–43.

ACADEMICA SINICA, [1957 and 1958], On the general circulation over eastern Asia; *Tellus*, 9, 432–446; 10, 58–75 and 299–312.

ALLEN, R. A., FLETCHER, J., HOLMBOE, J., NAMIAS, J. and WILLETT, H. C., [1940], Report on an experiment in five-day weather forecasting; *Papers in Physical Oceanography and Meteorology*, 8, No. 3 (Massachusetts Institute of Technology and Woods Hole), 94 pp.

ANDERSON, R., BOVILLE, B. W. and MCCLELLAN, D. E., [1955], An operational frontal contour-analysis model; *Quarterly Journal of the Royal Meteorological Society*, 81, 588–599.

BARRY, R. G., [1960], A note on the synoptic climatology of Labrador-Ungava; *Quarterly Journal of the Royal Meteorological Society*, 86, 557–565.

BARRY, R. G., [1963], Aspects of the synoptic climatology of central southern England; *Meteorological Magazine*, 92, 300–308.

BAUR, F., [1951], Extended-range weather forecasting; in *Compendium of Meteorology*; (Ed., T. F. Malone), American Meteorological Society, (Boston, Mass.), 814–833.

BELASCO, J. E., [1952], Characteristics of air masses over the British Isles; *Geophysical Memoirs, Meteorological Office*, 11 (No. 87), 34 pp.

BERGERON, T., [1928], Uber die dreidimensional verknüpfende Wetteranalyse; *Geofysiske Publikationer*, 5, No. 6 (Oslo), 111 pp.

BERGERON, T., [1930], Richtlinien einer dynamischen Klimatologie; *Meteorologische Zeitschrift*, 47, 246–262.

BERGERON, T., [1954], The problem of tropical hurricanes; *Quarterly Journal of the Royal Meteorological Society*, 80, 131–164.

BERGERON, T., [1959], Methods in scientific weather analysis and forecasting; in *The atmosphere and the sea in motion*; (Ed., B. Bolin), (Rockefeller Institute Press, New York), 440–470.

BJERKNES, J. and SOLBERG, H., [1922], Life cycle of cyclones and the polar front theory of atmospheric circulation; *Geofysiske Publikationer*, 3, No. 1 (Oslo), 18 pp.

BLAIR, T. A., [1942], *Climatology, general and regional*, (New York), 484 pp.

BOLIN, B., [1950], On the influence of the earth's orography on the general characteristics of the westerlies, *Tellus*, 2, 184–195.

BOLIN, B., [1952], Studies of the general circulation of the atmosphere; *Advances in Geophysics*, 1, 87–118.

BOLIN, B., [1955], Numerical forecasting with the barotropic model; *Tellus*, 7, 27–49.

BOVILLE, B. W. and KWIZAK, M., [1959], *Fourier analysis applied to hemispheric waves of the atmosphere*; CIR-3155, TEC-292, Meteorological Branch, Department of Transport, Canada, 21 pp.

BOYDEN, C. J., [1963], Jet streams in relation to fronts and the flow at low levels; *Meteorological Magazine*, 92, 319–328.

BRADBURY, D. L., [1958], On the behaviour of cyclones and anticyclones as related to zonal index; *Bulletin of the American Meteorological Society*, 39, 149–151.

BROWNING, K. A. and LUDLAM, F. H., [1962], Airflow in convective storms; *Quarterly Journal of the Royal Meteorological Society*, 88, 117–135.

BUTZER, K. W., [1957A], Mediterranean pluvials and the general circulation of the Pleistocene; *Geografiska Annaler*; 39, 48–53.

BUTZER, K. W., [1957B], The recent climatic fluctuation in lower latitudes and the general circulation of the Pleistocene; *Geografiska Annaler*; 39, 105–113.

BUTZER, K. W., [1958], Quaternary stratigraphy and climate in the Near East; *Bonner Geographische Abhandlungen*, Nr. 24, 157 pp.

BUTZER, K. W., [1961], Climatic change in arid regions since the Pliocene; in *A history of land use in arid regions*; (Ed., L. D. Stamp), Arid Zone Research, UNESCO, (Paris), 31–56.

BYERS, H. R., [1959], *General Meteorology*, (New York), 3rd Edn., 540 pp.

CHANG CHIA-CH'ENG, [1960], Some views on the nature of the China monsoon; *Trudy Glavnoi Geofizicheskoi Observatorii, Leningrad*, Vyp. 90, 24–42 (translated 1961 by Office of Technical Services, Washington, D.C.).

CHARNEY, J. G. and ELIASSEN, A., [1949], A numerical method for predicting the perturbations of the middle westerlies; *Tellus*, 1 (2), 38–54.

CLAPP, P. F., [1961], Normal heat sources and sinks in the lower troposphere in winter; *Monthly Weather Review*, 89, 147–162.

COURT, A., [1957], Climatology: complex, dynamic and synoptic; *Annals of the Association of American Geographers*, 47, 125–136.

CRADDOCK, J. M., [1964], The analysis of time series for use in forecasting; *The Statistician*, 15, 167–190.

CROWE, P. R., [1949], The trade-wind circulation of the world; *Transactions of the Institute of British Geographers*, No. 15, 37–56.

CROWE, P. R., [1950], The seasonal variation in the strength of the trades; *Transactions of the Institute of British Geographers*, No. 16, 23–47.

CROWE, P. R., [1965], The geographer and the atmosphere; *Transactions of the Institute of British Geographers*, No. 36, 1–19.

CURRY, L., [1962], Climatic change as a random series; *Annals of the Association of American Geographers*, 52, 21–31.

DONN, W. L. and SHAW, D. M., [1966], The heat budgets of an ice-free and an ice-covered Arctic Ocean; *Journal of Geophysical Research*, 71, 1087–93.

DÖÖS, B. R., [1962], The influence of exchange of sensible heat with the earth's surface on the planetary flow; *Tellus*, 14, 133–147.

DUNN, G. E., [1940], Cyclogenesis in the tropical Atlantic; *Bulletin of the American Meteorological Society*, 21, 215–229.

EADY, E. T. and SAWYER, J. S., [1951], Dynamics of flow patterns in extra-tropical regions; *Quarterly Journal of the Royal Meteorological Society*, 77, 531–551.

ELDRIDGE, R. H., [1957], A synoptic study of West African disturbance lines; *Quarterly Journal of the Royal Meteorological Society*, 83, 303–314.

ELIASSEN, A., [1956], Instability theories of cyclone formation; Numerical forecasting; Chapters 15 and 18 in *Weather Analysis and Forecasting* Volume 1, S. Petterssen, (New York), 305–319 and 371–387.

ELIASSEN, A., [1958], A study of the atmospheric waves on the basis of zonal harmonic analysis; *Tellus*, 10, 206–215.

ERICKSON, C. O., [1963], An incipient hurricane near the West African coast; *Monthly Weather Review*, 91, 61–68.

ESTOQUE, M. A., [1964], Formation and structure of tropical hurricanes; in *Proceedings of the Symposium on Tropical Meteorology, Rotorua, New Zealand*, 1963, (Ed., J. W. Hutchings), (Wellington, New Zealand), 595–613.

FERREL, W., [1856], An essay on the winds and currents of the ocean; *Nashville Journal of Medicine and Surgery*, 11(4), 287–301.

FITZROY, R., [1863], *The weather book. A manual of practical meteorology*, (London), 480 pp.

FLEAGLE, R. G., [1957], On the dynamics of the general circulation; *Quarterly Journal of the Royal Meteorological Society*, 83, 1–20.

FLEAGLE, R. G., [1960], The general circulation; *Science Progress*, 48, 72–81.

FLETCHER, J. O., [1965], *Climate and the heat budget of the central Arctic*; The Rand Corporation, (Santa Monica, California), 22 pp.

FLINT, R. F. and DORSEY, JR., H. G., [1945], Iowan and Tazewell drifts and the North American ice sheet; *American Journal of Science*, 243, 615–635.

FLOHN, H., [1960A], Recent investigations on the mechanism of the 'summer monsoon' of southern and eastern Asia; in *Monsoons of the World*, India Meteorological Department, (Delhi), 75–88.

FLOHN, H., [1960B], Monsoon winds and the general circulation; in *Monsoons of the World*, India Meteorological Department, Delhi, 65–74.

FLOHN, H., [1960C], The structure of the inter-tropical convergence zone; in *Tropical Meteorology in Africa*, (Ed., D. J. Bargman), Munitalp Foundation, (Nairobi), 244–251.

FLOHN, H., [1960D], Equatorial westerlies over Africa, their extension and significance; in *Tropical Meteorology in Africa*, (Ed., D. J. Bargman), Munitalp Foundation, (Nairobi), 253–264.

FORSDYKE, A. G., [1951], On zonal and other indices; *Meteorological Magazine*, 80, 151–160.

FORSDYKE, A. G., [1960], Synoptic models of the tropics; in *Tropical Meteorology in Africa*, (Ed., D. J. Bargman), Munitalp Foundation, (Nairobi), 14–19.

FREEMAN JR., J. C., [1948], An analogy between the equatorial easterlies and supersonic gas flow; *Journal of Meteorology*, 5, 138–146.

FREEMAN JR.,. J. C., [1951], The solution of nonlinear meteorological problems by the method of characteristics; in *Compendium of Meteorology*, (Ed., T. F. Malone), American Meteorological Society, (Boston, Mass.), 421–433.

FRENZEN, P., [1955], Westerly flow past an obstacle in a rotating hemispheric shell; *Bulletin of the American Meteorological Society*, 36, 204–210.

FRISBY, E. M. and GREEN, F. W. H., [1949], Further notes on comparative regional climatology; *Transactions of the Institute of British Geographers*, No. 15, 143–151.

FROST, R. and STEPHENSON, P. H., [1965], Mean streamlines and isotachs at standard pressure levels over the Indian and West Pacific Oceans and adjacent land areas; *Geophysical Memoirs, Meteorological Office*, 14 (No. 109), 24 pp.

FUJITA, T., [1955], Results of detailed synoptic studies of squall-lines; *Tellus*, 7, 405–436.

FULTZ, D., [1951A], Non-dimensional equations and modelling criteria for the atmosphere; *Journal of Meteorology*, 8, 262–267.

FULTZ, D., [1951B], Experimental analogies to atmospheric motions; in *Compendium of Meteorology*, (Ed., T. F. Malone), American Meteorological Society, (Boston, Mass.), 1235–1248.

FULTZ, D., [1960], Experimental models of rotating fluids and possible avenues for future research; in *Dynamics of Climate*, (Ed., R. L. Pfeffer), (Pergamon Press), 71–77.

FULTZ, D., [1961], Developments in controlled experiments on larger scale geophysical problems; *Advances in Geophysics*, 7, 1–103.

GALLOWAY, J. L., [1958A], The three-front model; its philosophy, nature, construction and use; *Weather*, 13, 3–10.

GALLOWAY, J. L., [1958B], The three-front model, the tropopause and the jet stream; *Weather*, 13, 395–403.

GARBELL, H. A., [1947], *Tropical and equatorial meteorology*, (London), 237 pp.

GILCHRIST, B., [1954], The seasonal phase changes of thermally produced perturbations in the westerlies; *Proceedings of the Toronto Meteorological Conference 1953*, 129–131.

GODSON, W. L., [1950], The structure of North American weather systems; *Centenary Proceedings of the Royal Meteorological Society*, 89–106.

GRIMMER, M., [1963], The space-filtering of monthly surface anomaly data in terms of pattern, using empirical orthogonal functions; *Quarterly Journal of the Royal Meteorological Society*, 89, 395–408.

HADLEY, G., [1735], Concerning the cause of the general tradewinds; *Philosophical Transactions, London*, 29, 58–62.

HALTINER, G. J. and MARTIN, F. L., [1957], *Dynamical and physical meteorology*, (New York), 470 pp.

HARE, F. K., [1958], The quantitative representation of the north polar pressure field; in *The Polar Atmosphere Symposium*, Part I, (Ed., R. C. Sutcliffe), (Pergamon Press), 137–150.

HARE, F. K., [1960], The westerlies; *Geographical Review*; 50, 345–367.

HARE, F. K., [1962], The stratosphere; *Geographical Review*; 52, 525–547.

HARE, F. K., [1965], Energy exchanges and the general circulation; *Geography*, 50, 229–241.

HESS, S. L., [1959], *Introduction to theoretical meteorology*, (New York), 362 pp.

HIDE, R., [1953], Some experiments on thermal convection in a rotating liquid; *Quarterly Journal of the Royal Meteorological Society*, 79, 161.

HOLMBOE, J. and BJERKNES, J., [1944], On the theory of cyclones; *Journal of Meteorology*, 1, 1–22.

HUSCHKE, R. E. (Ed.), [1959], *Glossary of meteorology*; American Meteorological Society, (Boston, Mass.), 638 pp.

JACOBS, W. C., [1946], Synoptic climatology; *Bulletin of the American Meteorological Society*, 27, 306–311.

JEFFREYS, H., [1926], On the dynamics of geostrophic winds; *Quarterly Journal of the Royal Meteorological Society*, 52, 85–104.

JOHNSON, D. H., [1962], Tropical meteorology; hurricane and typhoons; *Science Progress*, 50, 403–419.

JOHNSON, D. H., [1963], Tropical meteorology; other weather systems; *Science Progress*, 51, 587–601.

JOHNSON, D. H. and DANIELS, S. M., [1954], Rainfall in relation to the jet stream; *Quarterly Journal of the Royal Meteorological Society*, 80, 212–217.

JOHNSON, D. H. and MÖRTH, H. T., [1960], Forecasting research in East Africa; in *Tropical meteorology in Africa*, (Ed., D. J. Bargman), Munitalp Foundation, (Nairobi), 56–132.

KOTESWARAM, P., [1958], The easterly jet stream in the tropics; *Tellus*, 10, 43–57.

KLEIN, W. H., [1965], Synoptic climatological models for the United States; *Weatherwise*, 18, 252–259.

KNIGHTING, E., [1958], Numerical weather forecasting; *Weather*, 13, 39–50.

KRISHNAMURTI, T. N., [1961], On the role of the subtropical jet stream of winter in the atmospheric general circulation; *Journal of Meteorology*, 18, 657–670.

LAMB, H. H., [1950], Types and spells of weather around the year in the British Isles: Annual trends, seasonal structure of the year, singularities; *Quarterly Journal of the Royal Meteorological Society*, 76, 393–429.

LAMB, H. H., [1959], The southern westerlies: a preliminary survey; main characteristics and apparent associations; *Quarterly Journal of the Royal Meteorological Society*, 85, 1–23.

LA SEUR, N. E., [1954], On the asymmetry of the middle-latitude circumpolar current; *Journal of Meteorology*, 11, 43–57.

LA SEUR, N. E., [1960], Tropical synoptic models; in *Tropical meteorology in Africa*, (Ed., D. J. Bargman), Munitalp Foundation, (Nairobi), 47–54.

LEIGHLY, J., [1949], On continentality and glaciation; *Geografiska Annaler*, 31, 133–145.

LOCKWOOD, J. G., [1965], The Indian monsoon – a review; *Weather*, 20, 2–8.

LONG, R. R., [1952], The flow of a liquid past a barrier in a rotating spherical shell; *Journal of Meteorology*, 9, 187–199.

LUDLAM, F. H., [1961], The hailstorm; *Weather*, 16, 152–162.

LUND, I. A., [1963], Map-pattern classification by statistical methods; *Journal of Applied Meteorology*, 2, 50–65.

LYLE, H. H. and BRYSON, R. A., [1960], Harmonic analysis of the annual march of precipitation over the United States; *Annals of the Association of American Geographers*, 50, 157–171.

MALKUS, J. S., [1958], Tropical weather disturbances: why do so few become hurricanes? *Weather*, 13, 75–89.

MALKUS, J. S., [1962], Inter-change of properties between sea and air. Large-scale interactions; in *The Sea*, volume 1, (Ed., M. N. Hill), (Interscience Publishers, New York), 88–294.

MALKUS, J. S. and RIEHL, H., [1964], Cloud structure and distributions over the tropical Pacific Ocean; *Tellus*, 16, 275–287.

MANABE, S., SMAGORINSKY, J. and STRICKLER, R. F., [1965], Simulated climatology of a general circulation model with a hydrologic cycle; *Monthly Weather Review*, 93, 769–798.

MATHER, J. R., [1954], The present climatic fluctuation and its bearing on a reconstruction of Pleistocene climatic conditions; *Tellus*, 6, 287–301.

MERRITT, E. S., [1964], Easterly waves and perturbations, a reappraisal; *Journal of Applied Meteorology*, 3, 367–382.

MILES, M. K., [1962], Wind, temperature and humidity distribution at some cold fronts over S.E. England; *Quarterly Journal of the Royal Meteorological Society*, 88, 286–300.

MÖRTH, H. T., [1964], Primary factors governing tropospheric circulations in tropical and subtropical latitudes; in *Proceedings of the Symposium on Tropical Meteorology, Rotorua, New Zealand*, 1963, (Ed., J. W. Hutchings), (Wellington, New Zealand), 31–41.

MURRAY, R. and JOHNSON, D. H., [1952], Structure of the upper westerlies; a study of the wind field in the eastern Atlantic and western Europe in September 1950; *Quarterly Journal of the Meteorological Society*, 78, 186–199.

NAMIAS, J., [1950], The index cycle and its role in the general circulation; *Journal of Meteorology*, 17, 130–139.

NAMIAS, J., [1953], Thirty-day forecasting: A ten-year experiment; *Meteorological Monographs*, 2 (No. 6), 83 pp.

NAMIAS, J., [1963], Surface-atmosphere interactions as fundamental causes of drought and other climatic fluctuations; in *Changes of Climate*, Arid Zone Research XX, UNESCO, (Paris), 345–359.

NAMIAS, J. and CLAPP, P. F., [1951], Observational studies of general circulation patterns; in *Compendium of Meteorology*, (Ed., T. F. Malone), American Meteorological Society, (Boston, Mass.), 551–567.

NEWTON, C. W., [1959], Synoptic comparisons of jet stream and Gulf Stream systems; in *The atmosphere and the sea in motion*, (Ed., B. Bolin), (Rockefeller Institute Press, New York), 288–304.

NEWTON, C. W. and PERSSON, A. V., [1962], Structural characteristics of the subtropical jet stream and certain lower stratospheric wind systems; *Tellus*, 14, 221–241.

PALMÉN, E., [1951A], The aerology of extratropical disturbances; in *Compendium of Meteorology* (Ed., T. F. Malone), American Meteorological Society, (Boston, Mass.), 599–620.

PALMÉN, E., [1951B], The role of atmospheric disturbances in the general circulation; *Quarterly Journal of the Royal Meteorological Society*, 77, 337–354.

PALMÉN, E., [1954], General circulation of the tropics; in *Proceedings of the Symposium on Tropical Meteorology, Rotorua, New Zealand*, 1963 (Ed., J. W. Hutchings), (Wellington, New Zealand), 3–30.

PALMÉN, E. and VUORELA, L. A., [1963], On the mean meridional circulation in the northern hemisphere during the winter season; *Quarterly Journal of the Royal Meteorological Society*, 89, 131–138.

PALMER, C. E., [1951], Tropical meterology; in *Compendium of Meteorology* (Ed., T. F. Malone), American Meteorological Society, (Boston, Mass.), 859–880.

PALMER, C. E., [1952], Tropical meteorology; *Quarterly Journal of the Royal Meteorological Society*, 78, 126–164.

PÉDELABORDE, P., [1958], *The monsoon*; (translated 1963), (London), 196 pp.

PEDGLEY, D. E., [1962], *A course in elementary meteorology*; (H.M.S.O., London), 189 pp.

PENNER, C. M., [1955], A three-front model for synoptic analyses; *Quarterly Journal of the Royal Meteorological Society*, 81, 89–91.

PETTERSSEN, S., [1956], *Weather analysis and forecasting*, 2nd Edition, (New York), Vol. I 428 pp., Vol. II 266 pp.

PHILLIPS, N. A., [1956], The general circulation of the atmosphere: a numerical experiment; *Quarterly Journal of the Royal Meteorological Society*, 82, 123–164.

PISHAROTY, P. R. and ASNANI, G. C., [1960], Flow pattern over India and neighbourhood at 500 mb during the monsoon; in *Monsoons over the World*, India Meteorological Department, (Delhi), 112–117.

PRIESTLEY, C. H. B., [1949], Heat transport and zonal stress between latitudes; *Quarterly Journal of the Royal Meteorological Society*, 75, 28–40.

PRIESTLEY, C. H. B., [1951], Physical interactions between tropical and temperate latitudes; *Quarterly Journal of the Royal Meteorological Society*, 77, 200–214.

QUENEY, P., [1948], The problem of airflow over mountains: a summary of theoretical studies; *Bulletin of the American Meteorological Society*, 29, 16–26.

RAMAKRISHNAN, K. P., SREENIVASAIAH, B. N., and VENKITESHWARAN, S. P., [1958], Upper air climatology of India and neighbourhood in the monsoon seasons; in *Monsoons over the World*, India Meteorological Department, (Delhi), 3–34.

RAMANTHAN, K. R., [1958], Monsoons and the general circulation of the atmosphere – a review; in *Monsoons over the World*, India Meteorological Department, (Delhi), 53–64.

RAMASWAMY, C., [1962], Breaks in the Indian summer monsoon as a phenomenon of interaction between the easterly and the subtropical westerly jet streams; *Tellus*, 14, 337–349.

REITER, E. R., [1963], *Jet stream meteorology*; (Chicago), 515 pp.

REX, D. F., [1950], Blocking action in the middle troposphere and its effect on regional climate. Part II, The climatology of blocking action; *Tellus*, 2, 275–301.

RIEHL, H., [1951], Aerology of tropical storms; in *Compendium of Meteorology* (Ed., T. F. Malone), American Meteorological Society, (Boston, Mass.), 902–913.

RIEHL, H., [1954], *Tropical meteorology*, (New York), 392 pp.

RIEHL, H., [1962A], *Jet streams of the atmosphere*, Technical Paper No. 32, Department of Atmospheric Science, Colorado State University, (Fort Collins, Colorado), 117 pp.

RIEHL, H., [1962B], General atmospheric circulation of the tropics; *Science*, 135, 13–22.

RIEHL, H., and FULTZ, D., [1957], Jet stream and long waves in a steady rotating dishpan experiment: structure of the circulation; *Quarterly Journal of the Royal Meteorological Society*, 83, 215–231.

RIEHL, H., and FULTZ, D., [1958], The general circulation in a steady rotating dishpan experiment; *Quarterly Journal of the Royal Meteorological Society*, 84, 389–417.

RIEHL, H., YEH, T. C., MALKUS, J. C. and LA SEUR, N. E., [1951], The north-east trade of the Pacific Ocean; *Quarterly Journal of the Royal Meteorological Society*, 77, 598–626.

RIEHL, H. and Collaborators, [1952], Forecasting in middle latitudes; *Meteorological Monographs*, 1 (No. 5), 80 pp.

RICHARDSON, L. F., [1922], *Weather prediction by numerical process*, (Cambridge), 236 pp.

ROSSBY, C.-G., [1940], Planetary flow patterns in the atmosphere; *Quarterly Journal of the Royal Meteorological Society*, Supplement to 66, 68–87.

ROSSBY, C.-G., [1941], The scientific basis of modern meteorology; in *Climate and Man*, United States Department of Agriculture Yearbook, (Washington, D.C.), 599–655.

ROSSBY, C.-G., [1949], On the nature of the general circulation of the lower atmosphere; in *The atmospheres of the earth and planets* (Ed., G. P. Kuiper), (University of Chicago Press), 16–48.

ROSSBY, C.-G., and collaborators, [1939], Relations between variations in the intensity of the zonal circulation and displacements of the semi-permanent centres of action; *Journal of Marine Research*, 2, 38–55.

ROSSBY, C.-G. and WILLETT, H. C., [1948], The circulation of the upper troposphere and lower stratosphere; *Science*, 108, 643–652.

SABBAGH, M. E. and BRYSON, R. A., [1962], Aspects of the precipitation climatology of Canada investigated by the method of harmonic analysis; *Annals of the Association of American Geographers*, 52, 426–440.

SALTZMANN, B. and PEIXOTO, J. P., [1957], Harmonic analysis of the mean northern-hemisphere wind field for the year 1950; *Quarterly Journal of the Royal Meteorological Society*, 83, 360–364.

SANSOM, H. W., [1951], A study of cold fronts over the British Isles; *Quarterly Journal of the Royal Meteorological Society*, 77, 96–120.

SAWYER, J. S., [1958], Temperature, humidity and cloud near fronts in the middle and upper troposphere; *Quarterly Journal of the Royal Meteorological Society*, 375–388.

SAWYER, J. S., [1963], Notes on the response of the general circulation to changes in the solar constant; in *Changes of Climate*, Arid Zone Research XX, UNESCO, [Paris], 333–336.

SCORER, R. S., [1958], *Natural aerodynamics*, (Pergamon Press), 312 pp.

SCORER, R. S., [1966], Origin of cyclones; *Science Journal*, 2, 46–52.

SHAW, W. N. and LEMPFERT, R. K. G., [1906], *The life history of surface air currents*; Meteorological Committee, M.O. 174, (London), 107 pp.

SHEPPARD, P. A., [1958], The general circulation of the atmosphere; *Weather*, 13, 323–336.

SHEPPARD, P. A., [1962], Properties and processes at the earth's surface in relation to the general circulation; *Advances in Geophysics*, 9, 77–96.

SIMPSON, G. C., [1957], Further studies in world climate; *Quarterly Journal of the Royal Meteorological Society*, 83, 459–481.

SMAGORINSKY, J., [1953], The dynamic influence of large scale heat sources and sinks on the quasi-stationary mean motions of the atmosphere; *Quarterly Journal of the Royal Meteorological Society*, 79, 342–366.

SMAGORINSKY, J., [1963], General circulation experiments with the primitive equations; *Monthly Weather Review*, 91, 99–165.

SMAGORINSKY, J., [1964], Some aspects of the general circulation; *Quarterly Journal of the Royal Meteorological Society*, 90, 1–14.

STAFF MEMBERS, METEOROLOGY DEPARTMENT, UNIVERSITY OF CHICAGO; [1947], On the general circulation of the atmosphere in middle latitudes, *Bulletin of the American Meteorological Society*, 28, 255–280.

STARR, V. P., [1948], An essay on the general circulation of the earth's atmosphere; *Journal of Meteorology*, 5, 39–43.

STARR, V. P. and WHITE, R. M., [1951], A hemispherical study of the atmospheric angular-momentum balance; *Quarterly Journal of the Royal Meteorological Society*, 77, 215–225.

STARR, V. P. and WHITE, R. M., [1952], Schemes for the study of hemispheric exchange processes; *Quarterly Journal of the Royal Meteorological Society*, 78, 407–410.

STARR, V. P. and WHITE, R. M., [1954], *Balance requirements of the general circulation*; Geophysical Research Papers No. 35, Geophysics Research Directorate. (Cambridge, Mass.), 57 pp.

STEINER, D., [1965], A multivariate statistical approach to climatic regionalization and classification; *Tijdschrift van het Koninklij Nederlandsch Aardrijkskundig Genootschap*. 82 (4), 329–347.

SUTCLIFFE, R. C., [1947], A contribution to the problem of development; *Quarterly Journal of the Royal Meteorological Society*, 73, 370–383.

SUTCLIFFE, R. C., [1949], The general circulation – a problem in synoptic meteorology; *Quarterly Journal of the Royal Meteorological Society*, 75, 417–430.

SUTCLIFFE, R. C., [1950], Discussion on the general circulation; *Centenary Proceedings of the Royal Meteorological Society*, 180–183.

SUTCLIFFE, R. C., [1951], Mean upper contour patterns of the northern hemisphere – the thermal synoptic viewpoint; *Quarterly Journal of the Royal Meteorological Society*, 77, 435–440.

SUTCLIFFE, R. C., [1954], Cyclones and anticyclones – a comparative study; *Proceedings of the Toronto Meteorological Conference* 1953, 139–143.

SUTCLIFFE, R. C. and FORSDYKE, A. G., [1950], The theory and use of upper air thickness patterns in forecasting; *Quarterly Journal of the Royal Meteorological Society*, 76, 189–217.

SUTTON, O. G., [1960], *Understanding weather*, (Penguin Books), 215 pp.

SUTTON, O. G., [1962], *The challenge of the atmosphere*, (London), 227 pp.

TREWARTHA, G. T., [1961], *The earth's problem climates*, (University of Wisconsin Press, Madison), 334 pp.

TUCKER, G. B., [1959], Mean meridional circulation in the atmosphere; *Quarterly Journal of the Royal Meteorological Society*, 85, 209–224.

TUCKER, G. B., [1962], The general circulation of the atmosphere; *Weather*, 17, 320–340.

TUCKER, G. B., [1965], The equatorial tropospheric wind regime, *Quarterly Journal of the Royal Meteorological Society*, 91, 140–150.

VON HELMHOLTZ, H., [1888], Über atmosphärischer Bewegungen, trans. C. Abbé in The mechanics of the earth's atmosphere; *Smithsonian Institute Miscellaneous Collection*, 34 (843), 31–129.

WIIN-NIELSEN, A., BROWN, J. A. and DRAKE, M., [1964], Further studies of energy exchange between the zonal flow and the eddies; *Tellus*, 16, 168–180.

WILLETT, H. C., [1950], The general circulation at the last (Würm) glacial maximum; *Geografiska Annaler*, 32, 179–187.

WILLETT, H. C., [1951], The forecast problem; in *Compendium of Meteorology* (Ed., T. F. Malone), American Meteorological Society, (Boston, Mass.), 731–46.

WILLETT, H. C., [1961], The pattern of solar climatic relationships; *Annals of the New York Academy of Sciences*, 95, 89–106.

WILLETT, H. C. and SANDERS, F., [1959], *Descriptive meteorology*; 2nd Edn., (Academic Press, New York), 355 pp.

YANAI, M., [1964], Formation of tropical cyclones, *Reviews of Geophysics*, 2, 367–414.

YIN, M. T., [1949], A synoptic-aerological study of the onset of the summer monsoon over India and Burma; *Journal of Meteorology*, 6, 393–400.

CHAPTER FIVE

# Hydrological Models and Geography

ROSEMARY J. MORE

## INTRODUCTION

Hydrology is concerned with the science of water, its occurrence, circulation and distribution, its chemical and physical properties, its relation to the natural environment and its relation to living things, including man. The influence of water in moulding the landscape has been one of the principal concerns of physical geographers (e.g. Leopold, Wolman and Miller, 1964). The distribution of water has been important in the location of human settlements, and in the development of agriculture and industry. In Western Europe and North America the provision of adequate water supplies and the apportionment of water between competing demands are critical problems and ones which may set a limit to further economic developments.

Hydrological work has been traditionally dominated by engineers, whose scientific traditions have encouraged them to strive after generalizations and predictive schemes of a model character. Hence much of the present science of hydrology must fall naturally under the general heading of model building. One class of model associated with hydrology is that of hydraulic scale models of engineering projects, such as those built at the Wallingford Hydraulics Research Station or at the University of Delft's Hydraulics Laboratory. Although these hardware models are of immense practical value to the engineer (Allen, 1952), they are not the only models which have been used by hydrologists, for conceptual and mathematical models have also been applied to the appreciation and analysis of hydrological processes.

The basic conceptual model in hydrology is the idea of a cycle of water in its gaseous, liquid and solid form, and Figure 5.1 shows the principal components of this cycle. Water is evaporated from the oceans, stored as atmospheric moisture, and deposited as precipitation, which may be in the form of snow, sleet, hail, rain or dew. If the water resources of the earth are considered in terms of a budget, rainfall forms the largest item on the income side. However, much of the rain which falls does not immediately reach the river courses, but is lost by evaporation, transpiration, infiltration to soil moisture reserves, or deeper percolation to ground water in pervious rocks.

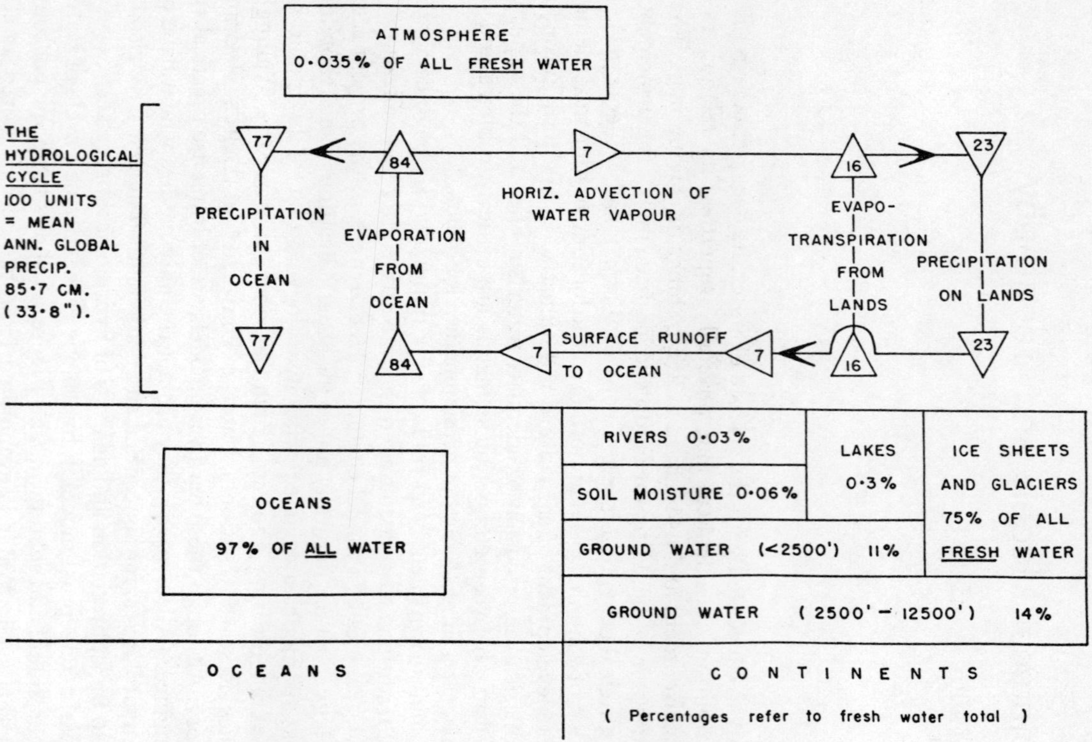


**5.1** The hydrological cycle and terrestrial water storage. The oceanic percentage relates to *all* terrestrial water; the percentages for continental and atmospheric water to all *fresh* water. The units in the hydrological cycle are related to an assumed 100 units of mean global precipitation (77 oceanic and 23 continental).

The water surplus to these needs and processes flows as direct runoff into the streams, lakes and major watercourses. One of the chief problems in hydrology is to define the relationship between rainfall 'input' and that part of 'output' represented by direct runoff, but because of the number of intervening factors, their complexity and the problems involved in measuring them, the nature of this relationship is still only very partially understood.

Lack of meteorological and hydrological data has hindered a greater understanding of hydrological processes. Large areas of the globe, notably over the oceans, have no recording stations and many existing stations have only limited instrumentation. This deficiency is being remedied by work done during the International Hydrological Decade (1965–1975) and, increasingly, by information remotely sensed from spacecraft (Carter *et al*, 1966). The increased volume of data received from these new sources is creating problems both of data processing and analysis, and conventional methods of manual computation (see, for example, Beard, 1962) are being supplemented by computer techniques (Johnson, 1965; Johnson and Lang, 1965). These techniques have not only made routine calculations less laborious, but have enabled new techniques to evolve which seem likely to contribute to our understanding of fundamental hydrological principles (O'Donnell, 1966).

In relation to the limited knowledge which we have regarding the energy and water budgets of the earth, the hydrological cycle itself may be regarded as a large-scale conceptual model which attempts to simplify a complex reality. Within the 'great cycle' there are numerous smaller cycles. For example, water is evaporated from the surface of lakes and falls again as precipitation on the surrounding hills, returning by streams and rivers into the lakes from which it was evaporated. The smaller cycles are all related to the 'great cycle' system, and all its components can be thought of as interlocking subsystems, so that, if any one of them varies, others will change in response. The quantities of water in the various subsystems of the cycle are continually changing, so that mathematical models (i.e. those used to quantify hydrological processes) which are made to simulate the cycle numerically must be dynamic or self-adjusting (George, 1965). Furthermore, variations in the significance of the component parts of the cycle vary from one climatic region to another, such that, for example, areas with different climates experience differences in amounts, duration, intensity and occurrence of rainfall and runoff.

For practical convenience, a more local hydrological cycle can be identified in catchment area units, which correspond to river valleys bounded by watersheds. The water resources of catchment areas are assessed in terms of water balances, which are prepared in a similar fashion to a bank balance, and which show the 'income', 'expenditure' and 'credit balance' of water. These studies are most commonly based on the model used by Penman (1950) to assess the water balance of the River Stour in Essex. The mathematical model which

Penman used to express the water balance for an area over any period was:

Initial storage + Rainfall = Evaporation + Runoff + Final storage.

By using this formula Penman was able to estimate monthly changes in storage for the Stour catchment over the period 1933–1948. The catchment was generally assumed to be at field capacity (i.e. the level of moisture content at which the maximum amount of undrainable water is held) at the end of March. The calculation then proceeded by subtracting evaporation loss from rainfall and making estimates of the gain to storage under different conditions of vegetational root cover and its location over the catchment. Figure 5.2 shows that there was a very good agreement between Penman's calculations regarding total storage and movements of rest water levels in wells in the chalk.

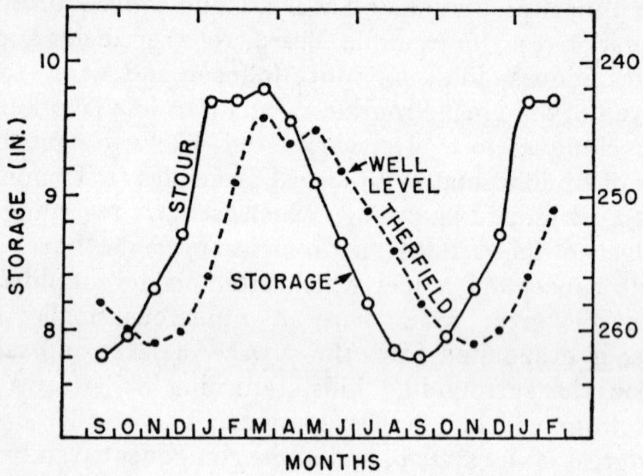


**5.2** Estimated mean monthly changes in storage for the Stour catchment, Essex (1933–1948) compared with observed mean monthly well levels at Therfield on the same chalk ridge some 30 miles away. The discrepancies in the peak levels can be attributed to differences in precipitation between the two areas (i.e. the Stour had wetter winters and drier summers during the period), and the two-month phase difference to the time for percolating water to reach the deep water table. (*Source: Penman, 1950, p. 465*).

## FUNCTION OF MODELS IN HYDROLOGY

Models are used for three main purposes in hydrology: (1) to simplify and generalize a complex reality, (2) to predict forthcoming hydrological events, and (3) to plan the future use of water resources.

## Models for simplification and generalization

The design of hydrometric networks is based on model theory, in so far as, for example, the real pattern of rainfall distribution is generally so variable in time and in space that it would require a very large number of rain gauges to give a detailed picture of the distribution. However, the cost of rain gauges, the problems of maintenance and the lack of observers often make it impossible to have as close a grid as would be desirable, and make it necessary to sample the rainfall distribution with a small number of judicially-spaced gauges. (See Krumbein and Graybill, 1965, for general considerations of areal sampling.)

Two main approaches to estimating from point samples the average depth of rainfall over an area have been used (Chow, 1964, pp. 9–26 to 9–49). The Thiessen method (1911) assumes that the amount at any station has fallen over that part of the catchment nearest to the station. It is applied by constructing a Thiessen polygon network, the polygons being formed by the perpendicular bisectors of the lines joining nearby stations. The area of each polygon is determined and its proportion to the total catchment area is used to weight the rainfall amount of the station in the centre of the polygon when finding the catchment average fall. It is necessary to change the polygons each time a station is added to or taken away from the network.

The isohyetal method consists of drawing lines of equal rainfall amount, using observed amounts at stations and any additional factors available to adjust or interpolate between observed stations. The average depth is then determined by computing the incremental volume between each pair of isohyets, adding these incremental amounts and dividing by the total area.

A recent approach to the sampling and generalization of hydrological conditions has been suggested by means of the use of experimental and representative basins (International Association of Scientific Hydrology, 1965; UNESCO, 1965). The philosophy behind the use of these 'model' basins is that a relatively small area, in which accurate measurements of precipitation, runoff, evaporation, groundwater, sediment, etc., are made, may be used to generalize and predict conditions over larger areas. A small catchment, or sub-area of a larger catchment, can be held to be typical in two senses: (1) the data of the different elements may provide 'sample' statistics to supply information about the 'population' of the larger area, or (2) the relationships derived between some of the elements, e.g. rainfall-runoff, rainfall-evaporation, may be assumed equivalent to the relationships for the same elements in the larger areas. The sampled results would be particularly useful where larger areas have complications which are otherwise impossible to assess (e.g. 'foreign' water intrusion due to the presence of towns).

Experimental basins have been defined as basins which have been instrumented to measure most, or all, of the major components of the water balance, and where some land-use change is to be made during the period of investigation. Representative basins are similarly instrumented, but no land-use change is contemplated. The British International Hydrological Decade Committee list twenty experimental catchments which are to be studied during the decade period (1965–1975), including the Grendon Underwood experimental catchment on the River Ray in the Thames Valley, which is being studied by the Hydrological Research Unit at Wallingford, and three representative basins wherein the present land use and hydrology relationships are being investigated during the decade period.

An interesting variant of the concept of representative hydrological basins is that of the 'Vigil Network' (Leopold, 1962; Slaymaker and Chorley, 1964; Leopold and Emmett, 1965; Emmett, 1965), under which small basins in supposedly representative lithologic and climatic regions are instrumented in an attempted to obtain sediment budgets in a manner similar to hydrological budgets. Thus soil and talus creep, surface slope erosion, stream bank erosion, and channel and reservoir erosion and sedimentation are measured by a variety of instrumental procedures varying from soil strain gauges to the surveying of painted debris movements. There are three main objectives of this Vigil Network Project: (1) the estimation of regional rates of erosion; (2) the association of individual erosional and climatic events; and (3) the investigation of the areal patterns of erosion and sedimentation in an attempt to understand the nature of the geometrical transformations of the earth's surface in the most direct manner. It is hoped that erosional models of basins under differing lithologic and climatic conditions will ultimately emerge, similar to the type of hydrological basin models resulting from representative hydrological basin investigations.

The above models are merely examples of some of the simplest attempts at areal generalizations in hydrology. Much of the remainder of this chapter is concerned with other kinds of models designed to simplify and generalize a complex hydrological reality.

## Models for hydrological prediction

Engineers responsible for the design of structures are naturally concerned with the magnitude and frequency of hydrological events experienced over long time periods. With the exception of records of discharges of the Nile and of some rivers in China, the majority of runoff records are for periods of less than a hundred years. This is a very short time span, and, within it, there is a great probability that extreme high and low values for water levels are not recorded. Yet, in the construction of flood-control dams, water-supply reservoirs and impounding works for irrigation schemes, it is important to know the 'recurrence intervals' for given magnitudes of floods and droughts. Thus there has arisen a practical need for predictive models. Wolf (1966A) has

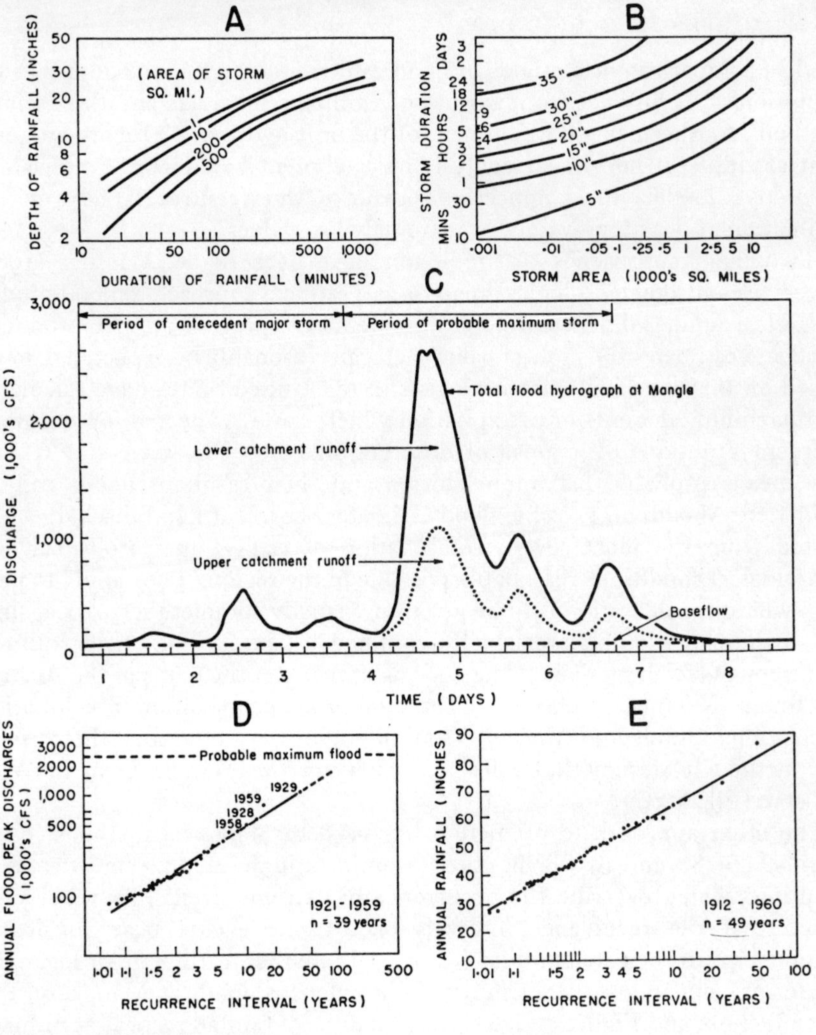


**5.3** The prediction of hydrological events.

A Enveloping depth/duration curves of maximum thunderstorm rainfall for small areas in the United States (*Source: After Berry, Bollay and Beers. Redrawn from Wiesner, 1964, p. 162*).

B Enveloping isohyets of greatest observed depths of rainfall in the United States (*Source: After Berry, Bollay and Beers. Redrawn from Wiesner, 1964, p. 162*).

C Probable maximum flood hydrograph for the Jhelum River at Mangla, West Pakistan, calculated with respect to the worst characteristics of 14 major historic storms combined to give an estimate of the Probable Maximum Precipitation in a 72-hour storm. This is assumed to follow 72 hours after the start of an antecedent major storm for which the actual flood of 4–7 August, 1958, was used as a model. The abnormal form of the flood hydrograph results from the unusual drainage basin geometry (*Source: Binnie and Mansell-Moullin, 1966, Fig. 15*).

D Recurrence intervals of flood peaks at Mangla (1921–1959) on a logarithmic Gumbel plot. (*Source: Binnie and Mansell-Moullin, 1966, Fig. 16*).

E Recurrence intervals of annual rainfall amounts of different magnitude for Sydney, Australia (1912–1960) on an arithmetic Gumbel plot (*Source: Dury, G. H., 1964, Some Results of a Magnitude-Frequency analysis of Precipitation*; Australian Geographical Studies, *2, p. 23*).

made a comparison of methods of flood estimation in which he describes the evolution of techniques from the use of empirical formulae and the 'rational method', to Sherman's development of the unit hydrograph theory and current attempts at non-linear analysis of catchment behaviour. Emphasis in predictive models for engineering purposes thus centres largely on the estimation of peak high values or extremely low values.

Two main approaches to prediction have been made. Wiesner (1964) advocates hydrometeorological analysis as a rational approach to flood prediction. He examines the meteorological elements and processes which produce a flood and estimates the highest value each can reasonably be expected to have. Based on this approach, meteorologists in the United States have calculated the maximum amounts of precipitation which can be expected for storms of different durations over areas of different size (see Fig. 5.3A and B). By this means probable maximum storms and their resultant floods can be estimated. Maximum possible flood estimates represent flood discharges expected from the most severe combination of critical meteorological and hydrological conditions reasonably possible in the region. They apply to projects where consideration is to be given to virtually complete security against risk to human life and are usually confined to the estimation of spillway requirements for high dams. The designs of many of the dams of the Australian Snowy Mountains scheme are based on probable maximum precipitation calculations. Binnie and Mansell-Moullin (1966) have described their use of this method to design the spillway capacity of the Mangla Dam in West Pakistan (Fig. 5.3C).

The other approach to prediction has been by statistical methods. E. J. Gumbel (1958A and 1958B) has been prominent in developing statistical distributions which describe the occurrence of extreme events. Gumbel suggested that the recurrence intervals of extreme events bear consistent relationships to their magnitudes (expressed in either arithmetic or logarithmic terms), and Figure 5.3D shows such a plot for floods at Mangla on the River Jhelum, and Figure 5.3E a similar analysis of rainfall records at Sidney, Australia (Dury, 1964). Gumbel proposed that the design of a structure should be based on the predicted flood discharge at a chosen recurrence interval (e.g. 50 years). While Gumbel's methods have great theoretical interest, they are not widely used at present for design purposes by practising engineers. Wiesner criticizes Gumbel's method in that he considers it statistically unjustifiable to extrapolate far beyond the period of record. He believes, for example, that the once-in-a-1,000-year flood, estimated from 50 years of stream-flow record, has confidence limits so wide that the results are of little value.

However, these two approaches to prediction are complementary rather than independent of one another in that statistical inferences are used to a considerable extent in hydrometeorology. Binnie and Mansell-Moullin

(1966), for example, use statistical frequency analysis to determine the recurrence intervals of flood peaks on the Jhelum (Fig. 5.3D). Both these approaches to flood and drought estimation are in their infancy and techniques are subject to constant revision, but this is a part of scientific hydrology in which the need to develop better techniques is critical and model theory plays a large part in their evolution.

### Models for resource planning

The third function of models in hydrology – that of resource planning – is closely related to their predictive function. The management of water-supply reservoirs, both during a single year and over a longer period, requires the use of stochastic methods, such as queueing theory, so that it becomes possible to apportion a varying input to storage over specified demands, and within specified risks of failure. On a broader scale, models are used to achieve the optimum exploitation of resources in catchment areas which have already been partially developed, and to plan water-resource utilization in undeveloped catchments. Eckstein (1958), Krutilla and Eckstein (1958), McKean (1958), Maass *et al* (1962), Kuiper (1965), Smith and Castle (1965), Schwab *et al* (1966) and Wolf (1966B) have discussed the principles of water-resource management. Such multiple-purpose schemes are in various stages of planning and execution in the Rhône and Durance valleys in France (Giguet, 1957), in the Tennessee and Central Californian valleys of the United States and in the Mekong valley in South-East Asia (White *et al.*, 1962). An opportunity has been given to River Authorities in England and Wales (under the terms of the Water Resources Act, 1963) to develop their rivers in the light of water-resource management techniques (Thorn, 1966). Further reference will be made to the resource-planning function of hydrological models in the section on hydro-economic models.

## PHYSICAL AND SYSTEMS HYDROLOGY

Classification is always a difficult and arbitrary operation, and this is particularly true of hydrological models at this stage of their development. The use of computers has given model theory greater scope, and ideas as to the nature and function of models are developing rapidly (O'Donnell, 1966). Dawdy and O'Donnell (1965) and Amorocho and Hart (1964) agree, in broad terms, in their analyses of current methods of approach to hydrological research. They distinguish two main schools of thought in model building, one of which may be described as Physical Science Research, or Physical Hydrology, and the other as System Synthesis Investigations.

Physical hydrology involves the pursuit of scientific research into the basic operation of each component of the hydrological cycle in order to gain a full understanding of the mechanisms and interactions involved. Although the immediate motivation of an individual researcher may not go beyond a specific phenomenon, it is implicit that a full synthesis of the hydrological cycle *in toto* is the ultimate goal.

The other approach seeks to achieve the comprehensive simulation of catchment behaviour in the form of gross catchment models. In these models catchment components are treated in a lumped form and their behaviour is simulated in an approximate way by largely empirical relationships. The construction of the component parts of the model and the parameters of the relationships are adjusted until known responses, within an acceptable tolerance, are achieved from known inputs. Subjective decisions must be made in the choice of components and in the specification of behavioural relationships. The adjustments of the parameters has generally also been subjective, but objective techniques have been used. The construction and manipulation of such models rely to a great extent on certain methods of systems engineering, whereby the relationship between input and output is synthesized in a model of the prototype process.

These two approaches to model building in hydrology are in some respects complementary to one another. As the investigations of the physical hydrologists provide more information on the details of the component parts of the hydrological cycle, so the comprehensive simulation of catchment behaviour, by means of overall models, will show where further detailed specification is needed.

## MODELS IN PHYSICAL HYDROLOGY

Because of the large number of interlocking component parts and the difficulties of scaling them down to manageable size, few attempts have yet been made to construct a complete working model of the physical hydrology of a catchment (e.g. Chery, 1966). However, much research has been carried out in an attempt to obtain a better understanding of each of the component parts with the long-term object of synthesizing this knowledge into a complete model. Figure 5.4A shows schematically the disposition of the main storm rainfall components (minus evaporation and transpiration) with time for an ideal drainage basin, assuming a constant rainfall intensity.

### Precipitation simulation: A hardware scale model in hydrology

One of the basic requirements for many laboratory models is the ability to manufacture precipitation of the required intensity and duration, for, ideally,

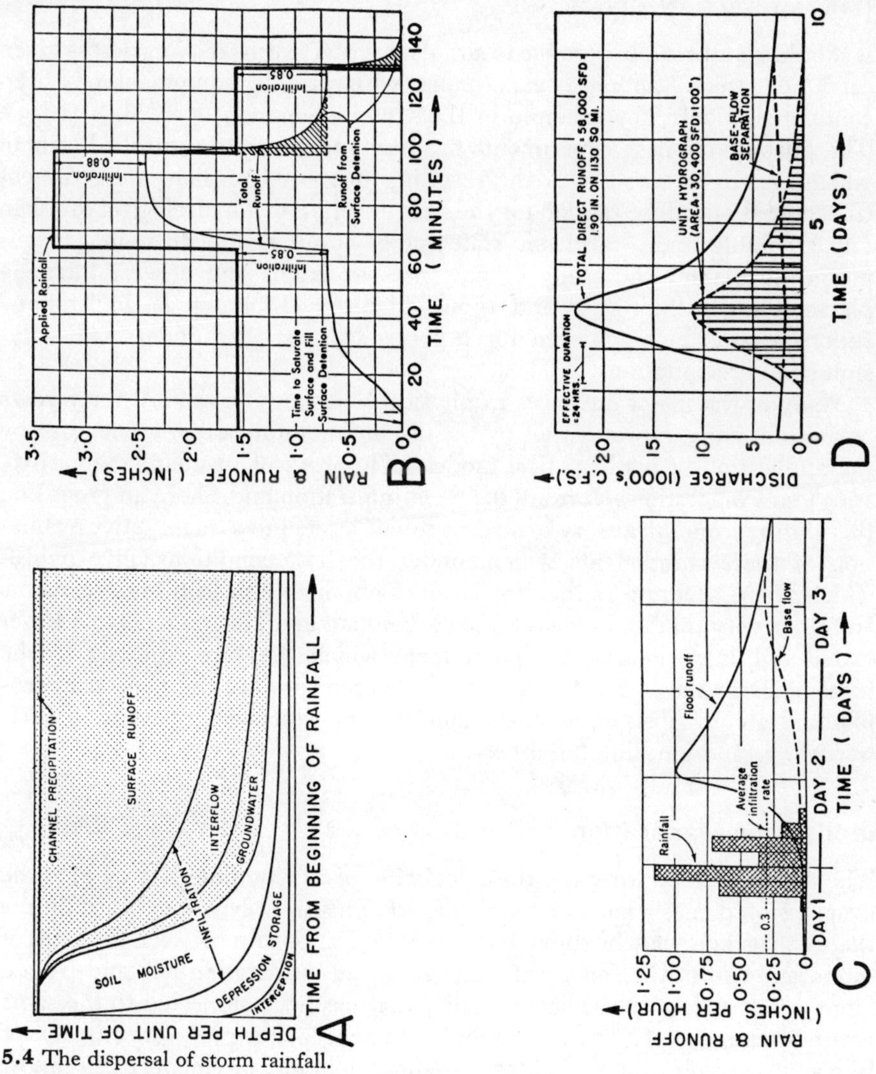


**5.4** The dispersal of storm rainfall.

A Schematic diagram of the dispersal of storm rainfall through time. Evapo-transpiration is ignored, and rainfall is assumed to be constant and continuous. The dotted area represents that proportion which eventually becomes streamflow (*Source: Linsley, Kohler and Paulhus, 1949, p. 410*).

B Runoff on a plot resulting from artificially-applied precipitations of 1·55 and 3·30 inches per hour (*Source: After Sharp and Holton. From Foster*, Rainfall and Runoff, *1949, p. 309*).

C Hydrograph of Sugar Creek, Ohio (310 square miles), resulting from a storm of 6.3 inches lasting 14 hours. The immediate stream discharge was 3·0 inches – the rest evaporating, being taken up by vegetation, or going into soil or ground-water storage (*Source: After Hoyt and Langbein. From Strahler*, Introduction to Physical Geography, 1965, *p. 274*).

D A simple derivation of a unit hydrograph for an isolated storm of 24-hours duration, yielding 1·9 inches over a basin of 1,130 square miles (*Source: Linsley, Kohler and Paulhus, 1949, p. 446*).

it should be ultimately possible to simulate actual storms moving across given catchment areas. Laboratory experiments with sprinkler equipment have been conducted widely, for example in the United Kingdom by Childs (1953). The purpose of these experiments has been: to generate controlled rainfall which can be equated with the resulting measured erosion and sediment transport; to produce rainfall for the simplification of a hydrological problem into a soluble hydro-mechanic state; or to apply rainfall to homogeneous porous media for the study of flow nets to drains and other infiltration phenomena (Luthin, 1957 and 1966). Figure 5.4B shows the infiltration/runoff relationships achieved for a controlled test plot under artificially-simulated precipitation.

Many of the major advances in physical hydrology in the last forty years have been made, however, not by the use of laboratory experiments, but by the application of mathematical models. The work of Penman (1948, 1956, 1963) on evaporation, Horton (1933) on infiltration and Sherman (1932) on the unit hydrograph theory of stream runoff has led to a quantitative evaluation of these components which, under the best conditions, give figures which are as accurate as the direct values obtainable in field measurements and do it very much more easily. Both Penman and Horton, and to a lesser extent Sherman, look at the component which they are studying as the focal point in the hydrological cycle. All three simplify the natural hydrological state by abstracting the major factors which are relevant to their hypothesis and ignoring the others.

## Models for evaporation

The most direct approach to the calculation of evaporation is to employ the evaporation pan as a hardware scale model. This work has been based on the theory that there may be some determinable relationship between the rates of evaporation from the small water surface of an evaporation pan and from a large water surface in the same locality that has been subjected to the same natural influences. The analogy is difficult to check, however, for estimates of lake and reservoir evaporation losses require field measurements of all inflow of moisture from the atmosphere, and runoff and ground water measurements, such that the difference between measured liquid inflow and measured liquid outflow can be attributed to evaporation. This technique reflects the limited accuracy of hydraulic measurement and all observational errors show up as evaporation, so that the inherent error is probably not less than 20 per cent for large bodies of water. To add to the difficulties of this method a variety of evaporation pans, with different pan coefficients, are in common use, so that it is difficult to make wide-ranging comparisons between evaporation values obtained in this way. Another major difficulty is that most natural catchment evaporation losses take place from vegetated surfaces, rather than from standing water. It is in this latter respect that some of the following mathematical models are especially significant.

Penman has based such a mathematical model of evaporation on a combination of the energy balance theory and the aerodynamic approach. During summer the amount of energy used in converting water to vapour is the largest single term in the net exchange of energy between atmosphere and ground. Because of this, it is possible to draw up a balance sheet of energy income and expenditure in which evaporation is the only important unknown quantity. Penman formulated this in the equation:

$$H = E + K = R_C(1-r) - R_B$$

Where: $H$ is the heat budget, or the essential equality between the energy used and the energy received.
$E$ is the energy available for evaporation.
$K$ is the energy used to heat the air.
$R_C$ is the energy from the sun.
$R_C \times r$ is the energy reflected by the earth.
$R_B$ is the energy radiated by the earth.

He then used the aerodynamic approach (i.e. that evaporation rates depend upon wind velocity and differences in vapour pressure of the air) to determine how the available energy is shared between $E$ and $K$, the ratio between which can be used in conjunction with the energy balance equation. In this combination of the aerodynamic and energy balance methods of approach, some meteorological elements that are difficult to measure directly are eliminated from the equations to give an expression for evaporation in terms of those meteorological elements that can be measured; viz. mean air temperature, mean air humidity, duration of bright sunshine, and mean wind speed.

Since his first paper on evaporation (1948) Penman's formula has been modified by further research, notably by Monteith (1959) on values of the reflection coefficient. In his 1963 publication Penman published a generalized expression for evaporation as:

$$E = \left(\frac{\Delta}{\gamma}H + E_a\right) \Big/ \left(\frac{\Delta}{\gamma} + x\right)$$

Where $E_a$ is an expression for the 'drying power' of the air, involving wind speed and saturation deficit.
$\Delta$ is a temperature-dependent constant (=the slope of the saturation vapour-pressure curve at mean air temperature, which is obtainable from standard tables).
$\gamma$ is the constant of the wet- and dry-bulb psychrometer equation.
$H$ is the heat budget.

The ratio $\frac{\Delta}{\gamma}$ is dimensionless, and is effectively a weighting factor in assessing the relative effects of energy supply and ventilation on evaporation.

$x$ is a factor dependent on stomatal geometry and day length.

Much practical value has been obtained from Penman's work on evaporation in that he has been able to show the appropriateness of his model in the estimation of water use by natural vegetation and in the calculation of irrigation need. Penman has shown that the water used by a plant is primarily dependent on atmospheric conditions, is largely independent of the soil or the type of crop, and, except for drainage (which virtually ceases when the soil moisture content falls below field capacity), water is only lost from the soil by evaporation from the surface and by transpiration from plants. The quantity of water removed from the soil for a given intake of solar energy can be thus stated simply in terms of air temperature, humidity and wind, such that the potential transpiration of a crop can be estimated from meteorological observations. Potential transpiration is defined as the amount of water transpired by a green crop of about the same colour as grass, which completely covers the ground, and which has an adequate supply of water. Potential transpiration is expressed as follows:

$$E_T = \left(\frac{\Delta}{\gamma} H_T + E_{aT}\right) \Big/ \left(\frac{\Delta}{\gamma} + 1\right)$$

Where $E_T$ is potential transpiration for a 30-day month.

$H_T$ is the heat budget.

$\Delta$ is a temperature-dependent constant.

$\gamma$ is the constant of the wet- and dry-bulb psychrometer equation.

$E_{aT}$ is an expression for the drying power of the air.

Penman's formula is limited in its applicability to British and Western European climatic conditions. Formulae based on similar theoretical assumptions have been formulated by Blaney and Criddle (1950) and by Lowry and Johnson (1942) for use in the arid western part of the United States; by Thornthwaite (1948 and 1954) for use particularly in the eastern and central parts of the USA; and by Olivier (1963), who uses a correlated latitude/radiation factor in a simple formula by which plant water needs may be calculated for any part of the world.

## A model of infiltration

R. E. Horton (1933) adopted a different approach to the relationship between rainfall, runoff and storage. He was able to express the characteristics of the infiltration curve (Fig. 5.5A) in terms of a mathematical model (Fig. 5.5B)

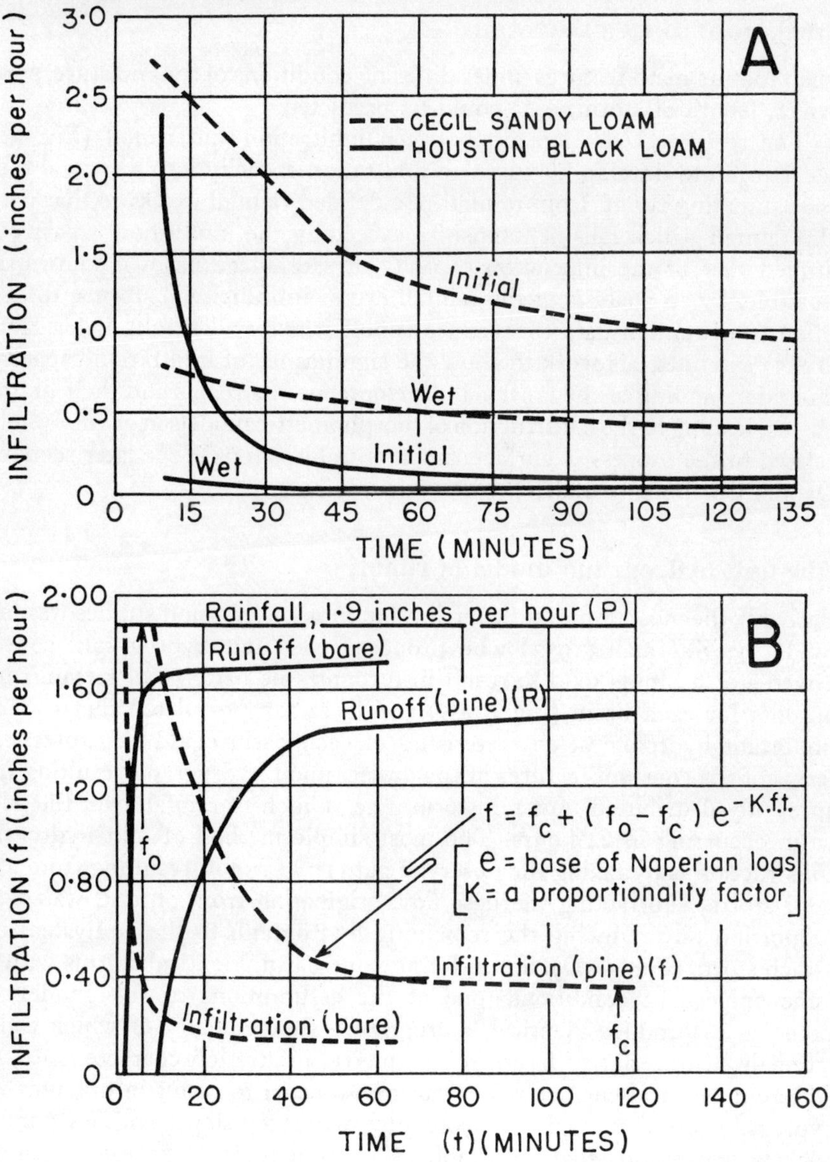


**5.5** The effects of time, soil composition, initial soil moisture conditions and vegetational cover on surface infiltration.

A Comparative infiltration curves for the relatively permeable Cecil and the less permeable Houston loams. In both cases the initial (dry) tests yield higher infiltration rates than the (subsequent) wet tests, with a constant applied rainfall. The curve for the wet run is not simply the last portion of the curve for the dry run, and the initial soil moisture conditions seem to influence the character of the whole infiltration curve (*Source: After Free, Browning and Musgrave, From Linsley, Kohler and Paulhus, 1949, p. 313*).

B Infiltration and surface runoff from forested land (55-year-old pines) and bare abandoned land in the Tallahatchie River basin, Mississippi under a precipitation of 1·9 inches per hour. The infiltration curve equation is that suggested by Horton (*Source: After Musgrave. From Meinzer, O. E., Ed.,* Hydrology; *1942, p. 250*).

such that its main features under differing conditions of soil moisture, surface cover, length of storm, etc., could be predicted.

The reciprocal relationship between infiltration and runoff (Fig. 5.5B) led Horton to develop his ideas on infiltration capacity into a general theory for estimating runoff from rainfall. He defined rainfall excess as that part of the rainfall which falls at intensities exceeding the infiltration capacity and argued that, if the infiltration capacity of a drainage basin is known, it is possible, by an analysis of the rainfall excess in individual storms, to determine from rainfall data the surface runoff which will result. Later Horton (1945) extended his work to show the significance of infiltration capacity in an understanding of the erosional development of streams and their drainage basins, leading to the construction of morphometric models in geomorphology related to the process of surface runoff (Strahler, 1964) (See later section on Overall Catchment Models).

## The unit hydrograph model of runoff

Probably the most important contribution to rainfall/runoff studies was made by L. K. Sherman (1932) who proposed the 'unit hydrograph' concept. Given storm inputs tend to result in recognizable patterns of stream runoff outputs for each basin (Fig. 5.4C) and Sherman postulated that the most important hydrological characteristics of each basin could be expressed in terms of the constant features of the direct runoff hydrograph resulting from an evenly-distributed unit rainstorm – i.e. 1 inch of rainfall over the whole basin occurring in 24 hours. The most simple method of unit-hydrograph construction is by taking the observed pattern of runoff resulting from a 24-hour storm, subtracting the base flow originating from ground water, and proportionately reducing the remaining hydrograph to the equivalent of a 1-inch storm (Fig. 5.4D). Thus, for any one basin, identical storms with the same antecedent conditions (one of the assumptions of this model) are expected to produce identical hydrographs, the geometry of which can be rationalized in terms of basin morphometry, infiltration characteristics, etc. Different physical characteristics can be expected to result in different unit hydrographs. For example, a topography with steep slopes and few pondage pockets gives a high, sharp peak and a short time period, whereas flat country with large pondage pockets gives a graph with a flat rounded peak and a long time period. Figure 5.6 shows how unit hydrographs for different watersheds reflect the variations in shape, size, morphometry, slope, pondage, etc.

The empirical unit hydrograph model has numerous advantages over other methods of determining runoff from given rainfalls, in that it is tailored to the characteristics of particular drainage basins such that patterns of runoff can, by adapting the basin unit hydrograph, be predicted for any rainfall input (O'Donnell, 1960; Dooge, 1959; Chow, 1964, pp. 14–13 to 14–35).

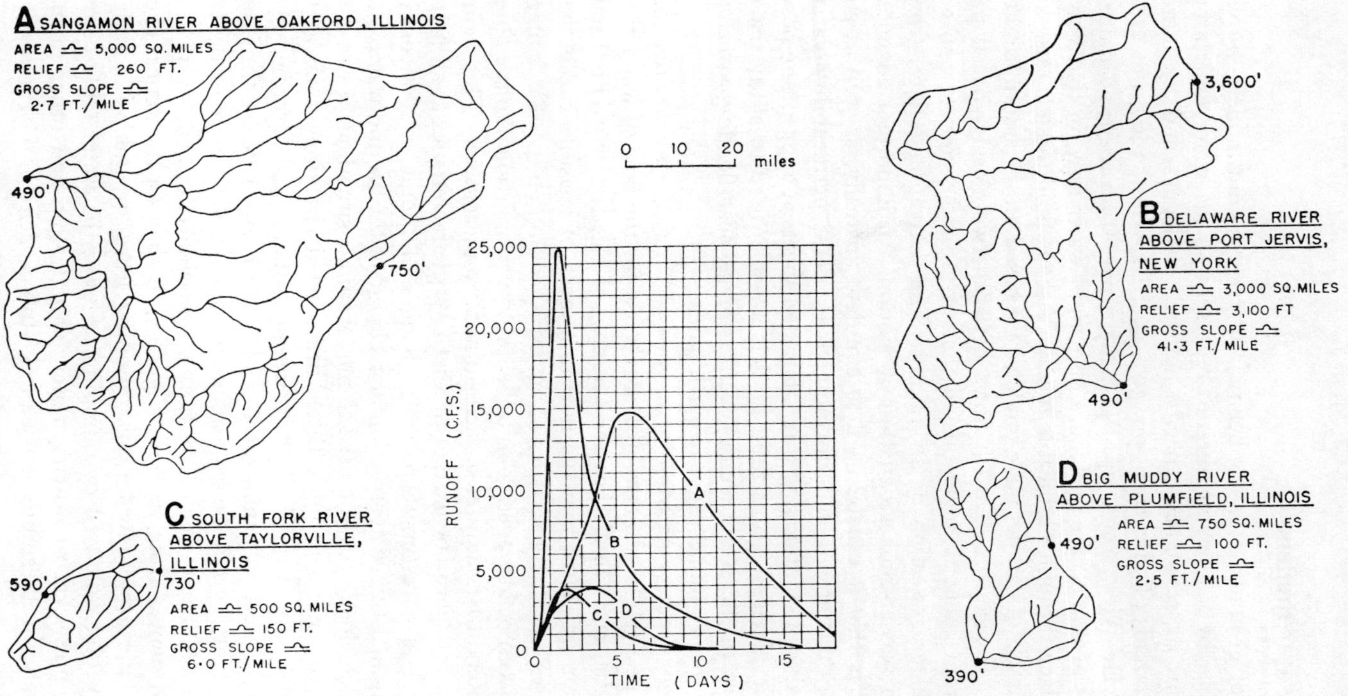


**5.6** Unit hydrographs for (A) the Sangamon River, Illinois, (B) the Delaware River, New York, (C) the South Fork River, Illinois, and (D) the Big Muddy River, Illinois. These show clearly the effect of area on the magnitude of the unit hydrograph, and the effect of basin slope on its form (*Source: Partly from Sherman, 1932*).

## Ground water models

The flow of ground water has been simulated by sand models, by viscous fluid models, by rubber membranes representing the water table around a well system, by resistance-capacity network analogues and by numerical analysis methods (Todd, 1959, pp. 307–325; De Wiest, 1965, pp. 318–348; Davis and De Wiest, 1966, pp. 241–256). Apart from the sand model, the other types of models are analogues of ground water flow, in that flow does not take place through permeable media. Viscous fluid models (also known as Hele-Shaw and parallel plate models) are based on the assumption that if a liquid such as oil or glycerin flows between two closely-spaced parallel plates its movement is analogous to that of ground water flow in a two-dimensional cross-section of an aquifer. The derivation follows from the generalized Navier-Stokes equation of motion. An important advantage of this model is that it can be used to simulate unsteady, as well as steady, flows in both confined and unconfined aquifers; permeability variations being introduced by attaching thin laminated sheets at appropriate positions between the plates. Viscous fluid models are also used to study sea water intrusion (as in the Netherlands), bank storage near flooding streams and the movement of ground water in earth dams.

Resistance capacity-network analogues are based on the relationship which can be postulated between Darcy's law for the flow of water through porous media, and Ohm's law for the flow of electrical current through a resistor (Fig. 5.7), such that the flow in a mesh of resistors can be held to represent the flow characteristics of an entire aquifer (Fig. 5.8). Voltages at mesh nodes can be arranged so as to be directly proportional to the observed head of water in the aquifer as determined by piezometer. This being so, the analogue model can be used to predict flow characteristics at all points in the aquifer and to estimate possible changes of ground water conditions attendant upon artificial abstraction or recharge of the ground water.

Electrical analogues can be used to model simple situations of two-dimensional flow, or more complex three-dimensional situations such as those investigated by the United States Geological Survey and described by Skibitzke (1963). Electrical analogue models have been made by the Thames Conservancy for the Lambourn Valley and the Kennet Valley in Berkshire, one of the functions of which is to study the possible effects of pumping wells at varying distances from streams. Figure 5.9B shows a theoretical cycle of recharge of the ground water, by pumping from streams with high discharges in winter, followed by pumping to lower the water table below its natural summer level to provide water for irrigation and other purposes. Thus the water in the aquifer is used to its full capacity in the same way as a surface reservoir might be managed for optimum supply.

Figure 5.9A shows a modification of this theoretical cycle which is to be

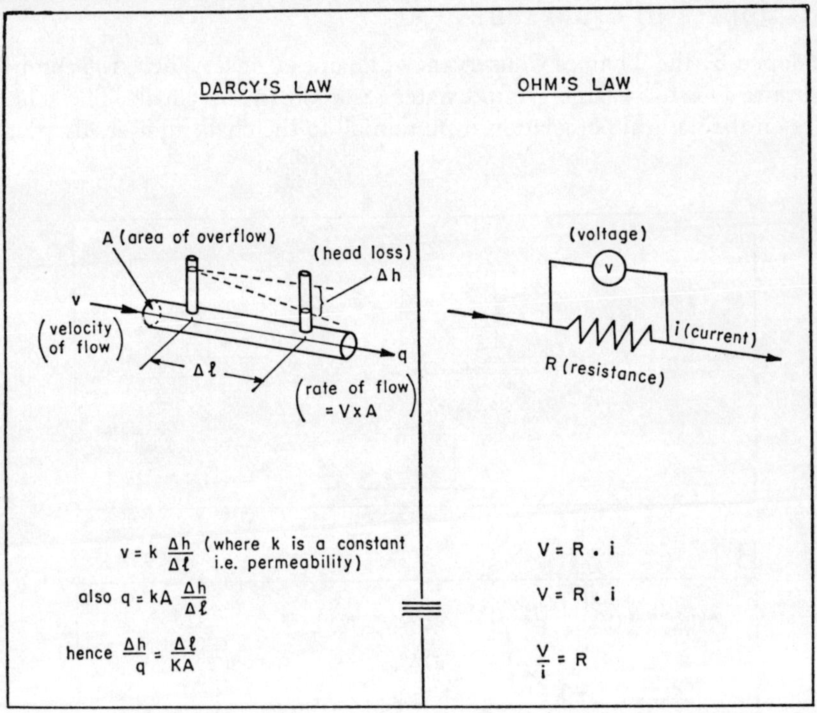


**5.7** The analogy between Darcy's Law, for the flow of water through porous media, and Ohm's Law, for the flow of current through a resistor, which forms the basis of the ground water electrical analogue (*Source: Water Research Association, TH/H4*).

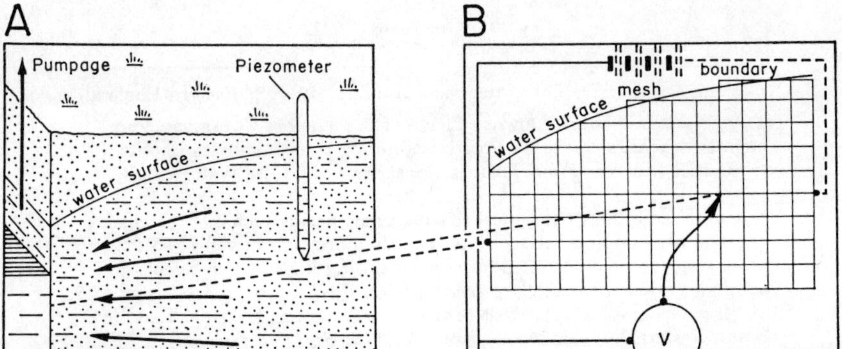


**5.8** Electrical analogue for ground water flow. The flow of water in an underground aquifer (A) can be simulated by the flow of electricity through a network of electrical resistances (B). In this simple case of water pumpage from a ditch, the voltage at a mesh node is directly proportional to the water pressure in the aquifer, as determined by a piezometer (*Source: Water Research Association, Technical Paper TP 32*).

developed by the Thames Conservancy (Thames Conservancy, 1965) for the Berkshire Downs, a large ground water reservoir in the Chalk. The scheme relies on the natural percolation from rainfall to the chalk aquifer. By pump-

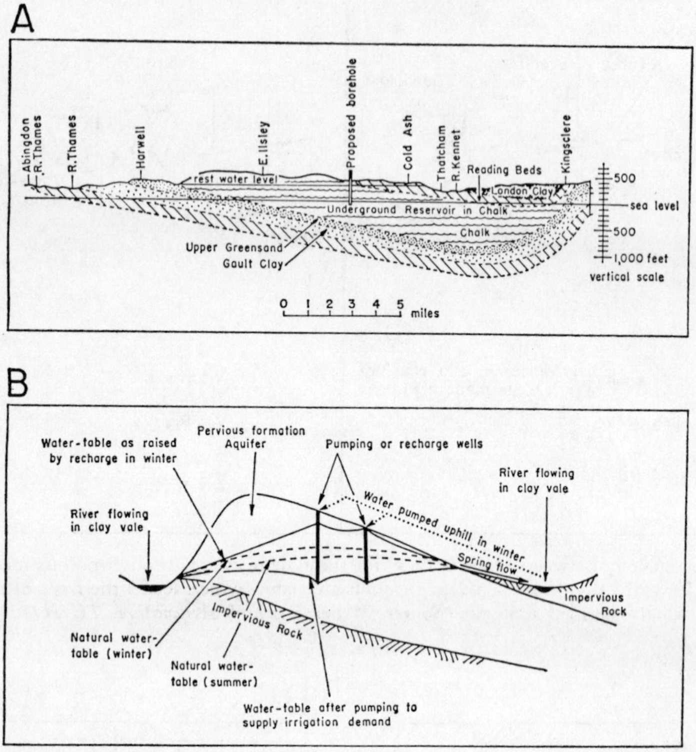


**5.9** Increased utilization of the chalk aquifers of South-Eastern England.

A Proposed Lambourn Downs scheme of the Thames Conservancy to augment the flow of the Thames by pumping from boreholes into its tributaries (*Source: From* Thames Conservancy and the Water Crisis, *1965*).

B A more ambitious theoretical scheme to artificially recharge the chalk aquifers by pumping surface water *into* the boreholes in winter so that more water is available for pumping *out* during the water-deficient summer period (*Source: Suggested by P. O. Wolf. Illustrated in More, R. J., 1965*, A Geographical Analysis of the Distribution of Irrigated Land in South Eastern England; *unpublished Ph.D. thesis, Department of Geography, Liverpool University*).

ing from the underground reservoir, flows in the river may be augmented during dry periods in the summer. In the winter, the pumps are rested and the aquifer recharges by natural percolation from the winter rainfall. The Thames Conservancy contemplate the eventual construction of a network of boreholes, mainly in Berkshire and the Cotswolds, which may be capable of supplementing Thames flows by up to 270 million gallons per day, at an estimated cost of £8,000,000. It is hoped that the cost of pumping water from underground into the rivers will work out at about 3*d.* per 1,000 gallons, which compares very favourably with 3*s.* to 4*s.* per 1,000 gallons for many present-day reservoir schemes and much higher costs for desalinization. During a pilot scheme (1966–69) nine boreholes are to be sunk in the Lower Lambourn Valley in the area between Welford and Newbury, and the effects of pumping studied in detail. It is estimated that the output from the nine boreholes will be 16 million gallons per day, which will be pumped into the Lambourn, which joins the Kennet at Newbury, which in turn joins the Thames at Reading.

Recharging the chalk of the Berkshire Downs, by pumping from adjoining streams, may be a logical development at a much later date, if certain technical difficulties are overcome. Recharge of the chalk in the Lee Valley north of London has already been undertaken by the Metropolitan Water Board with encouraging results. (Boniface, 1959; Buchan, 1959 and 1963).

## OVERALL CATCHMENT MODELS

General catchment models in hydrology can also be conveniently divided into the physical and the systems approaches (Amorocho and Hart, 1964); the former attempting to understand each mechanism and interaction of the hydrological cycle in the search for a complete, rational synthesis, the latter concerned with the establishment of workable relationships between sub-systems of hydrological components such that hydrological events can be predicted from known climatic inputs. One of the most successful advocates of physical hydrology, Robert E. Horton, was much concerned with the interactions of hydrological events and drainage basin geometry ('morphometry') and, in an important paper (Horton, 1945), he showed both how a drainage system can be dissected into components of differing order having dynamic significance (later modified by Strahler, 1952 and 1964; see Fig. 5.10A) and how the characteristics of the network (e.g. the density of drainage lines) can be rationalized on the basis of infiltration runoff theory. It is interesting how, depending upon the time scale under consideration, drainage systems can be variously considered as both the effects and causes of runoff patterns. From the point of view of physical hydrology the latter case provides an interesting 'morphometric model' for hydrologists. Strahler (1964) has shown how the 'bifurcation ratio' $\left(R_b = \frac{N_u}{N_{u+1}}\right)$ may control the pattern of runoff from storms over individual basins (Fig. 5.10B); Leopold and Miller (1956) that there is a relationship between mean discharge and basin order (Fig. 5.10C); and Hack (1957) that there is a logarithmic relationship between

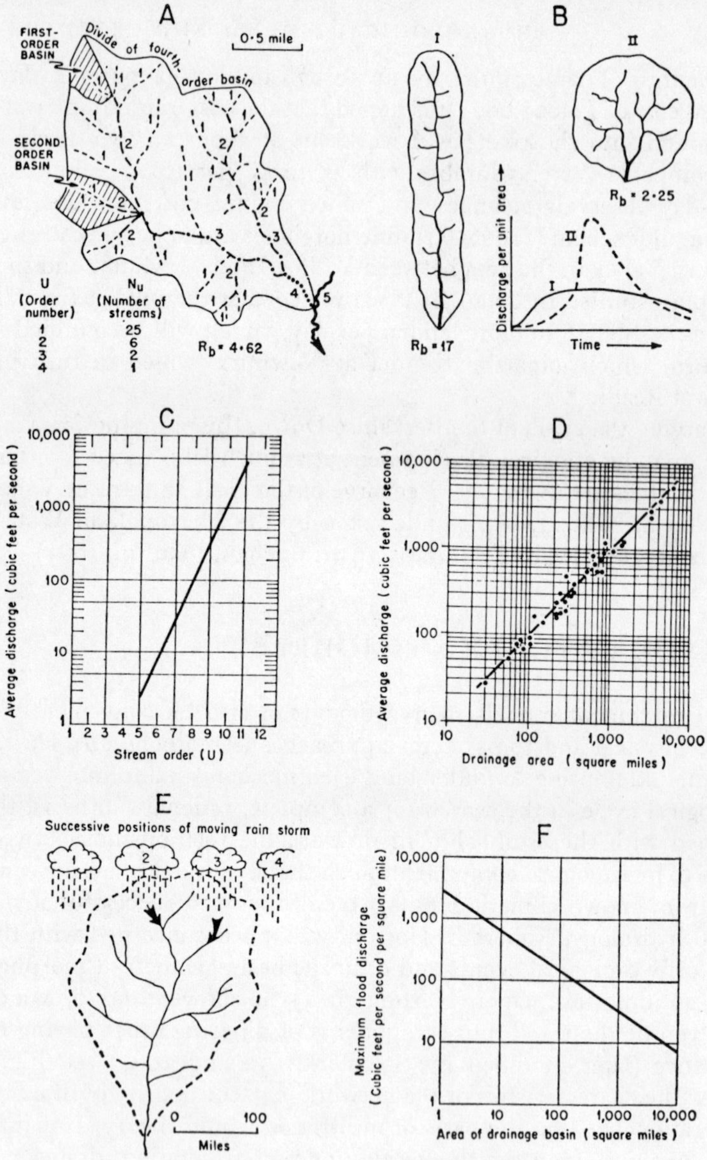


**5.10** The influence of basin morphometry on hydrological characteristics.

A A channel ordering system, showing the derivation of the bifurcation ratio ($R_b = N_u/N_{u+1}$) (*Source: From Strahler, 1964, p. 4–44*).

B Schematic hydrographs for basins having high (I) and low (II) bifurcation ratios (*Source: From Strahler, 1964, p. 4–44*).

C A general relationship of discharge and stream order for arroyos in New Mexico (*Source: From Leopold and Miller, 1956, p. 23*).

D Relationship of average discharge to drainage area for all gauging stations on the Potomac River (*Source: After Hack, 1957. From Strahler, 1964, p. 4–50*).

E and F The small area covered by the most intense storms means that maximum flood discharges per unit area are recorded for small basins. (*Source: After Follansbee and Sawyer. From Leet and Judson, 1954, pp. 122–123*).

mean discharge and drainage area (Fig. 5.10D). Considerations of basin morphometry also help to rationalize extreme hydrological events and Follansbee and Sawyer (1948) have shown how the areal localization of most intense precipitation (Fig. 5.10E) provides for the maximum flood discharges (per unit area) to occur in the smallest basins (Fig. 5.10F). In Great Britain, Nash (1960) has attempted to correlate unit hydrograph parameters with the morphometric characteristics of drainage basins.

If a quantitative overall hydrological catchment model is to be acceptably accurate in all particulars it must inevitably be complex, yet it must be feasible to operate. These conflicting requirements were not met until high-speed computers came into use, but, even so, the evaluation of such models is still circumscribed by the limitations of current knowledge and understanding of the processes being simulated, and by the capabilities of the computers and the computing techniques available. Ideally the data for the construction of an overall model should include a specification of the properties of, and the processes that occur in, all the relevant components of a catchment. Specification should be given in terms of physical parameters and should involve behavioural relationships for and between the catchment components stated in terms of these parameters. At the present time hydrological knowledge and techniques do not permit more than a coarse approximation to this ideal. Overall models have been developed to the stage of being an effective and acceptable engineering tool (Crawford and Linsley, 1966), but Amorocho and Hart (1964) caution against an excessive reliance on synthetic models of catchment behaviour, naming as perhaps the most important causes of unreliability errors in the recorded data, effects due to lumping of components, imperfections of the structure of any synthetic model and non-uniqueness of the processes of synthesizing an unknown system.

O'Donnell (1966) emphasizes that digital and analogue computers differ in their structure and mechanical capabilities and that the advantages of each type of computer should be used in solving hydrological problems to which the machine is best adapted. Digital computers are particularly suited to calculations involving matrix procedures, such as the prediction of flood runoff using the unit hydrograph technique, and the Tennessee Valley Authority has published a number of papers describing the use of digital computers in hydrological analyses (TVA, 1961; 1963A; 1963B). The most advanced and comprehensive digital computer evaluation of catchment behaviour is the study which has been made by Crawford and Linsley (1962 and 1966). This model aims to represent the whole of the land phase of the hydrological cycle and has been tested and successfully adjusted to reproduce within acceptable limits the runoff behaviour of more than thirty catchments. The model is a general one and its operation is controlled by the characteristics of its component storage and routing elements. These characteristics are

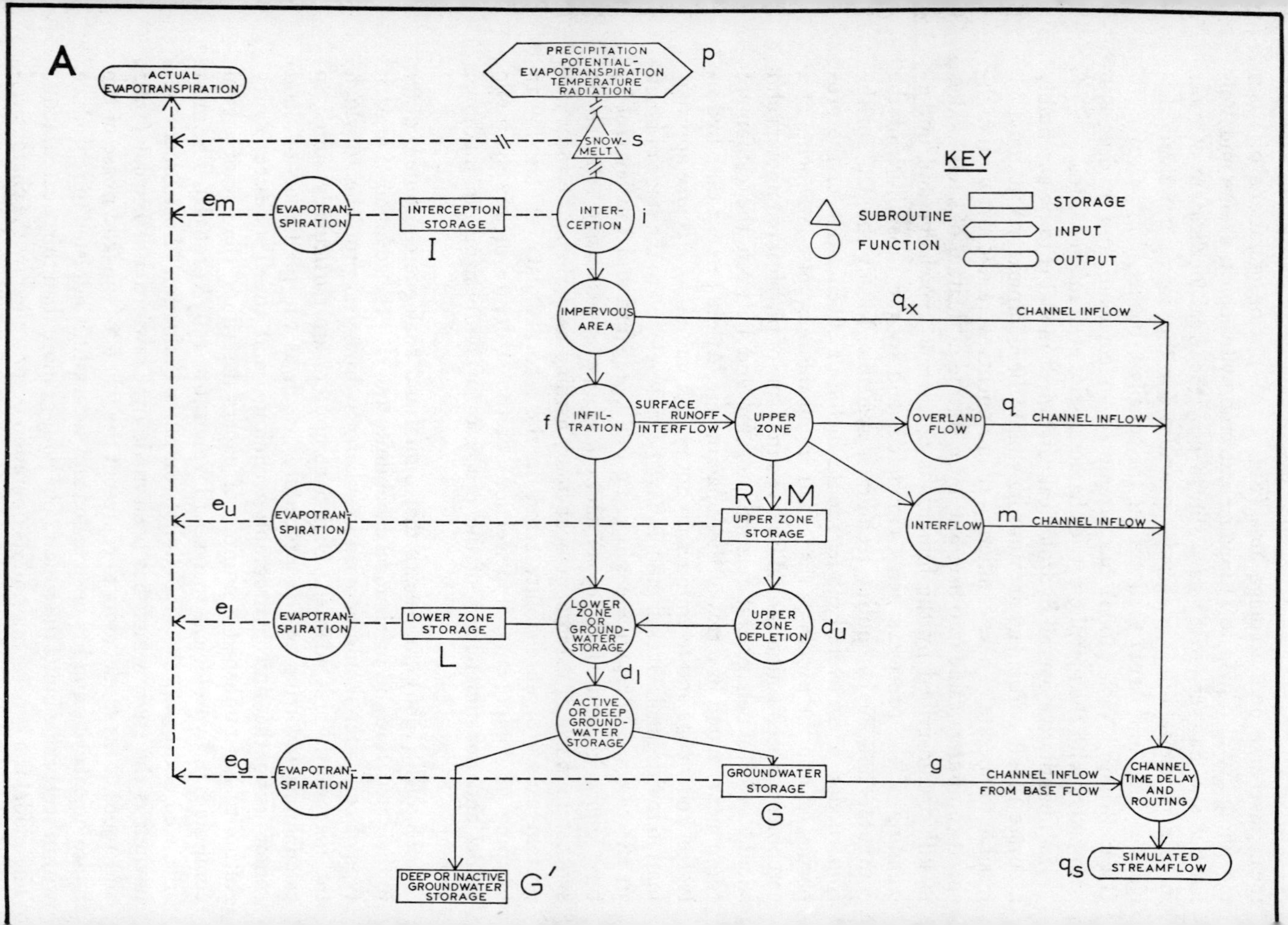
A
ACTUAL EVAPOTRANSPIRATION
PRECIPITATION
POTENTIAL-EVAPOTRANSPIRATION
TEMPERATURE
RADIATION
p
SNOW-MELT
s
KEY
SUBROUTINE
FUNCTION
STORAGE
INPUT
OUTPUT
em
EVAPOTRANSPIRATION
INTERCEPTION STORAGE
I
INTERCEPTION
i
IMPERVIOUS AREA
qx
CHANNEL INFLOW
f
INFILTRATION
SURFACE RUNOFF
INTERFLOW
UPPER ZONE
OVERLAND FLOW
q
CHANNEL INFLOW
R
M
eu
EVAPOTRANSPIRATION
UPPER ZONE STORAGE
INTERFLOW
m
CHANNEL INFLOW
el
EVAPOTRANSPIRATION
LOWER ZONE STORAGE
L
LOWER ZONE OR GROUND-WATER STORAGE
UPPER ZONE DEPLETION
du
dl
ACTIVE OR DEEP GROUND-WATER STORAGE
eg
EVAPOTRANSPIRATION
GROUNDWATER STORAGE
G
g
CHANNEL INFLOW FROM BASE FLOW
CHANNEL TIME DELAY AND ROUTING
qs
SIMULATED STREAMFLOW
DEEP OR INACTIVE GROUNDWATER STORAGE
G′

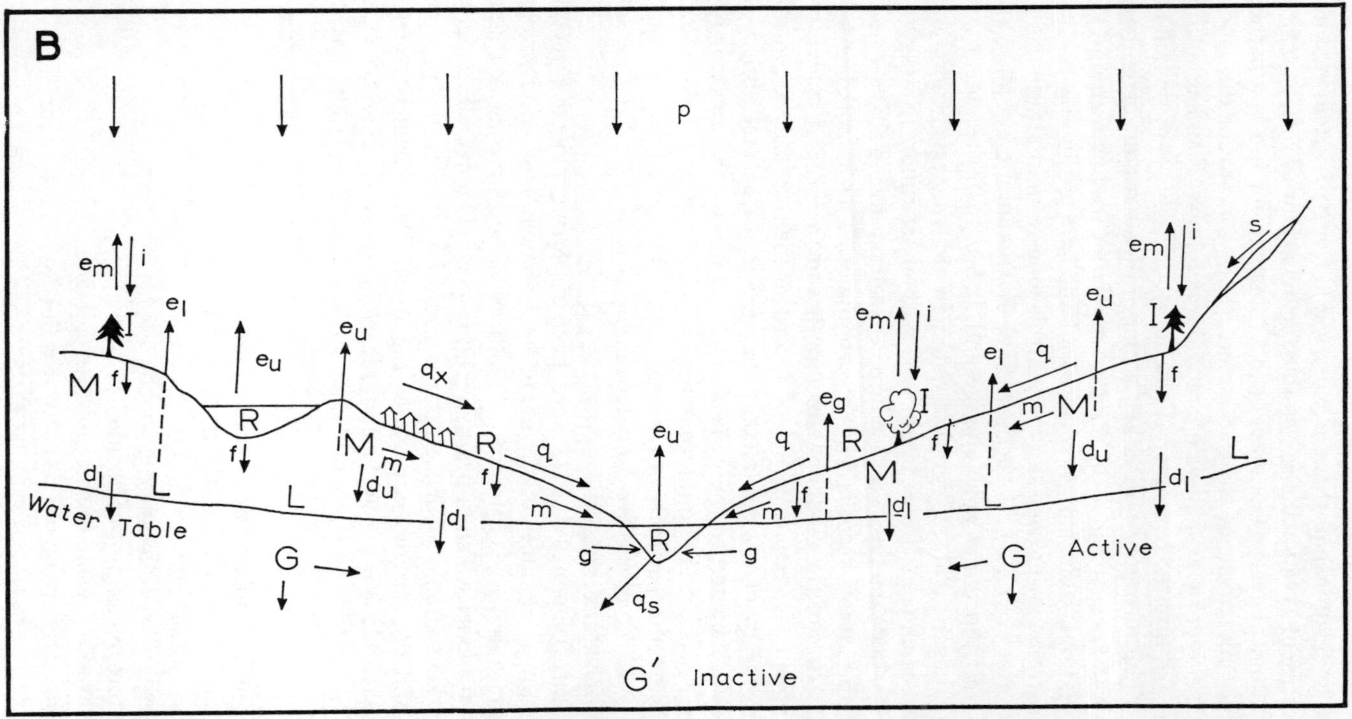


**5.11** The Stanford Digital Computer Watershed Model IV.

A The flowchart for the model based on 5 storage elements (I, R and M, L, G, G′) which are augmented or depleted by 13 transfers (*Source: From Crawford and Linsley, 1966*, Fig. 4.1).

B Schematic hydrologic cycle for a basin, employing the same nomenclature as A (*Source: Adapted from Linsley, Kohler and Paulhus, 1949*).

determined by relationships that are expressed in terms of certain parameters and that represent as rationally as possible the behaviour of various segments of the hydrological cycle. The model can be adapted for use in a particular basin by the adjustment of the parameters. The structure of the model is illustrated in Figure 5.11, and its operation is as follows. Estimates of initial storages are fed into the computer. Then hourly increments of rainfall enter the model. The incoming rain either becomes direct runoff, or is detailed into upper and lower soil moisture storage, the latter feeding a ground water storage. The upper zone storage absorbs a large part of the first few hours of rain in a storm; lower zone storage controls long-term infiltration and the ground water store influences the base flow. Evaporation takes place at the potential rate from upper zone storage and at less than the potential rate from the lower zone and from ground water. In applying this general model to a specific catchment (for example, one with short runoff records which are to be extended) the typical procedure is to select a five to six year portion of rainfall and runoff records for the catchment. This period is used to develop estimates of the model parameters that fit the general model to the given catchment. A second period of record is then used as a control to check the accuracy of the parameters obtained from the first period. The parameters are adjusted to the general model either by the operator or by the computer itself, using an internal looping routine of successive approximations. When the particular basin has been thus fitted by the general model, the working of the fitted model can be used to extend the period of record within certain confidence limits.

Dawdy and O'Donnell (1965) have been exploring objective methods of finding numerical values of the parameters of such synthetic hydrologic models using automatic optimization methods. They have used a simple mathematical model which appears in Figure 5.12. The overall model was deliberately kept simple so that emphasis could be given to the parameter sensitivity and optimization aspects of the work. The model is restricted to four storage elements – surface storage ($R$), channel storage ($S$), soil moisture storage ($M$) and ground water storage ($G$). The model has nine parameters:

| | |
|---|---|
| $R^*$ | the threshold value of surface storage |
| $f_o$ | maximum infiltration rate |
| $f_c$ | minimum infiltration rate |
| $k$ | exponential die-away exponent |
| $K_s$ | storage constant of channel storage |
| $M^*$ | threshold value of soil moisture storage |
| $G^*$ | threshold value of ground water storage |
| $c_{\max}$ | maximum rate of capilliary rise |
| $K_G$ | storage constant of ground water storage. |

At the start of the first time interval during the running of the model trial volumes in each of the four storage elements and the potential infiltration rate have to be specified, and a set of trial values has to be given to the nine parameters. Thereafter the computations for the first interval, and for each suc-

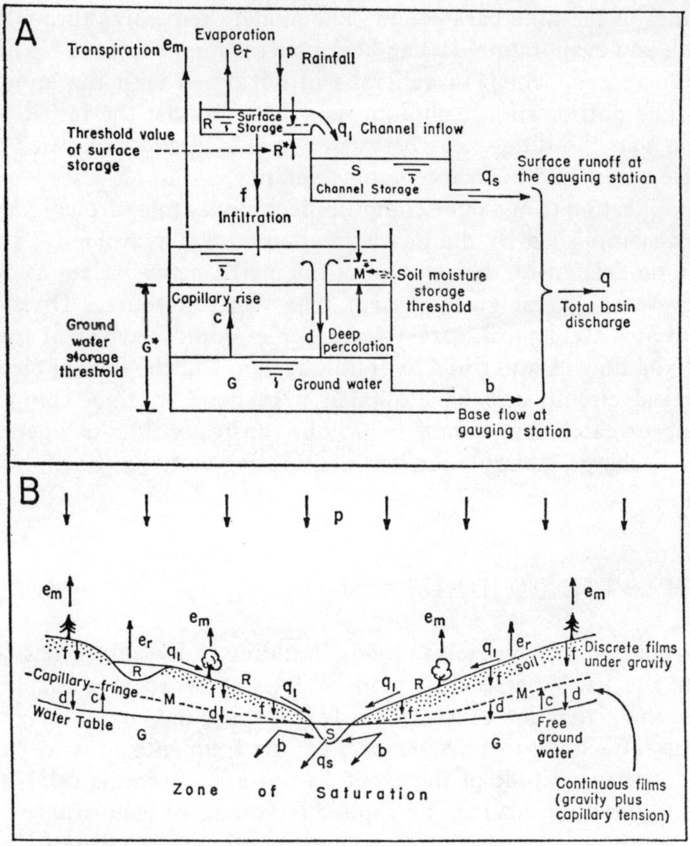


**5.12** The basin hydrological cycle.
A A basin model based on four storage elements (R, S, M and G), three of which operate with respect to thresholds (R*, M* and G*), augmented and depleted by a number of transfers (p, $e_m$, $e_r$, $q_1$, f, c, d, $q_s$, b and q) (*Source: From Dawdy and O'Donnell*, 1965).
B Schematic basin hydrological cycle, employing the same nomenclature as A (*Source: Adapted from Linsley, Kohler and Paulhus, 1949*).

cessive interval, yield values for the four storage volumes and the potential infiltration rate for the start of the next interval. A completely general optimization would include the start-of-synthesis values of each of the four storage volumes and the potential infiltration rate, as well as the nine parameter values, but by postulating a long dry period before the start of a rainfall runoff synthesis it seems reasonably accurate to set all four initial storages to zero and to assume that the starting potential infiltration rate has recovered to a maximum value of $f_o$, so that, in practice, there are only nine input items. The input data to the model consist of precipitation, potential evaporation and runoff data for each of the intervals of a known record, and the initial trial values of the nine parameters. The model then works through the precipitation and evaporation data and calculates a runoff volume for each interval of the record, which in general will not agree with the known runoff values. The optimization technique sets out to adjust the initial parameter values so that the differences between the known and calculated values of runoff are eliminated to an acceptable tolerance.

The application of analogue computer techniques to hydrological problems may be accomplished by the direct simulation of a hydrological system (an analogue model), or by the solution of the mathematical relations describing such a system (an analogue analyser). The Water Resources Division of the United States Geological Survey have, for example, developed the analogy between the flow of water in a hydraulic system and the flow of electricity in an electrical circuit, and have applied it to flood routing, comprehensive modelling of catchment runoff behaviour, unsteady flow in open channels and the discharge computation for slope-rating stations (Shen, 1963A and 1963B).

## STOCHASTIC MODELS

Another method of approach to model building in hydrology has been provided by the use of statistical models. It is assumed that each event involved in the generation of a series of hydrological data has a certain level of probability attached to its occurrence, i.e. the frequency of its occurrence is related to the magnitude of the event. If this assumption is valid, statistical probability distributions may be applied to lengths of historical hydrological records, and it may be possible to generate synthetic data from the statistical characteristics of the historical record. The major advantage of synthetic generation is to create records longer than the historical record and is particularly valuable in the study of reservoir operations and in the design of complex water-resource systems. (See earlier section on Models for Hydrological Prediction.)

Thus the hydrologic data series can be treated as a time series by considering it as a queue or waiting line (Cox and Smith, 1963; Duckworth, 1965). A queue involves arriving items or 'customers' that wait to be served at the facility which provides the service they seek. In a queueing system the customers arrive at the system, wait for service, receive service and then leave the system. The analogy between queueing theory and storage of runoff was first recognized by Moran and analysed theoretically by him and others to develop the theory of storage (Gani and Moran, 1955; Kendall, 1957; Moran, 1959; Harris, 1965).

Two principal attitudes towards synthetic data sequences have been adopted in the solution of hydrologic problems. Firstly, it may be assumed that hydrologic data are purely random. In this case Monte Carlo methods are used (Hammersley and Handscomb, 1964). The characteristics of the distribution of the recorded data are studied first. Then an appropriate simulation generation technique is employed (frequently involving the use of tables of random numbers) to produce artificial sequences of record having the same statistical structure as the measured historical record. By means of a table of random numbers Brittan (1961), for example, simulated stream flows in the Colorado River by selecting 100 random samples of 5 each corresponding to a 5-year runoff sequence. She was able to develop hydrologic records at Lees Ferry, Arizona, by determination of the probability distribution of mean flows in relation to the range.

The assumption of the randomness and independence of hydrologic events is not entirely realistic, however, particularly in regard to runoff where antecedent conditions often have a close connection with subsequent discharges. These 'carry-over' effects must be minimized by proper grouping of the data. For example, if 3-year carry-over effects are detectable in the historical record, the generation for Monte Carlo methods must be based on the total output for periods not shorter than 3 years, so that the period totals may approach statistical independence. However, the second statistical approach, the Markov chain, proceeds on the assumption that the outcome of any 'trial' depends on the knowledge of the state of the system at the immediately preceding time. Julian (1961) used a first-order Markov process for generating annual flows in the Colorado River at Lees Ferry whereby:

$$x_t = r x_{t-1} + \Sigma(y)_t$$

where $x_t$ is the annual runoff at year $t$

$x_{t-1}$ is the runoff at the preceding or the $(t-1)$st year

$r$ is the first-order serial correlation coefficient for the runoff, or a Markov chain coefficient

$\Sigma(y)$ is a random uncorrelated component due to annual rainfall.

This equation indicates that the runoff at a given year is equal to a constant times the runoff of the preceding year plus a random component. Much more elaborate multivariate chains have been used by Brittan (1961) and Fiering (1962), but the underlying principle common to all these techniques is that the system may be characterized largely by the statistical structure of the historical record. The relative success or failure of any simulation process depends ultimately on the validity of this premise.

Ven Te Chow (1964), in reviewing the significance of stochastic hydrology, admits that this field of study requires further research, but feels that potentially it has a promising future, particularly in water-resources development and management. For example, Hurst, Black and Simaika (1965) have used statistical modelling techniques to investigate the volume of storage required to make the greatest possible use of the Nile waters. A recent work by Bowden (1965) applied a Monte Carlo innovation-type model based on an azimuthal grid to simulate the spread of irrigation in part of the high plains of Colorado up to 1962, and to predict the rate and areal disposition of the spread of irrigation in the future.

## HYDRO-ECONOMIC MODELS

Model theory, as a whole, is becoming increasingly important in water-resource management, where the economic and social implications of projects have to be considered, as well as those of physical hydrology. Hydro-economic models attempt to simplify and simulate the combined hydrological and economic aspects of a project.

Cost-benefit analysis may be regarded as the classic procedure which has been used as a practical way of assessing the desirability of projects, where it is important to take a wide and long-term view (Grant and Ireson, 1964; Prest and Turvey, 1965). This technique involves the enumeration and evaluation of all the relevant costs and benefits, with the aim of maximizing the present value of all benefits, less that of all costs, subject to specified constraints. Although the technique has been discussed by economists for the last hundred years, its practical value has come to increasing prominence since Pigou's (1932) classic on welfare economics – the concept of social costs and benefits expressed in monetary terms. Cost-benefit analysis has been used by the United States Federal Government to evaluate the economic viability of a variety of public works engineering projects. 'The Green Book' published by the Inter-Agency River Basin Committee (1950), attempts to codify and agree the general principles of such evaluation. Another reason for the increasing interest by economists in cost-benefit analysis has been the rapid development in recent years of such techniques as operations research and systems analysis (McKean, 1958; Maass *et al.*, 1962) which prompt linked consideration of assemblages of factors involved in a broad investigation.

Cost-benefit analysis has two important limitations. Basically it is only a technique for taking decisions within a framework which has to be decided upon in advance and which involves a wide range of considerations, many of them of a political or social character. Also, cost-benefit techniques, as so far developed, are least relevant and serviceable for large-size investment decisions, where finance and investment on a national scale are involved. There are also considerable difficulties in the enumeration of all the cost and benefit factors involved and in their evaluation, particularly where the benefits are hard to express in monetary terms (e.g. social benefits). Some attempts have recently been made to express the less tangible benefits (e.g. recreational benefits) in quantitative terms, as Foster and Beesley (1963) have done for benefits to travellers on the projected London Underground Victoria Line. The great advantage of cost-benefit studies is that they force those responsible for planning to quantify costs and benefits as far as possible rather than to be content with vague qualitative judgements or personal 'hunches'.

Gilbert White and his associates at the University of Chicago have given considerable attention to economic studies of flood-plain management (White, 1945; White *et al.*, 1958, 1961 and 1962). Burton (1962), for example, has described various types of agricultural occupance of flood plains in the United States, suggesting a classification of the main types of agricultural land use in relation to the physical geography of flood plains, and Kates (1962) has undertaken an investigation of the awareness of flood-plain inhabitants to the hazards of flooding. The latter made case studies of six towns in the United States which have experienced flooding and used questionnaires to assess the attitude of the people to the flood problem. His study of the psychological and sociological attitude to floods reveals that there is remarkably little conscious human adjustment to this danger and that, even in places where action is taken to reduce the flood hazard, it may be 'casual, improvized, ineffective and far from optimal'.

White's studies of applied hydrology have extended beyond flood plain management to integrated river-basin development, as in the economic and social study of the Lower Mekong development (1963) to which he was a contributor. The purpose of this study was to find out as much as possible about the physical, economic and social background into which engineering structures would have to be fitted. It was an attempt to understand a cultural scene, completely different to American or European experience, and to attempt to adjust Western technology to it.

Many river valley developments are now multi-purpose schemes, and in trying to design for optimum development it is necessary to look at a number of possible combinations of water use. Calculations become more complicated and sometimes exceed the limits of feasibility by conventional methods. Here a systems approach to the design of hydro-economic models may become increasingly useful. By means of high-speed digital computers it is possible to simulate by simplified models the behaviour of relatively complex water-resource systems for periods of any desired length. It is now possible to perform the numerous and repetitive computations needed for many combinations of the system variables, and to evolve an optimal or near optimal design of the system.

Dorfman (1965) and Hufschmidt (1965) distinguish two main types of model used in a systems approach to water-resources studies – the simulation model and the analytical model. In a simulation model the time structure of the real-life project is faithfully preserved. The core of a simulation model is a set of differential equations that express the relationships between the various magnitudes describing the state and operation of the system during each small segment of time in the course of its history. The use of simulation analysis was begun by the U.S. Army Corps of Engineers on the Missouri River in 1953 (DATAmatic Corporation, 1957). In this analysis the operation of six reservoirs on the Missouri River was simulated on the Univac I computer to maximize power generation subject to constraints for navigation, flood control and irrigation specifications. Subsequently several studies have been made using model theory in the optimization of the operation of hydroelectric schemes (King and Peel, 1960; Rockwood, 1961; Lewis and Shoemaker, 1962; Rockwood and Nelson, 1966). In 1955 Morrice and Allan simulated the Nile Valley plan on an IBM 650 computer to determine the particular combination of reservoirs, control works and operating procedures that would maximize the use of irrigation water. In the Missouri River simulation and in those of the Columbia and the Nile only the purely hydrological conditions were optimized. In the Harvard Water Programme simulation analysis was applied to the economic optimization of water-resources system designs. Simulation was made of a simplified river basin system consisting of four reservoirs having three purposes. This system was purely hypothetical, but it created an adequate number of relationships to typify the complexities of an actual system. By sampling techniques, this analysis was able to evolve a design that yielded net benefits of $811 million against $724 million for the best design by conventional methods. Because of the enormous number of possible combinations of the system variables, the point of greatest net benefit cannot be easily determined, even with the use of high-speed computers. The only time-saving practical method of locating the point of optimality is to sample the variables and eliminate undesirable combinations from the computation. Here analytical models are used to simplify the real situation as far as possible to enable the resultant equations to be manipulated. Thus analytical models and the simulation approach are used in tandem. Initially the problem is analysed into a set of manageable mathematical relationships that can be solved for an approximation to a good or optimal design, and then a range of plausible variation around that tentative solution can be explored by a sequence of simulations.

Fiering (1965) gives an example of the use of the systems approach in the control of salinity in West Pakistan. The accumulation of salts in the soil is a by-product of a method of crop irrigation which has been employed with increasing intensity over a period of about 3,000 years. During the early part of the twentieth century British hydraulic engineers initiated the barrage system of irrigation and began to divert large quantities from the River Indus and its five tributaries. The drainage provisions were inadequate and the ground water level rose, as shown in Figure 5.13A. As the ground water table came within 5 ft. of the surface, evaporation from the soil and transpiration from the crops caused salts to rise in increasing quantities to the surface, thereby putting large acreages of once-fertile agricultural land out of

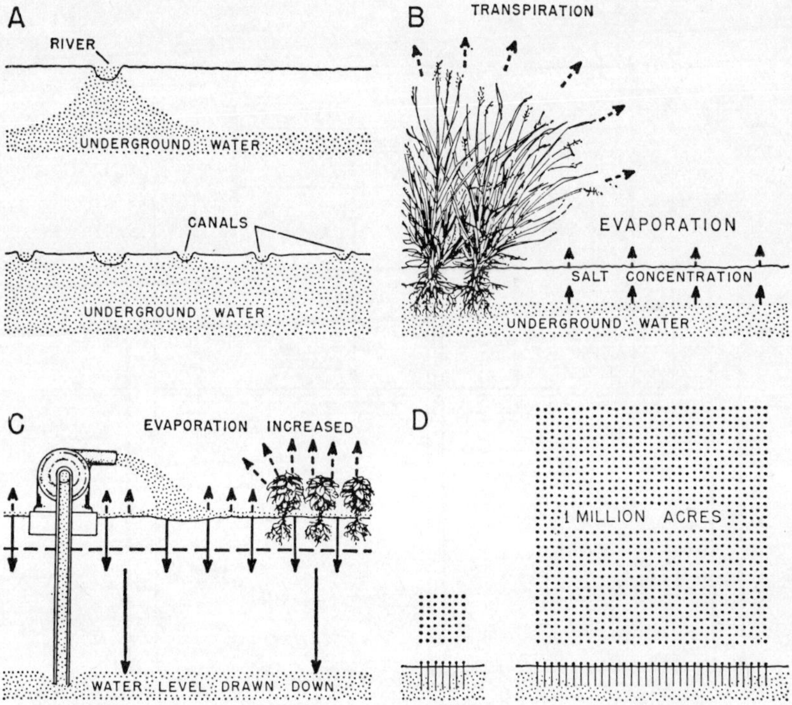


**5.13** Ground water relationships in irrigated areas of West Pakistan.

A The construction of leaky irrigation canals has allowed a general rise of the water table.

B Salt accumulates through the evaporation of saline water which rises into the topsoil under capillary action from the high water table.

C Waterlogging and soil salinity can be ameliorated by controlled cased tubewell pumping, which lowers the water table and supplies enough water at the surface to leach down soil salts before evaporation occurs.

D Large-scale tubewell pumping is required to negate the effects of lateral ground water seepage and to cause an appreciable lowering of the water table (*Source: Redrawn from Revell, R., 1963, Water;* Scientific American, *209 (3), p. 100*).

cultivation (Fig. 5.13B). Several engineering consultants have studied this problem recently (White House – Interior Panel, 1964) and have suggested remedial action by means of tube wells. Figure 5.13C shows the action of one tube well in pumping out the ground water, so as to bring the water table to a deeper level. At the same time drainage should be greatly improved, so as to leach out and carry away the concentration of salts from the top layers of the soil. A certain amount of irrigation of salt-resistant crops could be carried on over the period of time needed to draw down the water table and leach the top soil. The well field would have to be dense and extensive to prevent saline ground water from seeping into the pumping area from surrounding areas (Fig. 5.13D). A simulation model of this ground water problem was con-

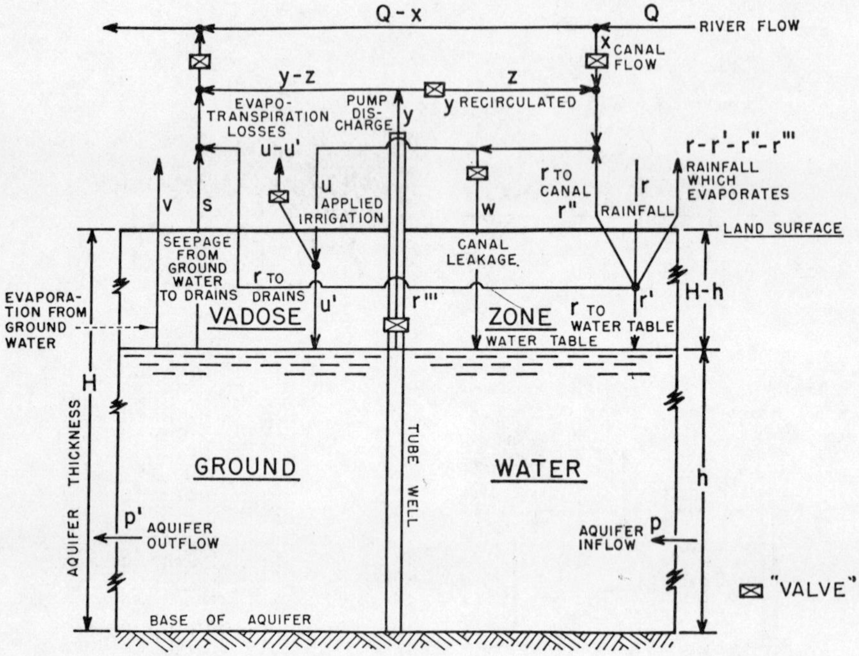


**5.14** Schematic diagram of the basic hydrological model to simulate the action of a tubewell and appurtenant hydraulic control devices for combating waterlogging and salinity in the irrigated regions of West Pakistan. The 'valves' are economic decision points – e.g. the setting of the valve at W represents the efficacy of the canal lining. (*Source: From Fiering, 1965, p. 43*).

structed at the Harvard Water Resources Centre (Fig. 5.14). By use of this model attempts were made to formulate answers to the following questions:

1 What is the best well spacing and pump capacity?
2 What proportion of the effluent should be re-applied to the land?
3 How much lining should be installed in the canal system?
4 How fast will the ground water drop?
5 What is the best cropping pattern that can be devised?
6 How much surface drainage should be provided?

The 'valves' in Figure 5.14 represent critical points which can be controlled in the operation of the system. Statistical sampling was used to derive a workable number of combinations of these 'valve settings'. The various combinations were then compared to find an optimal solution in terms of cost, agricultural production and engineering feasibility.

This salinity model is the most ambitious systems analysis yet attempted, but both Dorfman (1965) and Fiering (1965) see this type of study as a beginning leading to a new range of analyses made possible by computing techniques.

## HYDROLOGICAL MODELS AND GEOGRAPHY

Although there are large areas of overlap between hydrology and geography, especially in morphometry, hydraulic geometry and in the more economic aspects of hydrological planning, the two disciplines have developed quite separately. The work of Horton (1933 and 1945) in relating hydrology to geomorphology is a rare example of work which combines the two disciplines and shows how fruitful attempts to relate them can be.

Much of the research work in pure hydrology has been conducted by engineers, who have traditionally used model building in the solution of a variety of engineering problems. It is natural, therefore, that this approach should have extended into other types of hydrological investigations and that model building should be used as a tool throughout the whole field of hydrology. The generation of many of these models is a very specialist matter, but because the hydrologist deals so often with water in relation to man's activities, geographers should be extremely interested in them, in so far as the results of the hydrologists' work may very well be susceptible to remodelling into a geographically-orientated synthesis. The majority of the work described in this chapter is not geographical in its inception, and has not been done by geographers, but this does not mean that geographers can ignore the implications of this basic and fast-developing earth science, which possesses, as does geography, a strong bias towards the activity and welfare of man.

## REFERENCES

ALLEN, J., [1952], *Scale Models in Hydraulic Engineering*, (London), 407 pp. .

AMOROCHO, J., [1963], Measures of the linearity of hydrologic systems; *Journal of Geophysical Research*, 68, 2237–2249.

AMOROCHO, J. and HART, W. E., [1964], A critique of current methods in hydrologic systems investigation; *Transactions of the American Geophysical Union*, 45, 307–321.

AMOROCHO, J. and HART, W. E., [1965], The use of laboratory catchments in the study of hydrologic systems; *Journal of Hydrology*, 3, 106–123.

APPLEBY, F. V., [1954], Run-off dynamics; A heat conduction analogue of storage flow in channel networks; *International Union of Geodesy and Geophysics* (Rome).

BEARD, L. R., [1962], *Statistical Methods in Hydrology*, U.S. Army Engineer District Corps of Engineers, Sacramento, California, Project CW-151, 62 pp.

BINNIE, G. M. and MANSELL-MOULLIN, M., [1966], The estimated probable maximum storm and flood on the Jhelum River – a tributary of the Indus *The Institution of Civil Engineers, Proceedings of the Symposium on River Flood Hydrology*, 189–210.

BLANEY, H. F. and CRIDDLE, W. D., [1950], *Determining Water Requirements in Irrigated Areas from Climatological and Irrigation Data*; United States Department of Agriculture, Soil Conservation Service, 96 pp.

BONIFACE, E. S., [1959], Some experiments in artificial recharge in the Lower Lee Valley; *Proceedings of the Institution of Civil Engineers*, 14, 325–338.

BOWDEN, L. W., [1965], *Diffusion of the Decision to Irrigate*; The University of Chicago, Department of Geography, Research Paper No. 97, 146 pp.

BRITTAN, M. R., [1961], Probability analysis applied to the development of synthetic hydrology for Colorado River; part 4 of *Past and Probable Future Variations in Stream Flow in the Upper Colorado River*, University of Colorado, Bureau of Economic Research, (Boulder, Colorado).

BUCHAN, S., [1959], Artificial replenishment of aquifers; *Journal of the Institution of Water Engineers*, 9, 111–163.

BUCHAN, S., [1963], Conservation by integrated use of surface and ground water; *The Institution of Civil Engineers, Proceedings of the Symposium on Water Conservation*, 181–185.

BURTON, I., [1962], *Types of Agricultural Occupance of Flood Plains in the United States*; The University of Chicago, Department of Geography, Research Paper No. 75, 167 pp.

CARTER, D. B. *et al.*, [1966], Energy and water budget, In *Spacecraft in Geographic Research*, National Academy of Sciences, National Research Council, Publication 1353, (Washington, D.C.), 23–47.

CHERY, D. L., [1966], Design and tests of a physical watershed model; *Journal of Hydrology*, 4, 224–235.

CHILDS, E. C., [1953], A new laboratory for the study of the flow of fluids in porous beds; *Proceedings of the Institution of Civil Engineers*, 2 (3), 134–141.

CHILDS, E. C., [1957], Physics of land drainage; In Luthin, J. N., (Ed.), *Drainage of Agricultural Lands*, (Madison, Wisconsin), 1–66.

CHOW, VEN TE. (Ed.), [1964], *Handbook of Applied Hydrology*, (New York).

CIRIACY-WANTRUP, S. V., [1963], *Resource Conservation, Economics and Policies*, (Berkeley, California), 395 pp.

COX, D. R. and SMITH, W. L., [1963], *Queues*, (London), 180 pp.

CRAWFORD, N. H. and LINSLEY, R. K., [1962], *Synthesis of continuous streamflow hydrographs on a digital computer*; Department of Civil Engineering, Stanford University, Technical Report No. 12.

CRAWFORD, N. H. and LINSLEY, R. K., [1966], *Digital simulation in hydrology: Stanford Watershed Model IV*; Department of Civil Engineering, Stanford University, Technical Report No. 39, 210 pp.

DATAMATIC CORPORATION AND RAYTHEON MANUFACTURING COMPANY for U.S. Army Corps of Engineers, [1957], *Report of Use of Electronic Computers for Integrating Reservoir Operations.*

DAVIS, S. N. and DE WEIST, R. J. M., [1966], *Hydrogeology*, (New York), 463 pp.

DAWDY, D. R. and O'DONNELL, T., [1965], Mathematical models of catchment behavior; *Proceedings of the American Society of Civil Engineers, Journal of the Hydraulics Division*, 91, No. HY4, Part 1, 123–137.

DE WEIST, R. J. M., [1965], *Geohydrology*, (New York), 366 pp.

DOOGE, J. C. I., [1959], A general theory of the unit hydrograph; *Journal of Geophysical Research*, 64, 241–256.

DORFMAN, R., [1965], Formal models in the design of water resource systems; *Water Resources Research*, 1, 329–336.

DUCKWORTH, W. E., [1965], *A Guide to Operational Research*, (London), 153 pp.

ECKSTEIN, O., [1958], *Water Resource Development: The economics of project evaluation*, (Cambridge, Mass.).

EMMETT, W. W., [1965], The Vigil Network: methods of measurement and a sampling of data collected; *International Association of Scientific Hydrology, Symposium of Budapest*, Publication No. 66, 89–106.

FIERING, M. B., [1962], Queueing theory and simulation in reservoir design; *Transactions of the American Society of Civil Engineers*, 127, 1114–1144.

FIERING, M. B., [1965], Revitalizing a fertile plain; *Water Resources Research*, 1, 41–61.

FOLLANSBEE, R. and SAWYER, L. R., [1948], Floods in Colorado; *United States Geological Survey, Water Supply Paper* 997.

FOSTER, C. D. and BEESLEY, M. E., [1963], Estimating the social benefit of constructing an underground railway in London; *Journal of the Royal Statistical Society*, Series A, 126, 46–78.

GANI, J. and MORAN, P. A. P., [1955], The solution of dam equations by Monte Carlo methods; *Australian Journal of Applied Science*, 6, 267–273.

GEORGE, F. H., [1965], *Cybernetics and Biology*, (Edinburgh), 138 pp.

GIGUET, R., [1957], A French dual-purpose scheme – the Durance project and the Serre Ponçon dam; *Proceedings of the Institution of Civil Engineers*, 6, 550–576.

GRANT, E. L. and IRESON, W. G., [1964], *Principles of Engineering Economy*, (New York), 574 pp.

GUMBEL, E. J., [1958A], *Statistics of Extremes*, (New York), 375 pp.

GUMBEL, E. J., [1958B], Statistical theory of floods and droughts; *Journal of the Institution of Water Engineers*, 12, 157–184.

HACK, J. T., [1957], Studies in longitudinal stream profiles in Virginia and Maryland; *United States Geological Survey, Professional Paper*, 294-B, 45–97.

HAMMERSLEY, J. M. and HANDSCOMB, D. C., [1964], *Monte Carlo Methods*, (London), 178 pp.

HARRIS, R. A., [1965], Probability of reservoir yield failure using Moran's Steady-State Probability Method and Gould's Probability Routing Method; *Journal of the Institution of Water Engineers*, 19, 302–328.

HORTON, R. E., [1933], The role of infiltration in the hydrologic cycle; *Transactions of the American Geophysical Union*, 14, 446–460.

HORTON, R. E., [1945], Erosional development of streams and their drainage basins:

Hydrophysical approach to quantitative morphology; *Bulletin of the Geological Society of America*, 56, 275–370.

HUFSCHMIDT, M. F., [1965], Field level planning of water resource systems; *Water Resources Research*, 1, 147–171.

HURST, H. E., BLACK, R. P. and SIMAIKA, Y M., [1965], *Long-Term Storage*, (London), 145 pp.

INTER-AGENCY RIVER BASIN COMMITTEE (Sub-Committee on Costs and Budgets), [1950], *Proposed Practices for Economic Analysis of River Basin Projects*, '*The Green Book*', (Washington, D.C.).

INTERNATIONAL ASSOCIATION OF SCIENTIFIC HYDROLOGY, [1965], *Symposium of Budapest, Representative and Experimental Areas*; Publication No. 66.

JOHNSON, A. I., [1965], Computer processing of hydrologic and geologic data; *Ground Water*, 3 (3), 9 pp.

JOHNSON, A. I. and LANG, S. M., [1965], Automated processing of water information; *Proceedings of the First Annual Meeting, American Water Resources Association*, (Chicago), 324–350.

JULIAN, P. R., [1961], A study of the statistical predictability of stream-runoff in the Upper Colorado River basin; part 2 of *Past and Probable Future Variations in Stream Flow in the Upper Colorado River*, University of Colorado, Bureau of Economic Research, (Boulder, Colorado).

KATES, R. W., [1962], *Hazard and Choice Perception in Flood Plain Management*; The University of Chicago, Department of Geography, Research Paper No. 78, 157 pp.

KENDALL, D. G., [1957], Some problems in the theory of dams; *Journal of the Royal Statistical Society*, Series B, 19, 207–212.

KING, P. F. and PEEL, D. A., [1960], An analysis of a hydro-electric system; *The Computer Journal*, 3, 161–163.

KRUMBEIN, W. C. and GRAYBILL, F. A., [1965], *An Introduction to Statistical Models in Geology*, (New York), 475 pp.

KRUTILLA, J. V. and ECKSTEIN, O., [1958], *Multiple Purpose River Development*, (Baltimore), 301 pp.

KUIPER, E., [1965], *Water Resources Development, Planning, Engineering and Economics*, (London), 483 pp.

LEET, L. D. and JUDSON, S., [1954], *Physical Geology*, 1st Edn., (New York), 466 pp.

LEOPOLD, L. B., [1962], The Vigil Network; *Publication of the International Association of Scientific Hydrology*, Year 7 (2), 5–9.

LEOPOLD, L. B. and EMMETT, W. W., [1965], Vigil Network sites: A sample of data for permanent filing; *Publication of the International Association of Scientific Hydrology*, Year 10 (3), 12–21.

LEOPOLD, L. B. and MILLER, J. P., [1956], Ephemeral streams – Hydraulic factors and their relation to the drainage net; *United States Geological Survey, Professional Paper* 282-A, 1–37.

LEOPOLD, L. B., WOLMAN, M. G. and MILLER, J. P., [1964], *Fluvial Processes in Geomorphology*, (San Francisco), 522 pp.

LEWIS, D. J. and SHOEMAKER, L. A., [1962], Hydro system power analysis by digital computer; *Journal of the Hydraulics Division, Proceedings of the American Society of Civil Engineers*, 88, 113–130.

LINSLEY, R. K., KOHLER, M. A. and PAULHUS, L. H., [1949], *Applied Hydrology*, (New York), 689 pp.

LOWRY, R. L. and JOHNSON, A. F., [1942], Consumptive use of water for agriculture; *Transactions of the American Society of Civil Engineers*, 107, 1243–1302.

LUTHIN, J. N. (Ed.), [1957], *Drainage of agricultural lands*, (Madison, Wisconsin), 620 pp.

LUTHIN, J. N., [1966], *Drainage Engineering*, (New York), 250 pp.

MAASS, A., HUFSCHMIDT, M. M., DORFMAN, R., THOMAS, H. A., MARGLIN, S. A. and FAIR, G. M., [1962], *Design of Water-Resource Systems*, (Cambridge, Mass.), 620 pp.

MCKEAN, R. N., [1958], *Efficiency in Government through Systems Analysis: With Emphasis on Water Resources Development*, (New York), 336 pp.

MONTEITH, J. L., [1959], The reflection of short-wave radiation by vegetation; *Quarterly Journal of the Royal Meteorological Society*, 85, 386–392.

MORAN, P. A. P., [1959], *The Theory of Storage*, (London), 110 pp.

MORRICE, H. A. W. and ALLAN, W. N., [1959], Planning for the ultimate hydraulic development of the Nile Valley; *Proceedings of the Institution of Civil Engineers*, 14, 101–156.

MURPHY, F. C., [1958], *Regulating Flood-Plain Development*; The University of Chicago, Department of Geography, Research Paper No. 56, 204 pp.

NASH, J. E., [1958], Determining run-off from rainfall; *Proceedings of the Institution of Civil Engineers*, 10, 163–184.

NASH, J. E., [1960], A unit hydrograph study, with particular reference to British catchments; *Proceedings of the Institution of Civil Engineers*, 17, 249–282.

O'DONNELL, T., [1960], Instantaneous unit hydrograph derivation by harmonic analysis; *Commission of Surface Waters, International Association of Scientific Hydrology*, Publication 51, 546–557.

O'DONNELL, T., [1966], Computer evaluation of catchment behaviour and parameters significant in flood hydrology; *The Institution of Civil Engineers, Proceedings of the Symposium on River Flood Hydrology*, 103–113.

O'KELLY, J. J., [1955], The employment of unit hydrographs to determine the flows of Irish arterial drainage channels; *Proceedings of the Institution of Civil Engineers*, 4, 365–445.

OLIVIER, H., [1961], *Irrigation and Climate*, (London), 250 pp.

PENMAN, H. L., [1948], Natural evaporation from open water, bare soil and grass; *Proceedings of the Royal Society, Series A*, 193, 120–145.

PENMAN, H. L., [1950], The water balance of the Stour catchment area; *Journal of the Institution of Water Engineers*, 4, 457–469.

PENMAN, H. L., [1956], Evaporation – an introductory survey; *Netherlands Journal of Agricultural Science*, 4, 9–29.

PENMAN, H. L., [1963], *Vegetation and Hydrology*; Commonwealth Bureau of Soils, Harpenden, Technical Communication No. 53, 124 pp.

PIGOU, A. C., [1932], *The Economics of Welfare*, (London), 837 pp.

PREST, A. R. and TURVEY, R., [1965], Cost-Benefit analysis: a survey; *The Economic Journal*, 75, 683–735.

ROCKWOOD, D. M., [1961], Columbia Basin streamflow routing by computer; *Transactions of the American Society of Civil Engineers*, 126, 32–56.

ROCKWOOD, D. M. and NELSON, M. L., [1966], Computer application to streamflow synthesis and reservoir regulation; *Proceedings of the Sixth Congress of the International Commission of Irrigation and Drainage*, (New Delhi, India).

RUTTAN, V. W., [1965], *The Economic Demand for Irrigated Acreage: New Methodology and Some Preliminary Projections* 1954–1980, (Baltimore), 154 pp.

SCHWAB, G. O., FREVERT, R. K., EDMINSTER, T. W. and BARNES, K. K., [1966], *Soil and Water Conservation Engineering*, 2nd Edn., (New York), 683 pp.

SHEN, J., [1963A], Use of hydrologic models in the analysis of flood runoff; *Administrative Report, United States Geological Survey* (*Water Resources Division*).

SHEN, J., [1963B], The role of analogue in surface-water hydrological problems; *Administrative Report, United States Geological Survey* (*Water Resources Division*).

SHERMAN, L. K., [1932], Streamflow from rainfall by unit-graph method; *Engineering News-Record*, 108, 501–505.

SKIBITZKE, H. E., [1963], The use of analogue computers for studies in groundwater hydrology; *Journal of the Institution of Water Engineers*, 17, 216–230.

SLAYMAKER, H. O. and CHORLEY, R. J., [1964], The Vigil Network System; *Journal of Hydrology*, 2, 19–24.

SMITH, S. C. and CASTLE, E. N., [1965], *Economics and Public Policy in Water Resource Development*, (Ames, Iowa), 463 pp.

STRAHLER, A. N., [1952], Hypsometric (area-altitude) analysis of erosional topography; *Bulletin of the Geological Society of America*, 63, 1117–1142.

STRAHLER, A. N., [1964], Quantitative geomorphology of drainage basins and channel networks; Section 4-II in Chow, Ven Te, (Ed.), *Handbook of Applied Hydrology*, 4–39 to 4–76.

TENNESSEE VALLEY AUTHORITY, Office of Tributary Area Development, [1961], *Matrix operations in hydrograph computations*; Research Paper No. 1, (The Authority, Knoxville, Tennessee).

TENNESSEE VALLEY AUTHORITY, Office of Tributary Area Development, [1963A], *A water yield model for analysis of monthly runoff data*; Research Paper No. 2, (The Authority, Knoxville, Tennessee).

TENNESSEE VALLEY AUTHORITY, Office of Tributary Area Development, [1963B], *T.V.A. computer programmes for hydrologic analyses*; Research Paper No. 3, (The Authority, Knoxville, Tennessee).

THAMES CONSERVANCY, [1965], *Thames Conservancy and the Water Crisis*, 11 pp.

THIESSEN, A. H., [1911], Precipitation for large areas; *Monthly Weather Review*, 39, 1082–1084.

THORN, R. B. (Ed.), [1966], *River Engineering and Water Conservation Works*, (London), 520 pp.

TODD, D. K., [1959], *Ground Water Hydrology*, (New York), 336 pp.

THORNTHWAITE, C. W., [1948], An approach towards a rational classification of climate; *Geographical Review*, 38, 85–94.

THORNTHWAITE, C. W., [1954], A re-examination of the concept and measurement of potential transpiration; In Mather, J. R. (Ed.), *The Measurement of Potential Evapo-transpiration*, Problems in Climatology, (Seabrook, N.J.), 200–209.

UNESCO INTERNATIONAL HYDROLOGICAL DECADE CO-ORDINATING COUNCIL, [1965], *Final Report*, (Paris).

VAJDA, S., [1961], *The Theory of Games and Linear Programming*, (London), 106 pp.

WHITE, G. F., [1945], *Human Adjustment to Floods*; The University of Chicago, Department of Geography, Research Paper No. 29, 225 pp.

WHITE, G. F., CALEF, W. C., HUDSON, J. W., MAYER, H. M., SHEAFFER, J. R. and VOLK, D. J., [1958], *Changes in Urban Occupance of Flood Plains in the United States*; The University of Chicago, Department of Geography, Research Paper No. 57, 235 pp.

WHITE, G. F. (Ed.), [1961], *Papers on Flood Problems*; The University of Chicago, Department of Geography, Research Paper No. 70, 228 pp.

WHITE, G. F., DE VRIES, E., DUNKERLEY, H. B. and KRUTILLA, J. V., [1962], *Economic and Social Aspects of Lower Mekong Development*; A Report to the Committee for Co-ordination of Investigations of the Lower Mekong Basin, 106 pp.

WHITE, G. F., [1963], The Mekong River Plan; *Scientific American*, 208 (4), 49–59.

WHITE HOUSE-INTERIOR PANEL, [1964], *Report on Land and Water Development in the Indus Plain*.

WIESNER, C. J., [1964], Hydrometeorology and river flood estimation; *Proceedings of the Institution of Civil Engineers*, 27, 153–167.

WOLF, P. O., [1966A], Comparison of Methods of Flood Estimation; *The Institution of Civil Engineers, Proceedings of the Symposium on River Flood Hydrology*, 1–23.

WOLF, P. O., [1966B], Notes on the management of water resource systems; *Journal of the Institution of Water Engineers*, 20 (2), 95–105.

CHAPTER SIXTEEN

# Maps as Models

C. BOARD

*I don't believe in maps because it never looks like it says on the maps when you get there.* From an advertisement issued by the BREWERS' SOCIETY

*Maps, representing results of original surveys in visual form, constitute simple models of a 'real' world, . . .* KANSKY, 1963, p. 7.

*There is no such thing as a perfectly faithful model; only by being unfaithful in* some *respect can a model represent its original.*
BLACK, 1962, p. 220.

One suspects that the young ladies on the advertisement, who said they did not trust maps, were complaining more about the ability of their male companions to make something of the 'wiggly lines on the map'. Of course, no map can perfectly depict reality, but in *not* doing so it is all the more useful. The only perfectly faithful depiction would be an identical copy of reality itself. The reasons are not far to seek. Reduction in scale, the loss of the third dimension, human artifice in the creation of the representation, and the inability to read the depiction perfectly satisfactorily are the most important. Although some of the secrets of nature can be unravelled without maps, patterns over relatively large areas are often best detected and problems identified by careful studies of maps (Wooldridge and East, 1951, p. 65). The same authors observe, with envious sympathy, that 'a quick-witted mobile urchin may in one sense "know his geography" as he conducts us by tortuous by-ways from station to hotel, but neither he nor we have any adequate picture of the pattern of the town, without the benefit of maps' (1951, p. 65).

This chapter will regard maps as iconic or representational models and as conceptual models in a framework provided by the human being's struggle to communicate to his fellows something of the nature of the real world. There have been few previous attempts to generalize about maps in this way. Among the most notable are Schmidt-Falkenberg (1962), de Dainville's historical study (1964), Moles (1964) and Bunge's account of metacartography

(1962). Chorley (1964, p. 136) in a statement of the place of analogue models in geographical investigation, has pointed out that though none of them are completely successful, few are without some value. At the conclusion of a discussion which employed maps as conceptual models (Haggett, 1964, p. 380), Stamp expressed the hope that such models would be torn up if necessary, just as the representational maps would have been, once they had served a purpose or had been superseded. The map can so easily be the point of contact between 'the modern quantitative approach' and the traditional approach.

## THE MAP-MODEL CYCLE: THE ARGUMENT

It is comparatively easy to visualize maps as representational models of the real world, but it is important to realize that they are also conceptual models containing the essence of some generalization about reality. In that role, maps are useful analytical tools which help investigators to see the real world in a new light, or even to allow them an entirely new view of reality.

There are then two major stages to the cycle of map making. First, the real world is concentrated in model form; secondly the model is tested against reality. In practice the scientist who makes such maps has a new view of the real world. It is also axiomatic that the cycle may begin again with the revised view of the real world. For example, a series of journeys through an area for which there is only minimal map cover may suggest that there are interesting variations in land use pattern. The next and obvious step is to make (by some suitable method) a map which records the significant elements of that land use pattern. Once completed, the map of land use is taken into the field, or is compared with reality in some other way. Speculations on the relationships between land use and physical, economic and cultural factors may be tested. In many cases such tests will involve the design and construction of new maps of both trends and relationships in an attempt to unravel some of the complex patterns of the real world. Sometimes the process of investigation starts with a map, whose elements somehow give rise to some speculation in relation to the origin of, for example, a drainage pattern or some peculiarity in a maze of property boundaries. In this case the map, which is already a model of what it portrays, is dissected as is the landscape or the environment of the real world (e.g. Conzen, 1960, Chaps. 1 and 2). After such an investigation some of the results may well be presented in map form, thus entering yet another phase of the cycle. In both model making and model testing the principles underlying the methods of map making are strikingly similar. Though the processes of abstraction, model construction and model testing may very well continue without any marked breaks, for the purposes of this essay it is convenient to start with the real world and identify distinct steps in the processes in order that the relationship between maps and models may be more clearly seen. Figure 16.1 summarizes these steps and provides a 'map' of the ensuing account.

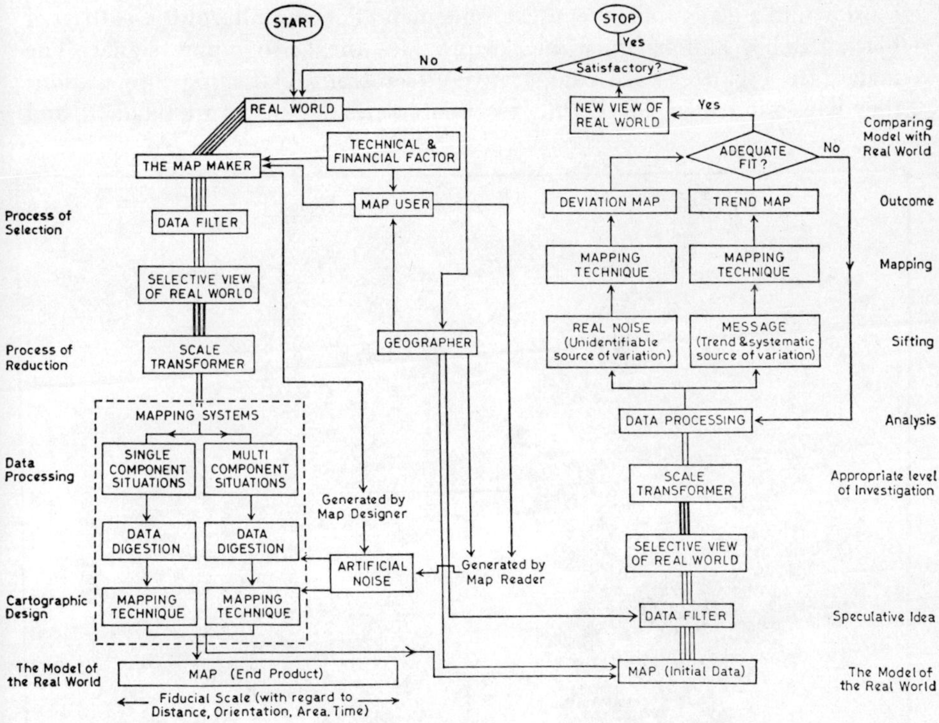


**16.1** The Map-Model Cycle.

It is a truism to assert that maps are vehicles for the flow of information. Some are better vehicles than others, but the functions they perform are nevertheless similar, irrespective of their quality. It is instructive to look at the role of maps in an adaptation of a general communications system. Figure 16.2 shows such a system. The source is comparable to the real world, or the stimulus which a real world situation gives to an investigator. For example, much of the surface of the Midland counties of England is characterized by a pattern of 'ridge and furrow' (Mead, 1954; Harrison, Mead and

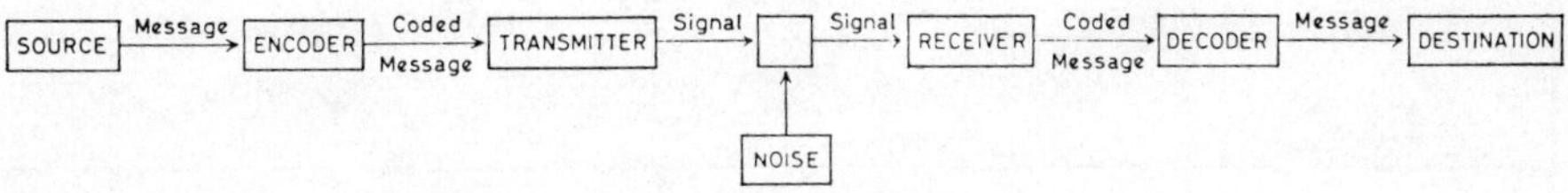


**16.2** Generalized communication system (*After Johnson and Klare, 1961, p. 15*).

Pannett, 1965). The curiosity of observers was excited by this phenomenon, which led to its being identified either on ground or on air photographs with a fair degree of certainty. This message generated by the cultural landscape has been simply coded (black indicates the presence of ridge and furrow) and set down in a geographical context, on a map. The stimulus of the pattern of black patches and irregular blocks provides the cartographic signal. The maps (see Fig. 16.3) are comparatively free from distracting information; they have a low noise level. Only essential names and insets are retained, and

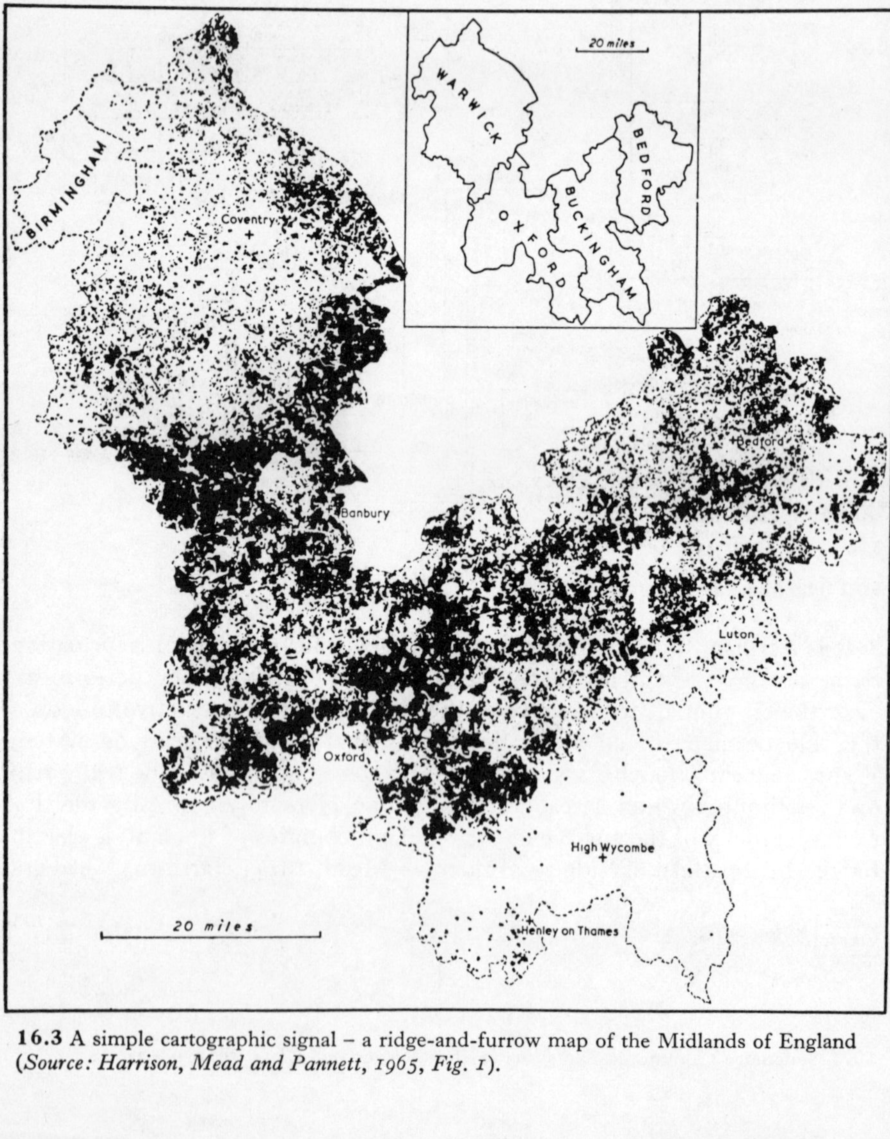


**16.3** A simple cartographic signal – a ridge-and-furrow map of the Midlands of England (*Source: Harrison, Mead and Pannett, 1965, Fig. 1*).

they do not to any extent blur the signal. The stimulus, once printed and circulated in the pages of the *Geographical Journal*, is taken up through the eyes (receiver) of the readers of that periodical. The black pattern and its complementary white matrix is decoded into a pattern of distribution. The shapes are deciphered and keyed into what is already known of that fragment of the English landscape.

In other cases, for instance where it is crucial that a representation of a landscape feature such as a road junction, be firmly placed in the mind of a traveller to help him to go in the right direction, simple maps containing a maximum of information and a minimum of noise are the most effective. The provision of superfluous detail only obstructs the transmission of the message to the recipient. Advertisements such as one used by French national railways to persuade Parisians to travel at week-ends are simple messages stressing only essentials (Fig. 16.4).

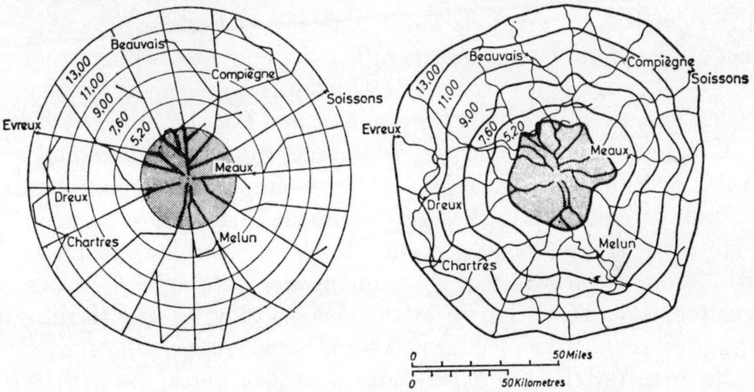


**16.4** A simple cartographic message (left) indicating the price of a second class week-end ticket to different stations around Paris adapted from a leaflet advertising special fares. Contrast (right) the actual pattern of lines and places and the configuration of the price zones on the left-hand topographic map (*Source: Leaflet published by Societé Nationale des Chemins de fer francais, 1965*).

## MAKING THE MODEL

### The map maker

It is a commonplace that cartography, the making of maps, combines the characteristics of both a science and an art. Even the maps produced as part of the output of computers (Tobler, 1965) require the input of instructions from a designer. In fact several different sets of instructions may be desirable if a series of trial maps from a body of data is being produced as preliminary to a fair-drawn map for publication (Monmonier, 1965, p. 13). Such maps minimize the influence that human factors may have upon the final design, but they do not remove it.

Harrison (1959, pp. 29–30) has pointed out that Eckert (1908) was not wholly correct in labelling the topographic map as precise, reproducing 'facts as they exist in nature', and the thematic or geographically abstract map as artistic. However, this criticism is answered by Eckert himself:

> As long as the scale allows the objects in nature to be represented in their true proportion on the map, technical skill alone is necessary. Where this possibility ends the art of the cartographer begins. With generalization art enters into the making of maps. (Eckert, 1908, pp. 346–347.)

No map can show the objects in nature in their true proportion. The largest conceivable scale may enable the exact width of a pavement and roadway to be shown, but nothing short of a facsimile will permit the true representation of all the details of a manhole cover with the maker's name included! The very act of selection of some details rather than others for portrayal involves a decision by the cartographer and one which introduces art to the map.

In what is perhaps the most thorough account of the subjective element in maps, Wright (1942) considers scientific integrity the fundamental quality in map makers. Sometimes cartographers are tempted to depict country 'with an intricacy of detail derived largely from the imagination', especially when the quantity of known detail is scanty. Such an attitude to map making may result in a loss of information by reducing the differences between regional patterns (see Fig. 16.5). However the reverse may be true in the case of a cartographer as skilled as Robert Dawson (1776–1860), who started work as a draughtsman for the Ordnance Survey at £54 per annum in 1794. (Dictionary of National Biography, p. 678). 'Drawing in its application to maps' was not for him 'confined to delineation only, but to the whole expression of form in respect of ground . . . it is *natural history* drawing of land in full perspective perception' requiring 'the ordinary graphic qualification of the artist combined with some knowledge of some physical geography and geology' (Dawson, 1854, quoted by Harris, 1959, p. 517).

## Bias in maps

The human element obtrudes further in the case of maps drawn for propaganda purposes. The aims of such persuasion may be commercial or political. Everyone is familiar with the tourist maps which are comfortably filled (or crowded) with details of the activities attractive to the potential tourist. But, it is not always so obvious that agencies, anxious to portray their territories to the best advantage, often adopt different standards for the inclusion of detail as between their territory and surrounding areas. This deliberate choice of

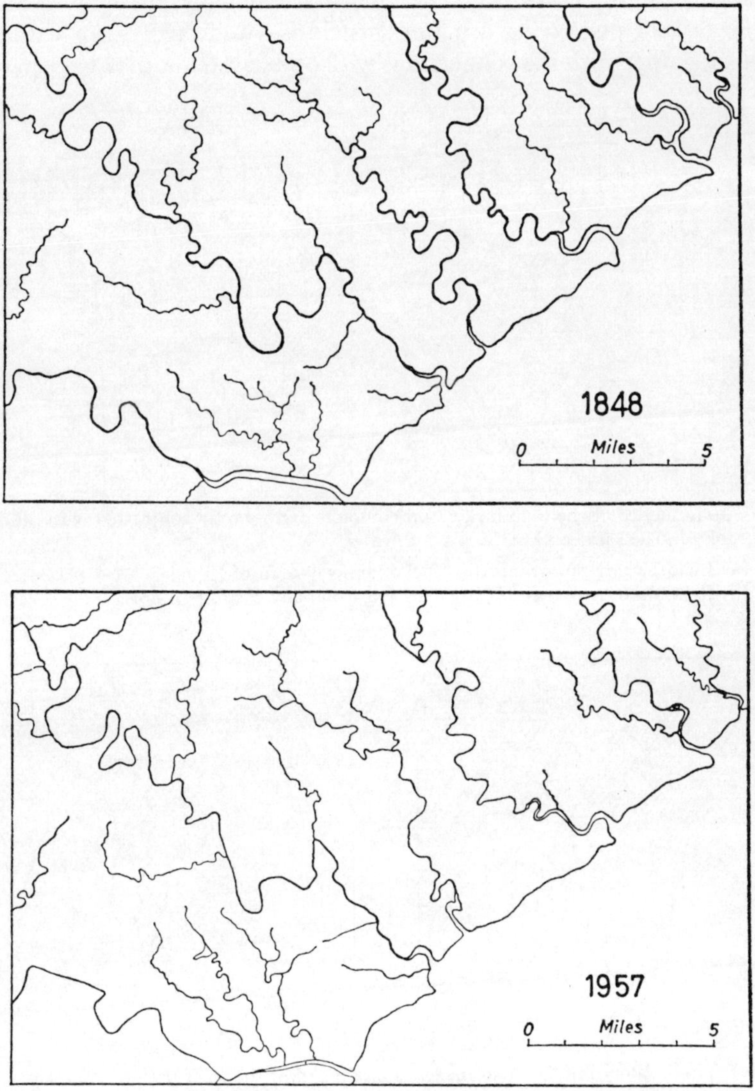


**16.5** Artificial intricacy contrasted with geographical 'reality'. Rivers on Jervois' map of British Kaffraria (Eastern Cape Province) in 1848 contrasted with the river pattern on a modern map (*Trigsurvey, Pretoria, 1957, In Board, 1962, volume of maps*).

detail can be even more misleading than the emphasis of detail in one area as opposed to another (Figs. 16.6 and 16.7). Occasionally, the map maker who has the task of portraying a pattern, such as a railway system with all its stations together with the connections with other systems, may have to distort

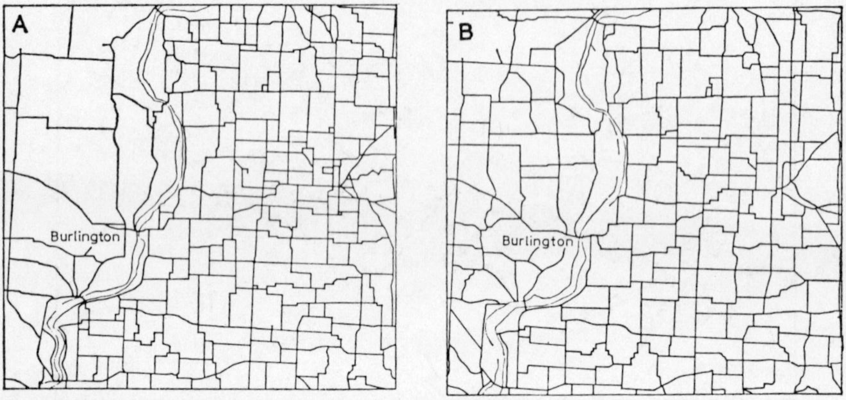


**16.6** The influence of the map maker: oil company cartography contrasted with official cartography in the same areas (Illinois – Iowa)

(A) Road pattern shown on the official highway map of Illinois.
(B) Road pattern shown on Standard Oil Company map of Illinois.

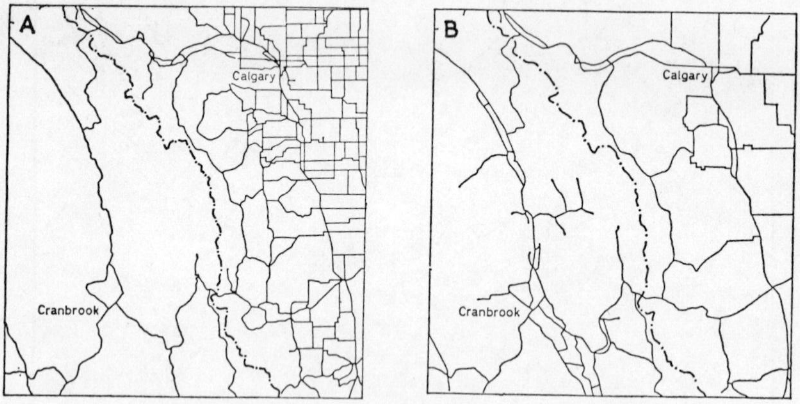


**16.7** The influence of the map maker: more detail of the road pattern is given in the territory belonging to the authority producing the map.

(A) Road pattern shown on official highway map of Alberta.
(B) Road pattern shown on official highway map of British Columbia.

the distances and directions of lines in order to make the best use of a rectangle.

Political motives affect cartography in two main ways. In many instances, there is an official position in relation to international boundary lines and

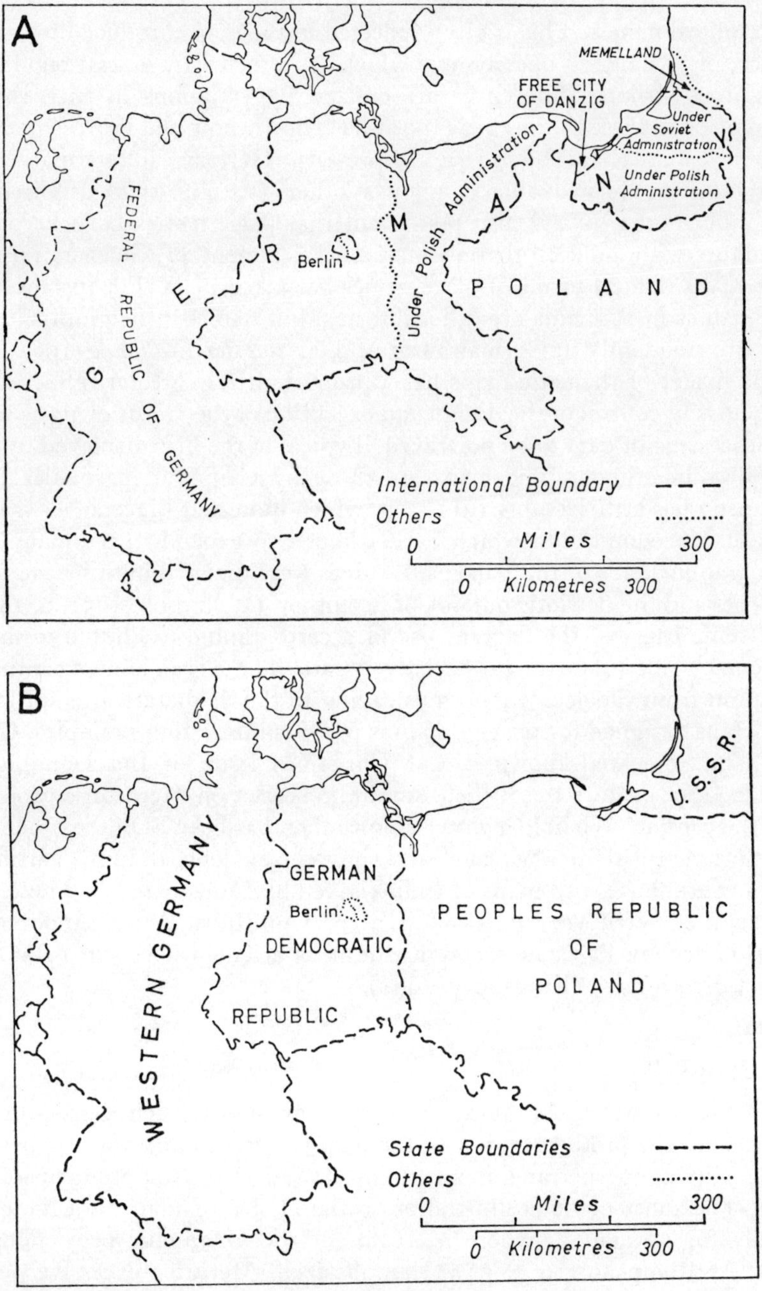


**16.8** The West German view (A) contrasted with the East German view (B) of political boundaries in Central Europe (*Source: Sinnhuber, 1964, Figs. 1 and 2*).

geographical names. This is often reflected in the maps produced by a single agency, or in maps in one country which are affected by official regulations. Sometimes striking differences are created by variations in such rules in different countries. Sinnhuber (1964) has shown how the representation of the political areas and boundaries of pre-1939 Germany differs markedly in West and East German atlases, amongst others (see Fig. 16.8). In discussion Sinnhuber (1964, p. 27) also points out that place names have become so mixed up with politics that inconsistent treatment of German forms as alternatives to local forms of place names has taken place. For instance most place names in Rumania are in local forms, but names in Belgium and Italy are more frequently in German form in the *Atlas der Erdkunde* (1962).

The maker of thematic maps has a proportionately greater influence, because he is in control of the design and execution of the fair drawing as well as the processing of data to be portrayed. Typical of the bias displayed by some compilers of ethnographic maps are those by Cvijić. In particular H. R. Wilkinson has criticized his 1913 map, which indicated Macedo-Slavs in the parts of Macedonia into which Serbia hoped to expand. 'Like many other ethnographic maps of the Balkans, its ideas were dictated both by the march of events and the patriotic outlook of its author' (Wilkinson, 1951, p. 180).

Extreme cases of the flagrant use of a cartographic technique to make a particular point are fortunately relatively rare. There is an almost continuous gradation from the accidentally misleading to the deliberate attempt to distort. Maps designed for postage stamps provide interesting examples. On the one hand, the well-known 2-cent Christmas issue of the Dominion of Canada (1898) shows the British Empire in scarlet on Mercator's projection with the caption 'We hold a vaster empire than has been'. On the other hand two more recent stamps which clearly show the evidence of human artifice in mapping are those with maps of India (issued by India in 1957). These show the disputed territories of Kashmir as part of India. A contemporaneous stamp issued by Pakistan shows Kashmir as a land whose 'final status (is) not yet determined' (Kingsbury, 1964).

## The map user

It would, however, be quite wrong to suggest that such aspects of map design were the product merely of the map maker's mind. Many departures from reality are perpetrated in an attempt to satisfy the requirements of map users. The most obvious illustration is the choice of map projections, especially for navigation, where Mercator or the Gnomonic projections are usual. As Robinson (1960, p. 71) emphasized, Mercator excessively exaggerates the size of the land masses in Northern latitudes. However its employment for small-scale world maps in countless atlases has been responsible for many misconceptions as to the relative sizes of different parts of the world.

Indeed Mackay (1954, p. 4) has shown that the usual aspect of Mercator does not even rate well with map users as against its oblique case when the map shape of Greenland is compared with its shape on the globe. Robinson's suggestion (1960, p. 75) that the cylindrical equal-area projection 'looks peculiar' to many people, when, provided standard parallels just below 30° are used, it has the least mean angular deformation of any equivalent world projection, demonstrates that mathematical balance may not be everything. Marschner (1943, p. 219) summarizes the position in the following manner. 'The structural property of maps on smaller scales, therefore, is a fundamental issue between the professional map users and the map makers'. But he also points out that the map users and map makers are frequently the same person. For geographical purposes Marschner (1944, p. 44) considers the property of equivalence to be most important, because so much depends on the measurement of area and the correlation of areal phenomena. Equivalence is all the more important because of the three elements (area, distance and angles) only area can be preserved as universally true on a map (Marschner, 1944, p. 45).

## Map purpose

No discussion of the role of the map designer could be complete without reference to the purposes for which maps are made. Some of the more tendentious maps are of course made with the deliberate intention to deceive. They may omit details which could prove useful to some enemy power, or insert details of towns and plots of land in places devoid of habitation in order to attract settlement and land purchasers. There is abundant evidence of the latter, for example, in tract maps filed in Los Angeles and adjacent counties for towns such as Sunset, Gladstone and Richland in the late 1880's (Dumke, 1963, Chap. 14). Although maps may be made for specific purposes there is naturally enough no guarantee that they will be used in the way intended. The tract maps referred to above are a valuable source of information for the historian of urban growth in southern California. Similarly topographic maps have often been made initially for military purposes, as for example the one-inch Ordnance Survey of Britain and the 1:75,000 Austrian Staff map. But just because they provide basic information about the country, they are frequently used by non-military personnel. Indeed it is customary to use such maps as base maps for overprinting specialist information such as geology, land use or population. Linton (1948) points out that the United States Geological Survey was charged with the task of both topographical and geological survey and was able to design a topographic map 'specifically as a base for geological and other overprints'. This had a very direct influence on the design of the 1:62,500 topographic base map, which was made simpler by omitting much man-made detail and finer by the use of narrower road symbols than those usually employed on maps of comparable scales. As a result USGS geological maps are very much clearer than many others.

Another large class of maps which show clear signs of the importance of purpose in design are those intended for navigation, whether by ship, aeroplane or motor car. Ideally and in fact frequently, such maps or charts carry such information that serves to speed individuals from one place to another. The sea charts of the Marshall Islanders incorporated dominant wave directions (Lyons, 1928) and modern Admiralty charts show lighthouses, wrecks and soundings. Aeronautical charts for high-speed, low-altitude flying require prominent features at roughly 70-mile intervals to serve as checkpoints for every 4 minutes of flying time at 1,000 miles per hour. Such features are set in an extremely simplified framework with shaded relief, outlines of towns and the main lines of the transport network (Davis, 1958).

Thus we can argue that if the map user is a specialist, who has appropriate technical and financial resources at his command, maps that he commissions will tend to be 'tailor-made' to his requirements. Maps made in great quantity, for large numbers of consumers, could perhaps command the same resources by virtue of large-scale sales. On the other hand, large numbers of consumers will probably have many different requirements to make of such maps and as a result the maps themselves will represent a compromise in design. Such a difference is discernible in the design and content of truly national atlases and special editions of atlases for certain parts of the world. National atlases, because of the prestige they carry, can usually command greater financial resources than would be warranted on the grounds of numbers sold and selling price. School atlases for specific countries or regions normally include in addition to the maps to be expected in a world atlas, maps depicting special aspects of the country or region concerned. Although they are drawn specifically for that atlas, they are normally derived from more detailed maps at a larger scale. National atlas maps are frequently compiled from raw data, in order to secure consistent treatment for different topics displayed.

We shall see later how the map user can, by virtue of inbuilt or acquired restrictions on his comprehension of maps, distort the flow of information directed at him. The map maker is of course also extremely liable to distort the information by means of the particular methods he chooses to use to depict segments of the real world. Such 'artificial noise' will be considered in more detail when mapping systems are being discussed. Information about the real world is taken by the map maker, prompted by the map user, and is transformed in a number of important ways before it is presented in idealized form as a model of the real world. This information is coded in symbolic form, indeed Robinson (1960, p. 136) goes so far as to say that 'the entire map is a symbol . . . and it is not quite correct to designate only certain components as symbols'. These symbols, including, conventional signs are the language of map making. Our ability to express ourselves in, or to understand this language contribute to the ease with which the cartographic message is transmitted and received. Very often trouble occurs when even the transmitter is confused about the exact nature of the message he is supposed to send. All too often in geographical texts we come across illustrations by regions of proportions of some measure related to a total amount. It is rarely clear whether the authors of pie-graph maps based on such data wish us to pick out regional patterns, estimate proportions, or estimate total figures for individual regions within the larger area. The pie-graph will not effectively do all these jobs simultaneously, judging by experiments carried out by the author.

## Data selection

Having accepted the necessity for making a map, let us begin to examine how the information output of the real world is further reduced by the processes of map making. We have already seen how the very intervention of the map user and the attitude of the map maker have both combined to reduce the flow of pure information. The map maker also decides how much information can be allowed to pass through to the map. Only a selection of the innumerable bits of information can be represented, unless the map were to be at the unlikely scale of 1:1. Usually the process of data selection starts by the selection of certain classes of information and the exclusion of others. Even the most detailed topographical map usually excludes information on the precipitation or occupations of the population. Some would argue that topographic might properly exclude man-made features on the earth's surface (Stamp, 1961, p. 454). But in cartographic usage most visible features together with place names included. At this stage therefore a distinction is drawn between maps which are called thematic because they are designed to emphasize particular features or concepts and topographic maps which are more catholic in scope. The distinction is one of degree rather than of kind, but it is a useful one in that it generally reflects a fundamental difference in design. Thematic maps, being designed for comparatively small numbers of users can make use of a more esoteric language of symbolization than can topographic maps, which are usually intended to be used by a multitude of users and readers of very different capabilities. However, by selecting only part of the real world for portrayal, the map designer automatically departs from a completely faithful representation of reality. What we observe in reality is not bounded by the map's neat line.

A clear parallel to this occurs in the field of art. Gombrich (1962, p. 78) in an extremely fascinating discussion of truth and the stereotype points out that drawings from nature are not correct because they contain more detail in them. Referring to the German painter Richter's experiences in copying the Tivoli in Rome, he argues

> those who understand the notation will derive *no false information* from the drawing – whether it gives the contour in a few lines or picks out 'every blade of grass' as Richter's friends wanted to do. The complete portrayal might be the one that gives as much correct information about the spot as we would obtain if we looked from the very spot where the artist stood. . . .

Gombrich concludes 'so complex is the information which reaches us from the visible world that no picture will ever embody it all'. Significantly, Maling (1963, p. 21), in a review of quantitative methods of generalization in cartography, points out that the empirical curve which in nature is the land/sea boundary is generalized even on an air photograph. Due to limitations imposed by film grain and lens resolution the boundary is about 10 microns in width. This zone of uncertainty is from 10 to 15 times wider on any map drawn at the same scale. Because of this, irregularities narrower than twice this width cannot be shown. What Lundquist (1963, p. 35) calls 'editorial generalization', selecting *which* discrete objects should appear on a map once the number of such objects has been decided according to technical generalization, connects our discussion of data selection with generalization. It is the purpose of the map which determines which objects are to be included. For instance, a railway passenger timetable map on a relatively small scale would not normally include lines used solely for freight, however important they were. To take an extreme view, any class of information can be depicted on a map at any scale, provided that the pattern of its distribution is appropriately generalized. The Atlas of the British Flora (1962) contains maps of minute plants, whose presence is indicated within 10 kilometre squares by a black symbol in that square. The general pattern of black squares represents at that level of generalization, the distribution of a particular species.

The decision to map a particular class of object or relationship is an editorial one taken early in the process of making a map. Once the classes of features have been determined, with the exception of the minimal topographic base, the map maker will concentrate on these to the exclusion of others. The map of restaurants in France at which a good meal may have been had for 10 New Francs in 1963 (Michelin Guide, 1963, pp. 30–31) is an interesting case. Paris appears only in red as the centre of a road network serving an area devoid of such restaurants. As would be expected, many important places such as Marseille do not appear. Such maps as these present very selective views of reality. Once the selection has been made to the satisfaction of the map maker, acting sometimes in concert with map users, the next decisions revolve around the question of an appropriate scale.

## Scale transformation

One seemingly uncomplicated (and as a result, overlooked) aspect of scale is the obvious one of how big an area is to be included in the map. With modern methods of printing it is perfectly possible for the whole world to be shown on a postage stamp. Such a map would not serve as a wall map to be used for teaching world political geography. There is of course an appropriate scale for any particular purpose, depending very largely upon the amount of detail the map maker wishes to include, but also upon the size of paper available. More severe restrictions operate in the case of working to a common format in an atlas or textbook, or even for a series of maps. One of the disadvantages of the Ordnance Survey's 'Third Edition' one-inch maps (published between 1901 and 1913) was the rigidity of the system of sheet lines producing a uniform size of sheet with no overlaps, taken over from earlier editions (Harley, 1962). In fact the Ordnance Survey as early as 1902 had been publishing combined sheets, where adjoining sheets contained very little land (Johnston, 1902, p. 5). Later sheet sizes were irregular, but the advantages of uniformity of format triumphed again with the Seventh Series (published from 1952 onwards). A system of overlaps between sheets ensures that map purchasers will usually get value for money. Figure 16.9 illustrates these differences for part of North Wales.

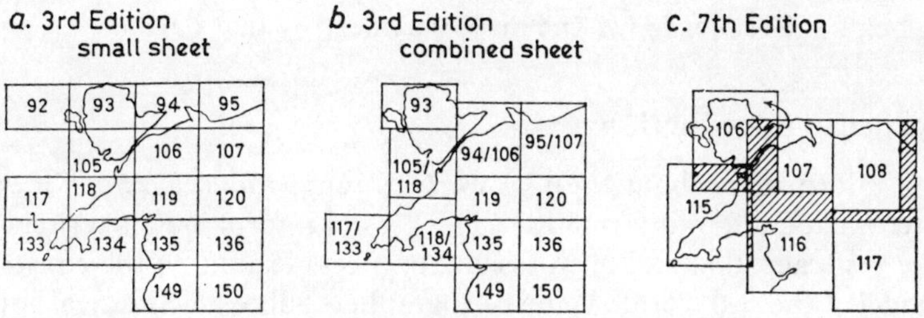


**16.9** Three different arrangements of land areas within a system of sheet lines, for the same area of North Wales.
(A) Sheets same size, no overlaps.
(B) Sheets different size, no overlaps
(C) Sheets same size with overlaps.
(*Source: Ordnance Survey indexes to One Inch to the Mile sheets*) (*Crown Copyright Reserved*).

At a rather different scale, there has been much experiment to overcome some of the cartographer's basic problem, the representation of the three-dimensional globe in two dimensions. Early map projections were mainly symmetrical. Even when projections began to be interrupted to emphasize the unity of the oceans or continents respectively, they were usually symmetrical (Dahlberg, 1962). More recently further variants of interrupted projections have been developed, including Goode's arrangement of the sinusoidal in seven staggered segments. A family of azimuthal projections centred on different places has helped our understanding of the spatial relationships with respect to those places. One intractable problem remaining is that of showing all places on the globe in their true spatial relationship on a flat surface. In some cases portions of the earth's surface have been depicted twice on one map, as in the map of ocean currents in Bartholomew's Atlas of the World's Commerce (1907). In order to show the Pacific system unbroken as the relationship between the southern Indian Ocean and the Australian Bight, the longitudinal extent of nearly the whole of Australia is repeated. Repetition, again on Mercator, of some 40 degrees of longitude including Britain and western Europe at the ends of the map (Philip's University Atlas, 1946, pp. 16–17) serves to emphasize the position of those parts in relation to Eurasia and to the Atlantic without breaking the contact by sea through the Pacific Ocean.

Quite apart from questions of functional utility, the arrangement of mapped area within the neat line is subject to the vagaries of artistic taste. Robinson (1952, Chap. VIII) agrees that a map should have visual unity, eschewing centrifugal tendencies such as those inherent in flow-lines leaving the map area; and a map ought to have visual balance with elements such as land masses, title, and references balanced about the optical centre or deliberately chosen centre of interest. The use of stretches of sea for information which helps the reader to use the map, may destroy that balance.

## The process of reduction

Passing information about the real world through the scale filter inevitably leads to a reduction in information. This loss of information is termed cartographic generalization and is an essential process leading to the construction of a model of the real world. Many cartographers adopt an empirical approach to generalization, using some rule of thumb such as tracing those elements of an outline which can be discerned after the original outline has been reduced to the drawing scale. The gauge of the line drawn itself determines the zone of uncertainty by obliterating indentations too small to figure independently. Maling (1963) has reviewed methods of quantitative generalization as applied to linear features, such as coasts and rivers. Imhof (1951, p. 99) points out that generalization has the effect of making the differences in patterns of features less noticeable. He also illustrates its operation through the suppression of sinuosity in rivers. In reality the Rhone from le Piz Badus to the sea is 1,320 kilometres long but measured on a map of scale 1:4 million it is about 1,000 kilometres. There is empirical evidence that the rate at which map length differs from real length is progressively reduced as map scale decreases. The Russian cartographer Volkov (Maling, 1963, p. 13) has fitted a parabola of form $y=a\sqrt{x}+b$ to Penck's measurements of the Adriatic coast on differnet maps. ($y$ is the length of coast and $x$ the denominator of the scale fraction) Sukhov's coefficient of general sinuosity measured for a stretch of Skärgärd coast (Maling, 1963, p. 12) also falls progressively slowly with decreasing map scale.

Lundquist (1959) in a preliminary survey of generalization sets forth several principles to be borne in mind by the map maker. These include being aware of the danger of excluding important features in areas where they may be scarce where one is applying a rigid scheme of quantitative reduction in numbers. The generalization of discrete features, such as towns and villages can be achieved in two ways. Their selection may be made according to importance until the map at the smaller scale is suitably filled with places. Or a set proportion of the places on the source map may be shown on the map of smaller scale, depending on the reduction factor. The latter is perhaps the more objective, except that the map maker still has to decide which of the

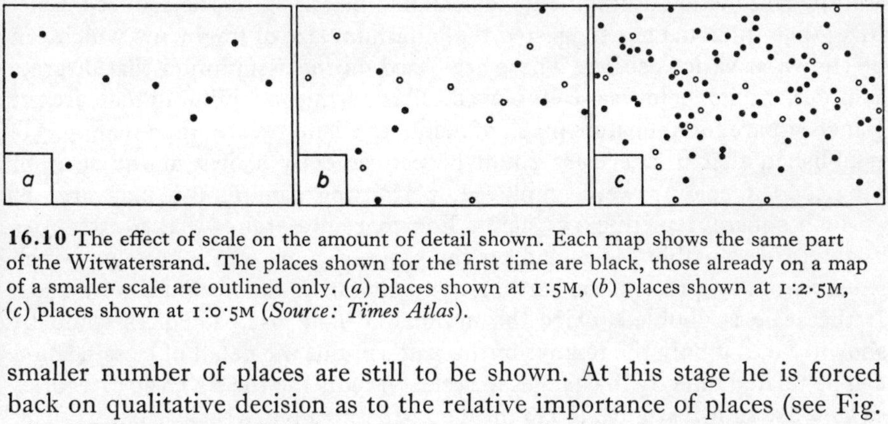


**16.10** The effect of scale on the amount of detail shown. Each map shows the same part of the Witwatersrand. The places shown for the first time are black, those already on a map of a smaller scale are outlined only. (*a*) places shown at 1:5M, (*b*) places shown at 1:2·5M, (*c*) places shown at 1:0·5M (*Source: Times Atlas*).

smaller number of places are still to be shown. At this stage he is forced back on qualitative decision as to the relative importance of places (see Fig. 16.10) Pillewizer and Töpfer (1964) have developed a formula from a study of well-designed maps for determining the number of symbols which should be shown on maps of smaller scales generalized from those of larger scales.

$$n_F=n_A C_B C_Z \frac{\sqrt{M_A}}{M_F}$$

Where $n_A$ is the number of symbols on the source map, of scale $1/M_A$ and $n_F$ is the number of symbols on the derived map, of scale $1/M_F$ and $C_B$ and $C_Z$ are constants respectively reflecting the significance and degree of coarseness of symbol character. Maling in explanatory notes on a paper by Töpfer and Pillewizer (1966) suggested that empirical data derived from examining a number of atlas maps of Scotland broadly fitted the form of Pillewizer and Töpfer's equation.

The preservation of a draft copy of the Ordnance Survey's outline map of Great Britain on the scale of 1:1,250,000 furnishes an interesting example of the process of generalizing the pattern of places on topographic maps. The earliest draft was a reduction of the Ten Mile map (1:625,000) with some features eliminated. This was redrawn to conform to the design for the smaller scale. Four sizes of town were recognized based on their population. They were distinguished by different kinds of symbols and lettering. In some parts of the map towns of 50,000 inhabitants had to be left off for fear of overcrowding it. In other areas where the density of population is low very small towns which had been left off at first were put onto the first published version (1946) on account of their local importance and others of importance were inserted. Some railway lines were also restored where they were deemed to be important through connections. A comparison of draft and published maps is given in Figure 16.11.

The generalization of symbols which cover areas, such as woodland or other land use type, involves the simplification of outline already discussed and the elimination of fragments too small to be shown at the reduced scale. Fox (1956, p. 26ff.) has suggested the minimum size of fragments which can be shown at various scales. These are based on the assumption that discrete symbols and not colours are employed. This permits a minimum map area of $\frac{1}{8}$-inch square for a separate patch of distinctive land use. Stamp (1948, p. 33) established that 6 categories could be conveniently shown at the scale of 1:633,600 if colours were employed, permitting a minimum map area of $\frac{1}{20}$-inch square depicting 155 acres. Fox therefore argues that 6 categories ought to be suitable for a map on the scale of 1:253,440 where an $\frac{1}{8}$-inch square would represent 160 acres, if symbols rather than colours were used. If the scale is doubled, twice the number of land use categories could be shown. So, not only the texture of the pattern, but the detail of classification can be related directly to changes in scale. In reducing the six basic categories of land use shown at 1:63,360 to the Ten Miles to an inch scale, Stamp (1948, p. 33) points out that although the shapes of fragments could not be preserved, the porportions of land in the different categories were maintained.

Lundquist (1963) discussing the generalization of networks of roads or railways emphasizes the importance of editorial generalization suggesting that there are relatively few problems in technical generalization. Such maps as these are usually tailored very much to the needs of particular groups of users so that the decision to retain or remove particular items of information is more critical. If a place is not sufficiently important to be shown, there is little point in inserting a road which leads to it alone.

Not only does the reduction in scale itself result in a loss of faithfulness with which reality is shown on the map, but the choice of the degree of generalization given what is possible with different techniques of printing is an important factor.

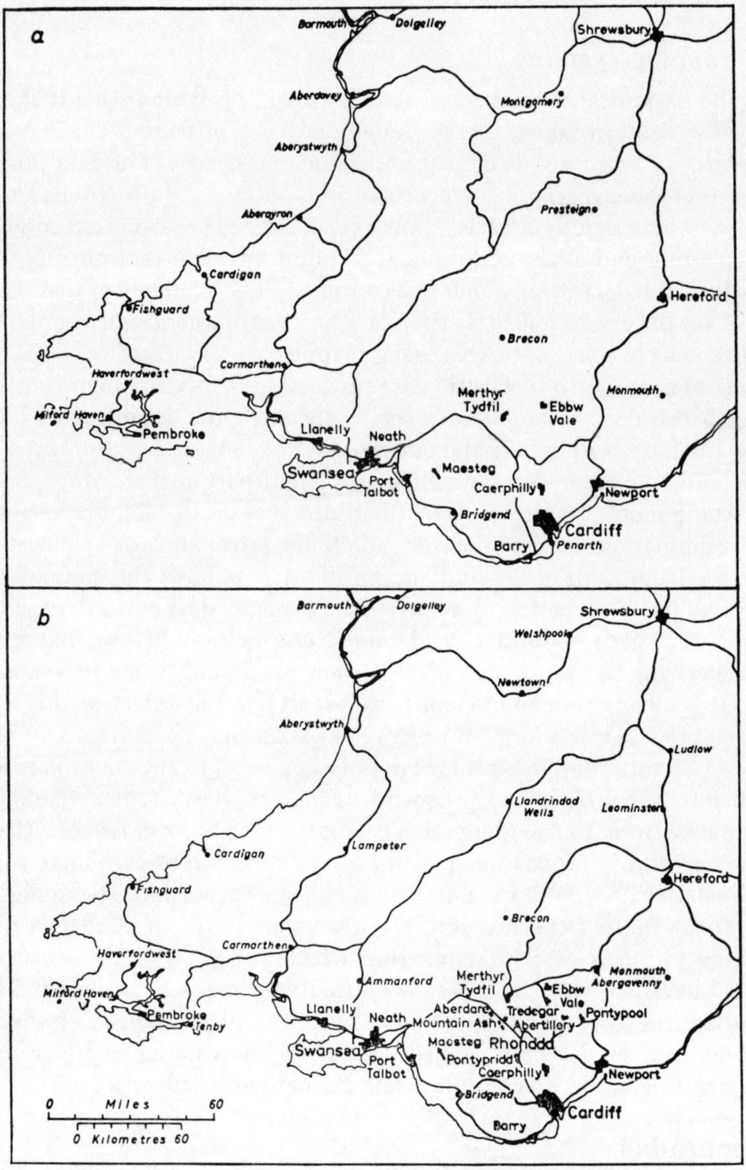


**16.11** The differences between the pattern of places and railways on a draft version and the earliest published version of a small-scale map.

(*a*) draft edition, 1943: has less detail of towns and leaves out some main railways; does not sufficiently differentiate the denser pattern of features in the South Wales Coalfield.

(*b*) published edition, 1946: possesses greater verisimilitude; regional differences between the coalfield and Mid-Wales preserved (*Source: Ordnance Survey outline map on the scale of 1:1,250,000. Crown Copyright reserved*).

## The mapping system

Once the basic decisions as to what to map and on which scale it should be mapped have been taken, a great range of choice of techniques is available. These are conveniently discussed under the two headings of data processing and cartographic method. Data processing, although quite vital to the end product, is not strictly a cartographic procedure. It is common to all types of description and analysis. Suffice it to point out here that no map can be better than the data from which it is compiled. The accuracy of that data may depend on observational bias, the scale of measurement and the density of measurement in relation to area being mapped.

Some maps require very little data processing before the map compilation begins. Most maps using nominal scale data and some dot maps are included in this category. Others are the end product of long and complicated calculations and are sometimes a rather insignificant part at that. Maps depicting multi-component situations clearly fall into this group. The crop-and-livestock combination maps of Weaver and others (1954 and 1956) look straightforward enough, but each enumeration district requires the computation of several sums of squares. Similarly facies maps developed by geologists (Forgotson, 1960) require considerable computation before mapping can begin. Perhaps the most complicated data processing prior to mapping is associated with the techniques of factor analysis. The effect of this form of analysis is to collapse a large number of variables upon which a series of areas is measured, for example aspects of population, income, production, mechanization, into a few basic and independent factors. Berry (1960) made choropleth maps from the loading of 95 countries on 5 components (factors). Imbrie and Purdy (1962) mapped the facies variation of carbonate rocks on the Great Bahama Bank by a similar classifying procedure. King and Henshall (1966) plotted the distribution of peasant farms in Barbados by dots according to four classes derived from a factor analysis of crops grown and kinds of livestock carried. These are essentially maps of farms of different enterprises, derived from a relatively objective procedure. A similar map, based however on subjectively defined criteria, showing enterprises of farms in Eastern England is given by Jackson, Barnard and Sturrock (1963).

## Symbolization

The flow of information transmitted by the real world and filtered in the ways suggested above is now ready to be mapped. An initial distinction must be made between topographic maps and thematic maps employing the nominal and ordinal scales of measurement, and the quantitative thematic maps at higher scales of measurement. The former convey their information by means of the presence and absence of features in particular positions or areas (Fig. 16.12). Maps employing ordinal scale measurement, indicate the rela-

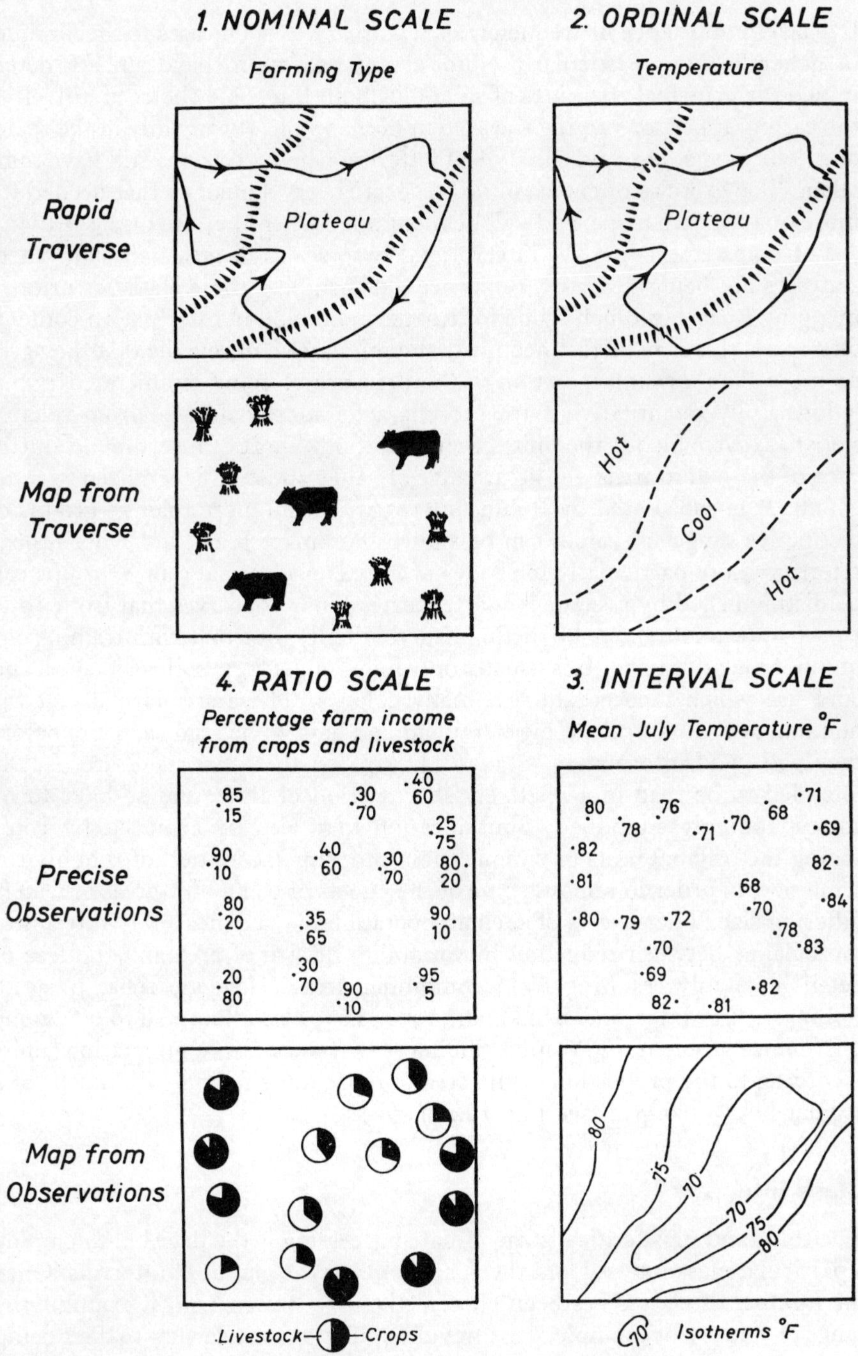


**16.12** Observations and maps appropriate to the four scales of measurement.

tive size, importance or frequency of features. For such cases the techniques of generalization, particularly editorial generalization, have already determined the principal attributes of symbolization. But some choice is still left to the cartographer for varying the system of mapping. It is mainly in the realm of colour or shading, and the style of lettering where decisions still have to be taken. The exercise of choice in *this* respect is very similar to that needed for more quantitative maps, and will be mentioned when they are discussed.

For maps employing the interval and ratio scales of measurement, where statistics are being depicted, the process of handling these statistics prior to mapping furnishes much scope for further reduction in information content. Of course, this may be justified if the statistics are so unreliable that they give no more than a rough indiction of the presence of some enumerated population. Fully quantitative maps are characteristically derived from reliable statistics, available for the units of enumeration. Short of inserting a value in each of these unit-areas, the great range of values must be generalized to some extent. It is considered by Robinson (1952) that no more than 10 groups of distinctive ranges of values can be shown on a map. Jenks and Knos (1963) referring more particularly to shades of grey consider that 7 or 8 groups can be distinguished by average readers. Keates (1962) observes that from 10 to 15 separate colours may be distinguished initially, but that with training the number may rise to 50. It is usually only maps of geology, soil, vegetation and land use which incorporate *that* many colours. These are normally at the nominal or ordinal scale of measurement and have to be read carefully before they will yield information. Statistical maps on the other hand are usually intended to be read in a relatively short period of time, and so have to be correspondingly simplified. Simplification may also be achieved by combining the original units of enumeration into 'super-counties' of roughly the same size, in order to remove some of the effects of using unit-areas of widely differing size. The success of such an operation may be measured in terms of the balance between reduction in variability of size of area and the loss of detail inevitably resulting from combining areas (Haggett, 1964, p. 371). Robinson, Lindman and Brinkman (1961) have also attempted to overcome this same problem by reapportioning the values for enumeration units according to the proportion of the area of those units falling within cells of a regular hexagonal grid (see Fig. 16.13).

## Class intervals

The third and most critical stage of data processing is the decision to employ a particular class interval and the base point for the gamut of intervals. Once the number of classes has been chosen, the map designer must examine the range of values (for example, densities of population) with a view to their being arranged to give adequate representation to the different parts of the range

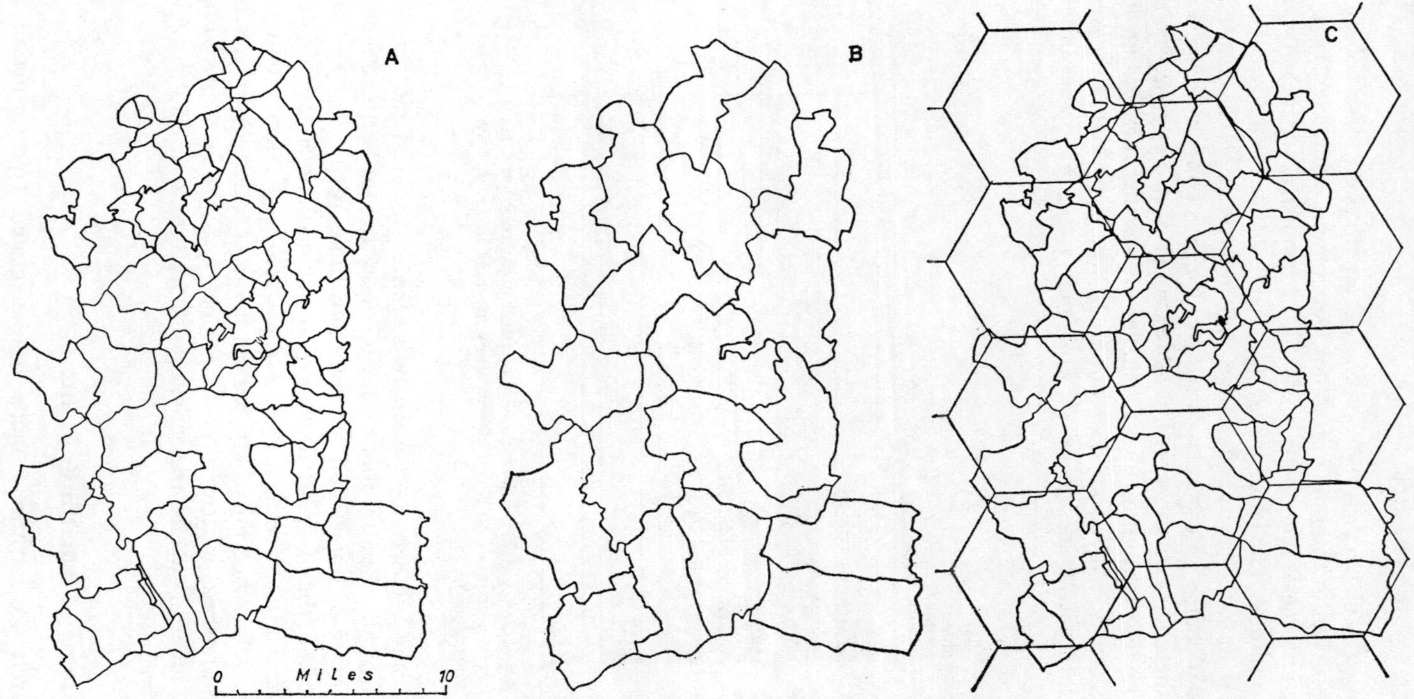


**16.13** The mesh of enumeration units and their modification, an example from Hertfordshire:

(A) Pattern of parishes in the period 1896–1898 – 69 units.
(B) Combinations of parishes in (A) (*After Coppock, 1960*) – 19 units.
(C) Parishes in (A) with pattern of hexagons superimposed – 137 units.

and to reproduce its characteristics. Common alternatives are steps chosen to coincide with breaks in the range of values, equal steps, steps increasing in magnitude by arithmetic increment, steps with a geometric (or logarithmic) increase in magnitude. The latter are arranged so that they normally concentrate the values in lower part of the range, and percentiles which ensure an equal number of values in each class whatever the range involved (see Fig. 16.14). Each method has its merits, but as Jenks (1963, p. 15) observes, the cartographer finds it more difficult to visualize an abstract distribution like population density and therefore does not know which method is the best. Many follow precedents set by others. One such procedure is set out by

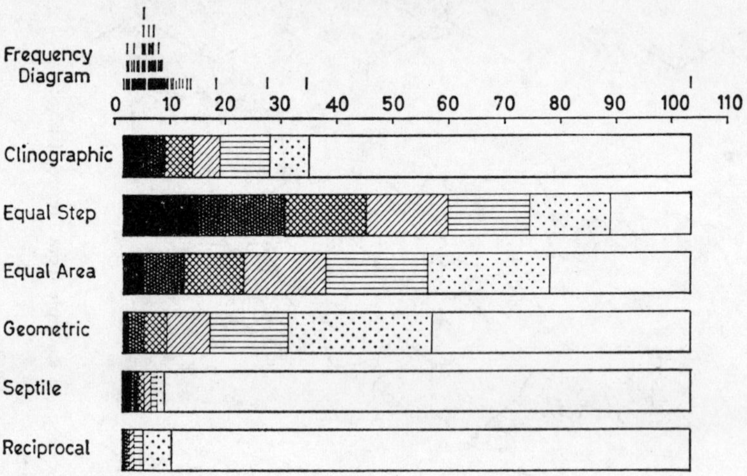


**16.14** A frequency distribution (density of rural population) and six possible arrangements into seven classes showing their size and range (*After Jenks, 1963*).

Mackay (1963) and is based upon an analysis of intervals and limits selected from many geographical publications. Having permitted the cartographer to choose the number of classes and the sizes and limits of the first and last classes in the range (the latter usually being open ended), an equation determines the spacing of intervening classes. Mackay also recommends rounding off the precise values of limits obtained from the equation. Although his work, like that of others, is strictly speaking concerned with isarithms rather than statistical maps in general, the conclusions and techniques apply equally to choropleth maps.

## Special problems of isarithmic maps

Further research on isarithmic maps has suggested transformation of original values in order to represent certain aspects of the distribution being mapped. For example, Krumbein (1957) discusses the advantages of transforming percentage values into angles by taking the square root of the percentage and finding the angle whose sine is equal to the square root. An arc tan. transformation can be similarly performed on ratio data giving exactly the same values (angular values) for ratios corresponding to percentages (e.g. 1:4=20%). Apart from the fact that such transformations tend to normalize the values concerned, when percentages in the middle range are bunched together on the map, those at the lower and upper ends of the scale, where percentage changes are more significant, are spread out. The opposite effect is achieved with ratio values. Such manipulations of data before mapping are applicable to multi-component situations, rather than to simple single-component cases concerning absolute values.

Blumenstock (1953) has shown that where values are derived from a sample, as for example in some agricultural and population censuses, but more commonly as meteorological observations are, observational error, sampling error and bias in some observations may all affect the reliability of isarithms. If a particular value has a relatively great chance of being incorrect by an amount which would not justify its being used to draw a detail in the isarithmic pattern, it is best ignored. Simpler, but more reliable maps result from the application of such corrections.

Another problem affecting the accuracy of isarithms based on units of enumeration is the influence of linear interpolation between the control points where the values are assumed to apply. In one case Porter (1958) showed that this procedure could result in 25 per cent of the enumeration units being misclassified, that is left on the wrong side of the isarithm. Porter was able to compare the degree of correspondence between isarithms drawn solely on a basis of statistics available for unit-areas and replotted values from supplementary information. His map shows that disagreements are related to anomalous areas of relatively small extent which the more generalized map ignored.

At the other extreme, if the units of enumeration happen to be square, so that the control points are arranged in a grid pattern, there may arise a

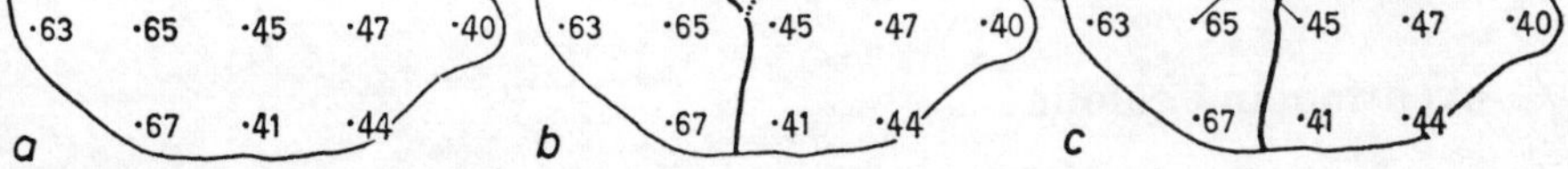


**16.15** Indeterminancy in compiling isarithmic maps and its solution.
(*a*) the raw data arranged in grid fashion
(*b*) the alternative choices
(*c*) the solution.

situation where two alternative isarithm patterns are equally valid. Interpolating between two high values arranged diagonally can give a result differing from an interpolation between two low values arranged on the diagonal intersecting the first (Fig. 16.15). A solution is found in averaging the two pairs of values to yield a fifth control point where the diagonals cross Mackay (1953). Uhorczak (1930) saw that if the units of enumeration were arranged like bricks in a wall, the control points would be arranged in a pattern of traingles, thus overcoming the problem of indeterminacy. Czekalski (1933) followed by Mackay (1953) also recommended hexagons, which also have a triangular pattern of central control points as units for mapping. These solutions involving a reorganization of the unit-areas remain little used because of the immense labour involved in the recalculation of areas and quantities before mapping can begin (see Fig. 16.13). Robinson and others (1961) have carried out such an exercise for a portion of the Great Plains of the United States, partly with the object of avoiding cases of indeterminacy.

## Point symbols

The third major class of quantitative maps are those employing a variety of point symbols. These range from the so-called dot map to the map showing quantities by proportional circles, squares, or spheres. Such maps rarely involve much preliminary data processing, and then mainly in relation to the value assigned to each symbol and the size of the symbol. Most of the calculation revolves around the values given to single dots, when it has already been decided that dots of a uniform size and value will be employed. The relative density of the pattern of dots in this case portrays the pattern of quantity over area. Robinson (1960, pp. 156–162) illustrates the effects of varying dot size and value on the appearance of the map.

With proportional symbols, such as circles, square roots are first calculated (cube roots for spheres) and a scalar applied to resulting values in order to obtain a reasonable size of symbol appropriate to the map. The representation of values by pictorial symbols (animals, men, sheaves of wheat) involves only the assignment of a range of values to a symbol of a certain size. It is unusual to find that more than four or five sizes of symbols of this kind on one map, since the map designer generally requires each size to be distinctly recognizable.

## Convention and colour

In the foregoing discussion some attention has already been paid to the various kinds of symbolization available to the map designer. Data processing and symbolization are so interrelated that it would be impossible to treat the two topics separately. Nevertheless there remain some aspects of map design that can play an important part in the propagation of the message. As maps become more thematic, more specialized and more quantitative, they become more abstract. The reader will readily recall the many examples of somewhat stylized pictograms used for tourist maps. The symbols normally employed on road maps used by the motorist are only a little less pictorial. Many of the conventional signs to be found on topographic maps fall into this category too. Bagrow (1964, p I and III) shows a Mexican map that uses a line of footprints for a track and trees with a central American cast as a symbol for woodland. The Flemish and Dutch cartographers, followed by the English and German made extensive use of stylized churches and house groups on topographic maps before the end of the eighteenth century. It appears to have been the decision of the French in the years of the First Republic (1802) to alter the conventional representation of objects shown in elevation by symbols based upon the plans of objects (De Dainville, 1964). That such a change was logical is beyond question, but the mixture of symbols in both plan and elevation characterizing British and American and most European maps (with the exception of French, Swiss, Swedish and Danish) by being more conventional than logical is perhaps the more easily interpreted.

The advent of colour printing about the middle of the nineteenth century, for the first time made it possible for maps to employ extensive washes of colour as a symbol. Such colours which could be hand-painted by artists, unemployed gentlewomen or apprentices, were normally reserved for political entities in atlases. A notable exception is provided by the cadastral maps, of which single or very few copies were ever made. Indeed in England at least the convention of showing arable land in brown, woodland in green and pasture in another shade of green is of some antiquity. Some of these features appear in an estate map of Wotton Underwood at the end of the sixteenth century (Schulz, 1939 and 1954). Colour printing was used for relief maps soon after its introduction (Lyons, 1914). Layer colouring apparently originating in Germany, was to be found mainly in atlases and on wall maps. Perhaps the system which has had the longest run of popularity is von Sydow's green, white and brown in upward succession. Towards the end of the century Bartholomew's layer coloured topographic maps employed a very similar system, which has also become conventional in Britain. Both these schemes make use of the idea that the higher land ought to be darker in tone.

In other respects the conventions established in the period of hand-colouring were transferred to colour-printed maps. Blue is now almost universally used for water, both fresh and salt on topographic maps (although this has not always been so). Arab maps a millennium ago commonly showed the Red Sea in pink. Red is a colour commonly associated with towns, and this too has been employed to distinguish urban areas on German school atlases, Ordnance Survey 6 inch to the mile town maps of the 1920's, and on the maps of the Land Utilization Survey of Britain.

On thematic maps colours are less conventional, but red, by association with warmth, is frequently used on maps of population density to show the high densities. By the same token blue, a cold colour, is conventionally associated with low densities or decreases. However for maps of rainfall the blue-water association is too strong so that dark blue is conventionally reserved for very wet areas and red for the driest areas. The association of ideas is probably responsible for the popularity of shades of red for igneous rocks on geological maps (Linton, 1948, p. 143), however the imitative colouring adopted by the designers of the early maps of the British Geological Survey is also worth noting. Here the brick reds used for Old and New Sandstone systems recall the common soil colour of those outcrops, but other colour choices are less consistent with these notions. Frequently a blue to red scheme of graded colours with an intermediate band of pale tones is employed for maps intended to show a great range of values for example annual rainfall, temperature, sometimes including decreases and increases of population or to differentiate the imports and exports per caput of a commodity like timber (Bartholomew, 1907). The association of blue, blue-green or green with decrease or low values and red shades with increases or high values is now widely accepted. It is well illustrated in the *Atlas de France* (1946) and in the Ordnance Survey's 1:625,000 population maps.

Colour symbolization is also frequently found in maps which are used a lot by the general public. Transport networks, such as the London Underground or Paris Metro employ a colour code for the individual lines. Here contrasting colours aid the identification of particular routes.

By employing principles and devices such as these, most of which have some rational or at least conventional basis, the map designer helps to convey the cartographic message. By generalizing the shapes and sizes (to say nothing of the population characteristics) of towns and cities by presenting them as a series of point symbols of clearly distinguishable type, the cartographer at once diminishes the amount of information the map will convey, but hopefully he makes possible a proportional gain in legibility. The message, although it then contains less information, has a much better chance of reaching its destination. The art of cartography lies in the balance and compromise involved in the choice range between the information level and the probability of its being comprehended.

## Noise

During the mapping process, both at the stage of data digestion, but rather more at the stage of mapping itself, interference of various kinds impinges upon the cartographic signal. Such unwanted disturbance is termed *noise*. Since it is generated by human agencies, it is here called *artificial noise*. Most of the *real noise* is eliminated at earlier stages in map making. It comprises information about the real world which is considered to be irrelevant to the purpose for which the map is being made. The elimination of all but the most obvious features from charts for air navigation, including the names of quite large towns, is a good illustration of the suppression of real noise. Were superfluous features to remain, the noise level of the map would be so high as to render it difficult to read in view of the high speeds of modern aircraft.

Artificial noise is of two kinds: that generated by the map designer by the methods he employs to put across the cartographic message, and that generated by the map reader who sees the elements of a map in a manner which sometimes differs from that which is intended by the designer.

## Designer noise

The map designer, by making choices about class intervals, symbol values and colour or shading schemes, inserts an element of interpretation into the map. His choice is subjective, and may depend on his desire to portray a distribution in a particular way. There is no really objective arrangement of class intervals in a choropleth map. Even the employment of percentiles involves the choice of a number of classes, and such a system of class division is also directly related to the number of separate areas providing values. Wright (1942, p. 541) has observed that when the map designer relates two distributions on one map, producing 'synthetic information' from, for example, isohyets of a certain value and percentage of farmland under a particular crop, he relies more on his judgment than on unsullied information derived directly from the real world. However, as we have seen, there is a continuous range of maps from the relatively objective to the propagandist or polemic. There are many ways in which designer noise can distort the cartographic signal. Apart from the more obvious graphical possibilities already suggested, the field of lettering and titling furnishes some interesting examples. The numbering of class intervals is sometimes done in such a way that the same value appears to occur in two adjacent classes. At other times, after transformation of values to a logarithmic scale, original values are retained on the map, thus hindering the reader who may need to interpolate values. The use of awkwardly uneven or unrounded values for class limits, especially when these are based on percentiles or on equal deviations from a mean value may irritate the map reader, but it will rarely seriously inhibit comprehension.

It would clearly be impossible to convey information by maps entirely by non-verbal symbols. Maps without lettering seem unfinished and disoriented (Imhof, 1951, p. 107). Cartographic lettering often follows conventions so that the reader, once he is familiar with them, is assisted to use the map. A distinction must however be made between the style and position of lettering, and the spelling of names. Riddiford (1952) has gone so far as to say that reaction to lettering may be so violent as to prevent the user reading the map. Two extreme views are held: that lettering should be subservient to the other map detail; that lettering should be bold, or sufficiently legible to be easily read. Most discussion has centred on the issue of style versus legibility, there being a strong school of thought in favour of inconspicuous lettering. Reeves (1929, p. 437) considered that it was regrettable that any names had to appear on a map. Winterbotham (1929, p. 436) argued convincingly that:

> the art of lettering is to make it so neat and unobstrusive as not to absorb attention. Names are given to signify something in the topographical sense. I do not think one would wish to make a J or K so distinctive as to draw the eye as, for example, would a man in Bond Street in pink trousers.

Ornate and antique lettering is now usually reserved for special features, such as archaeological sites. The positioning of names on maps is frequently regarded as a craft learnt only by long apprenticeship and practice. Imhof (1962) has set down in detail, with illustrations of both good and bad work, the ground rules for positioning names. Three main principles emerge from the distillation of his great experience. The first is that names should be conveniently read in the position which the map user usually takes up. Secondly, names should be so spaced and arranged that they can be read as whole names, not appearing as two separate fragments. Thirdly, names ought clearly and unambiguously to belong to the feature they refer to. Such principles should outweigh questions of convention, such as placing river names always on the northern side of the stream. (Balchin, 1952, p. 144) In the latter case, too rigid a rule can easily introduce artificial noise to distract the reader from his task. The relationship between names and the curves of the graticule is a noisy one at best: either names, being parallel to the lines are hard to read; or if names are placed parallel to the lower neat line, their varying angle to the lines of the graticule irritates the reader.

Since the question of spelling and transliteration is not restricted to maps, it would be out of place to dwell on it at this point. Suffice it to point out that countries which use two or more languages on maps magnify the problems of cartographic lettering. Such considerations apply mainly to the reference and marginal information, which may have to be repeated in each language. The thematic maps illustrating the Tomlinson Report on the Bantu areas of South Africa (1956) provide many examples of names and descriptions in both English and Afrikaans. It is customary, at least on individual maps to reduce the possibility of confusion by placing names in one language consistently above the other. Once one is used to the idea, one learns to read alternate lines almost automatically, a habit that can sometimes be awkward. Another device is to use English and Afrikaans forms on alternate sheets of a series, as with the provisional topographical field sheets of South Africa on the scale of 1:18,000.

To flout well-established conventions, such as the depiction of the sea in blue can be a way of introducing artificial noise into the map. The Reader's Digest Atlas of the British Isles shows the sea in shades of sea-green on the grounds that the British seas do not appear blue. It takes a little time to adjust to the unfamiliar colour. More distressing perhaps is the plastic relief model of the Oxford district (1964) which shows agricultural land use in shades of light (Cambridge!) blue. It is true that the intention is to emphasize the use of land in the urban areas by employing red, orange and grey there, but the unconventional use of colour for rural land use so offends the eye at first that the message is at once distorted.

## Reader noise

Cole (1964) has complained that our unwillingness to look at maps another way than with north at the top has led to 'North–South thinking', to an inability to perceive spatial relationships in other ways. Indeed one can argue that movements should be mapped so that they move away from the reader towards the top of the page. After all the coaching itineraries and motorists' routes are drawn in this fashion. Should not the expansion of settlement in the United States be plotted on a map with East at the bottom? But perhaps to follow this advice would create more noise than would be warranted in exchange for a more logical orientation, because most of us now accept that North is at the top.

The preferred orientation of maps with North at the top carries with it the corollary that they are best lit from the north-west corner. Capitalizing on this property, map designers have for some long time attempted to create an

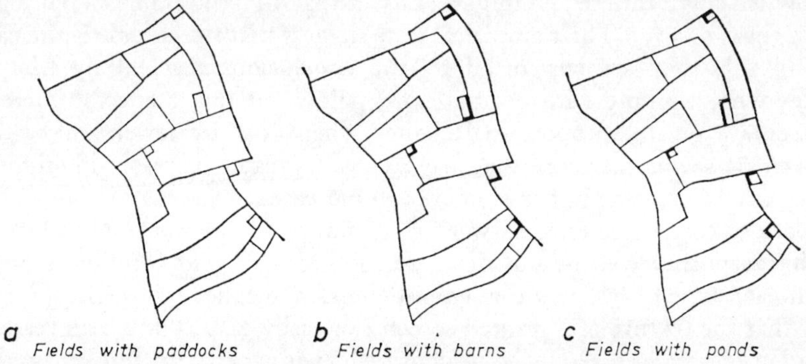


**16.16** An illustration of the effects of the conventional lighting of features from the North-West, or the top left-hand corner of the map.

illusion of the third dimension by thickening the boundaries of features on the shaded side, away from the north and west to indicate projections above the surface. This technique is widely employed in hill-shading. For depressions, such as ponds and pits the shaded sides lie on the north and west. In this way one can distinguish the two types of feature and identify ponds and pits with some certainty (Fig. 16.16). Reversing the map would of course destroy these illusions and would be rather like looking at air photographs the wrong way up (that is with the light not coming from the top). The discussion between Lewis (1959) and Sweeting (1958, 1959) on the interpretation of air photographs of Jamaica's Cockpit country underlines the possibility of confusion.

Another important source of noise in representational mapping is the failure of the map designer to choose shadings (or colours) to portray gradations in density or other value which look right. Map readers in some way instinctively know that one range of shadings is correct and another wrong. Skilled cartographers have for long been well aware of this phenomenon and have produced schemes of graded shades by trial and error. More recently cartographers, following the work of psychologists such as Thurstone, Ostwald and Stevens, have investigated the relationship between stimuli provided by the printed pattern and the sensation received by the map reader. Williams (1958) testing both patterns of dots and rulings found, when subjects were asked to select and place in order patterns from a wide range of choices in such a way that they gave an impression of equal gradations in density, that the percentage of white paper inked did not correspond with the Weber-Fechner Law. This states that sensation increases as the logarithm of the stimulus. This was true only of the lightly inked patterns, but for the major part of the range sensation increased arithmetically with stimulus. Similar results were obtained when using coloured patterns. Other investigators have argued that the Weber-Fechner Law implies 'that it is easier to distinguish slight differences in darker tones than in lighter tones' (Jenks and Knos, 1961, p. 323). Their empirical tests suggest that the reverse is the case. And in this they are supported by the conclusions reached by Mackay (1949), who, working with dot patterns, points out that a specific increase represents a greater proportional change when dots are fewer and farther between. It seems therefore that progressively greater increases in stimulus are required to give an impression of even increases in a sensation of density. The lack of correspondence between results of psychological tests and cartographic tests may well be put down to the very different conditions under which each type of test was carried out. Jenks and Knos (1961, p. 334) also show that the texture of a printed screen can easily disturb the map reader's sensation of a graded sequence of densities, even where the percentage area inked remains the same.

Less work has been done on the visual perception of *coloured* symbols on maps. This has not inhibited the discussion of the topic, but it is generally recognized that more research needs to be done. There are about as many views on the right system of colouring for layers on relief maps as there are colours in the spectrum. There appears to be a fundamental difference of opinion as to whether high altitudes should be lighter or darker. Imhof (1951, p. 94) suggests both are right. Starostin and Yanikov (1959) present several relief maps of the Kamchatka peninsula using different schemes of hypsographic tints. One of these is conventional in that the range passes from bright emerald green through brown to pink and white for progressively higher ground. Two others employ a green-brown range with dark brown on the mountain summits. Three make use of a range of browns, one having the dark colours for high ground and the others having light summits. The latter certainly stand out, but the lowlands, where one would expect more cultural detail, are so dark that lettering and line work there would be all but illegible. Experiments with test maps carried out by the author (1964) suggest that people who are familiar with maps tend to prefer monochrome colour schemes, with dark shades for the high points. Multi-coloured schemes are less favoured.

More research has been done on the perception of quantitative symbols of different colours. Williams (1956) found that when coloured symbols were compared with black ones of the same size and shape, only yellow ones were judged to appear more than 5 per cent larger. Considerably more is known of the way in which map readers perceive quantitative symbols printed in black. Such symbols as proportionate circles, squares, spheres and cubes are common in thematic maps used to illustrate texts on geographical topics. Williams (1956) was able to discover a general pattern from a series of experiments where students were asked to select a symbol from a large choice such that it bore some simple relationship to a standard one. For example, when asked to select a symbol which had twice the value of the standard one, the average response was to pick out one which was 1·8 times the diameter of the standard one, not one of twice the diameter, nor one which was twice the area of the standard symbol. Altogether, it seemed that for a symbol to appear $x$ times the size of a standard one its linear dimension ought to be $x^{\cdot 8}$ larger. For the area of a circular symbol it is now conventional to make the new circle $x^{\cdot 5}$ times larger if it is intended to represent $x$ times the value shown by the original one. Once again, it appears necessary to exaggerate the stimulus in order to convey an appropriate sensation of value. Further investigations have been carried out by Clarke (1959) and Ekman and others (1960 and 1961) to show the extent to which perception affects the interpretation of other quantitative, discrete symbols. Clarke extended work done by Croxton and Stein (1932), concluding that the 'difficulty in visual evaluation of proportional symbols increases with the number of dimensions of the symbol'. Ekman and others reworked Clarke's findings and compared them with their own experiments. They were concerned to investigate in more detail the psycho-physical 'law' that '*subjective intensity is a power function of the intensity of physical stimulation*'. Their studies confirmed that the map symbol reader operated on a scale which is a non-linear (exponential) function of the scales used in describing reality, or the variables being mapped. The values of the exponent varied from about 1 for line symbols to 0·9 for squares, 0·8 for drawn cubes, and 0·74 for drawn spheres. However the value of the exponent for Clarke's data on spheres lay between 0·5 and 0·6. Ekman and others (1961) hypothesized that if estimates of volume were essentially estimates of area for such symbols as spheres and cubes, the ratio between exponents for estimates of area and volume respectively should be 2 to 3. This was confirmed to their satisfaction because the exponent for estimates of volume were about 0·6, so that they were able to conclude that estimates of volume were almost exclusively estimates of perceived area which was in its turn not far from proportional to the geometric area covered by the symbols. Some experiments recently carried out by the author (1965) have suggested that the map reader's judgement of blacked-in segments of proportional circles, indicating the percentage of one item out of a varying total quantity, generally relate to the area blacked in rather than the angle of the segment at the centre of the circle. Because the circles varied in size, estimates of percentages were particularly affected. Since these tests were carried out in a real map situation with genuine data it would seem that the results of tests hitherto carried out only on diagrams divorced from maps *also* apply to maps. In experiments of the latter kind, von Huhn (1927, p. 34) observed that where circles of different sizes were employed, black segments were less effective for showing percentages because only angles and not arcs, chords or areas could be compared. In the author's map test general estimates of regional trend or pattern in the percentage values were difficult to make, presumably for the same reason. For relatively instantaneous impressions of spatially varying proportions the map reader needs more the angles at the centres of the circles. These difficulties are overcome only by showing proportions in circles of the same size.

## The model of the real world

Thus it may be safely concluded that only a fraction of the information in the real world eventually finds its way to the map reader from the map or representational model of the real world. The ways in which maps 'perform as a device in portraying spatial properties in competition with other devices, such as photographs, pictures, graphs, language and mathematics' are encompassed in what Bunge (1962, p. 38) calls metacartography. He isolates a group of devices other than maps and mathematics as *premaps*, but comes to the conclusion that they are perhaps a subset of maps (p. 71). Bunge outlines a number of traverses to establish the boundaries between premaps and maps, successively exaggerating spatial properties of different kinds. In this way he deals with scale, shape distortion, information content versus ab-

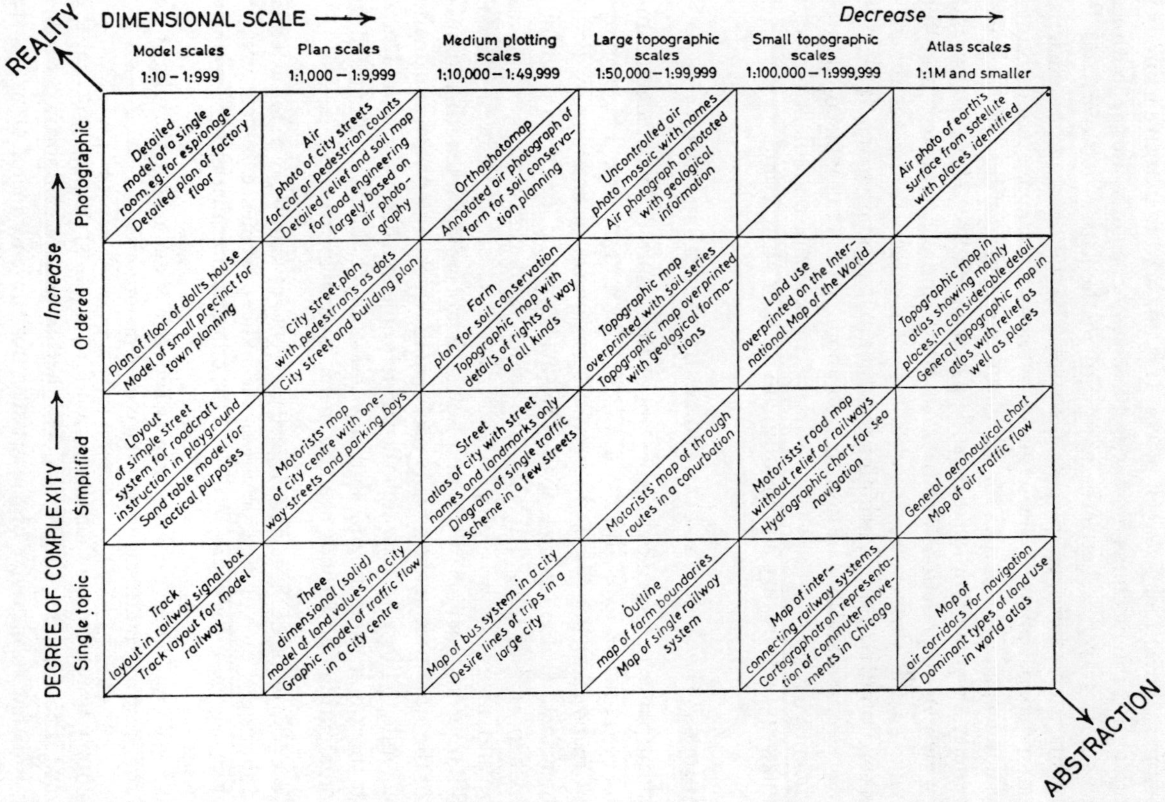


**16.17** The gradient between Reality and Abstraction indicating examples of types of maps at their appropriate level of abstraction.

straction, base map data, projection angle, correspondence to the earth's surface, psychological accuracy (apparent realism), projectional conventionality and connections between places. It is not hard to perceive a number of overlaps or intersections in these traverses. Secondly Bunge outlines another set of traverses between maps and mathematics. He discusses in turn: connections between places, distance, the number of dimensions, idealization, spatial analysis, depiction of overlaps. In a final section, Bunge reviews some spatial properties for which measures have still to be adopted: homogeneity, orientation, shape, and pattern. Figure 16.17 represents another view of the relationships between maps and other devices used for portraying spatial properties. It knits together many of the topics considered separately by Bunge, bringing together maps and premaps. It regards all maps as representations of information about spatially organized features and concepts relevant to the earth's surface, in terms of a gradient between an ultimate, infinite reality and an ultimate ideal or abstraction.

## Faithfulness: the gradient between reality and abstraction

In a recent paper on a general theory about the nature of knowledge, Bambrough (1964) makes extensive use of analogies with the map. The employment of maps as analogues to illustrate philosophical discussions by Bambrough, Kaplan (1964, pp. 284–285), Treisman (1966, p. 601) and Toulmin (1953, Chap. 4) suggests that they are widely and easily recognizable as models, and many of their properties furthermore are quite well known. Indeed Bambrough begins by quoting Lewis Carrol (1893) at length, reminding us of the attempt to make a really useful map which ended with one on the scale of a mile to the mile, to which the farmers objected because it would have covered the whole country and shut out the sunlight. Bambrough (1964, p. 102) commenting on methods used by metaphysicians writes:

> A direct description of the logical characteristics of our knowledge of the external world shows that each of these pictures gives undue prominence to some features of our knowledge and obscures or distorts the other features that the rival pictures emphasize. . . .
>
> Here again we can have either a map on the scale of one mile to the mile or we can have grasp and understanding at the cost of distortion.

It is salutary to remember that Gombrich (1962, p. 78) pointed out that the real world is so complex, so rich in detail that no picture can portray it completely and that subjectivity of the artist is not the only factor at work. He continues:

> But what matters to us is that the correct portrait, like the useful map, is an end product on a long road through schema and correction. It is not a faithful record of a visual experience but the faithful construction of a relational model.

> Neither the subjectivity of vision nor the sway of conventions need lead us to deny that such a model can be constructed to any required degree of accuracy. What is decisive here is clearly the word 'required'. The form of a representation cannot be divorced from its purpose and the requirements of the society in which the given visual language gains currency.

Once we are aware that the accuracy of a map is one thing and the way it replicates reality is another, there exist possibilities for general statements about all maps in relation to the real world. The less a map is like the real world, the more abstract it is, the more it is a model of that real world. Indeed it is perfectly easy to conceive of a scale of faithfulness between the one mile to one mile map and the directional road sign, which is not only extremely abstract, but is inaccurate in terms of angles, distance and area, but (usually) up to date. To some extent the property of scale also jibes with faithfulness, the smaller the scale the less faithful a map can be to reality. However, the very obvious variation that occurs in maps of the same scale indicate that although scale may set broad limits to the faithfulness of the map, in detail other properties may be more important. Reference to Figure 16.17 makes it clear that decreasing faithfulness, or increasing abstraction has two components which are the axes of the diagram – increasing dimensional scale and decreasing complexity. Conventionally air photographs are excluded from the category *maps*. They lie some way between them and reality. They show only landscape or visible features, but depict all such features, depending on the equipment used to take them. They, like maps are also distorted in terms of area and distance but are more faithful than maps in the sense that azimuths from the centre of vertical air photographs are true. They do at least furnish a record which contains all reproducable detail at one particular point of time. In spite of their being separated from maps by convention, they are more conveniently regarded as *pseudo-maps* because of their close affinities with them. Air photographs differ from them chiefly in respect of names and invisible features. In fact, air photographs are often used as if they were maps (Wilson, 1965; Langdale-Brown and Spooner, 1963, p. 1). In the first example traffic counts were made on them; in the second seemingly uniform areas of vegetation were delimited on mosaics of photographs. A link between air photographs and maps is provided by the orthophotomap (Pumpelly, 1964), which is a photographic presentation of an area over which symbols are printed, surrounded by the marginal information usually found on a map. Horizontal displacements are eliminated and the orthographically true detail is composed of the total photographic image in a subdued colour (pale khaki) with linear features printed in dark grey. This gives it the appearance of having a third dimension. Other experimental work on the conversion of air photographs to maps is discussed by Merriam (1965), who presents the results of screening a conventional 133-line half-tone air photograph so that only the texture of the image remains. By adding tones to that image some cartographic symbolism, such as hypsometric tints, can be incorporated without destroying the detail of the original photograph. Merriam points out that the overprinting of colours on a photographic image can obscure detail. Merriam calls documents such as these *substitute maps*.

## Faithfulness in terms of spatial properties

The spatial properties which underlie all others are distance, orientation and area. Any map may depart from reality in all, some or one of these characteristics. The selectivity or degree of completeness with which it represents reality which has been discussed already, is superimposed on these basic scalar qualities, but as seen in Figure 16.17, is highly dependent on them. Time furnishes a fourth factor affecting the information content of a map. Naturally the greater time lapse between survey and publication, the less

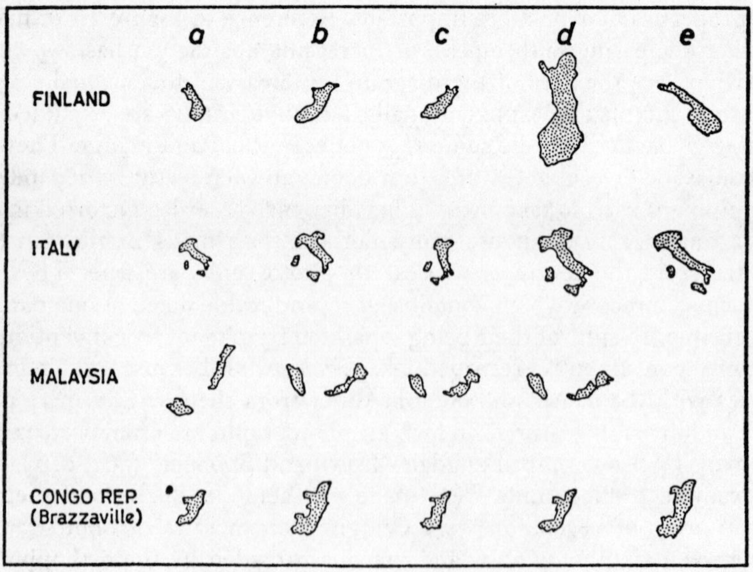


**16.18** The effects on the shape and area of four countries of approximately the same size produced by the employment of different map projections.

(*a*) Oblique Azimuthal Equidistant, centred on London and its antipodal point (*Source: Cassells New Atlas, 1961, plates 4 and 5*).
(*b*) Mollweide interrupted at 20° W. and 60° E. south of the equator (*Source: Philip's University Atlas, 1960, plates 8 and 9*).
(*c*) Zenithal Equal area with hemisphere centred at 70° E. and the equator (*Source: Oxford Atlas, 1958, plate 17*).
(*d*) Mercator (*Source:* The Times *Atlas, 1958, volume I, plate 6*).
(*e*) Mollweide with central meridian at 40° W. in order to show the Atlantic Ocean (*Source: Faber Atlas, 1964, plate 133*).

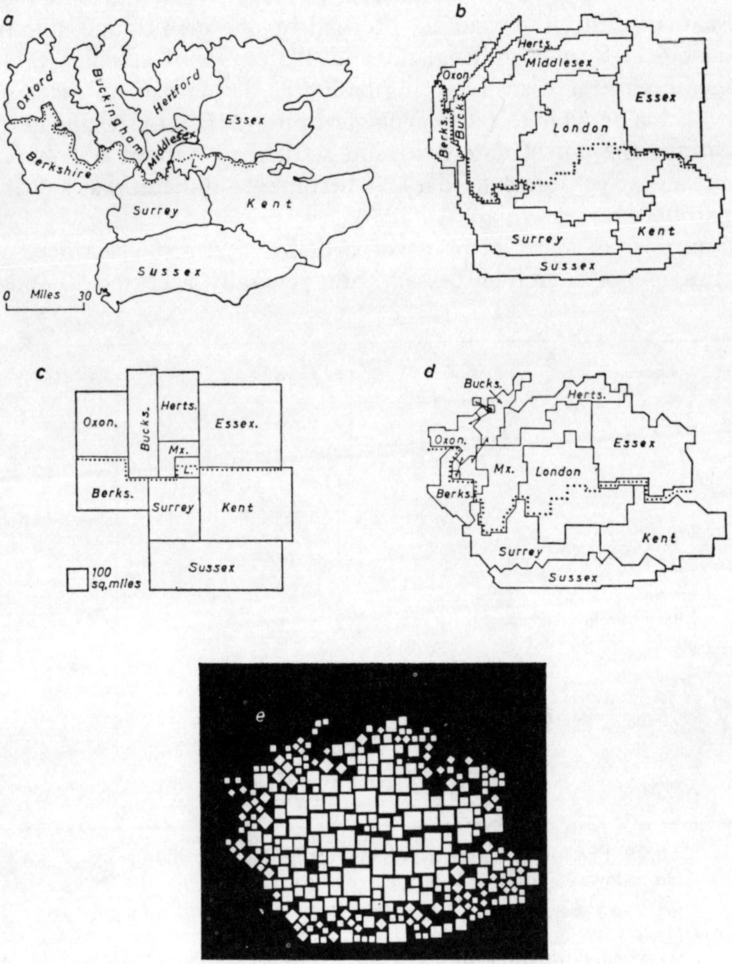


**16.19** Representations of counties in south-east England.

(*a*) Conventional, from a Transverse Mercator projection.
(*b*) From Hollingsworth's map of parliamentary constituencies in *The Times*, 19 October 1964. Roughly proportional to population.
(*c*) Counties shown as rectangles whose areas are proportional to their population.
(*d*) Approximate outline of counties derived from map (*e*).
(*e*) Representation of administrative areas in south-east England, with areas proportional to population. Squares – urban areas; diamonds – rural districts. (*An experimental map by the Ministry of Housing and Local Government, Crown Copyright reserved.*)

likely it is that the map will be faithful to reality. Sometimes this is not so important. Consider the example afforded by one inch to one mile maps of the Geological Survey of England and Wales. Until recently these were published on a progressively ageing base map dating back to the 1890's. As long as it was possible to relate geological information to specific geographical positions the out-of-date road and settlement pattern was immaterial. However a time comes when that task becomes so difficult that a change of a base map becomes essential.

The purpose of many maps is so particular that some sacrifices of basic properties is not merely necessary but generally accepted. Topographic

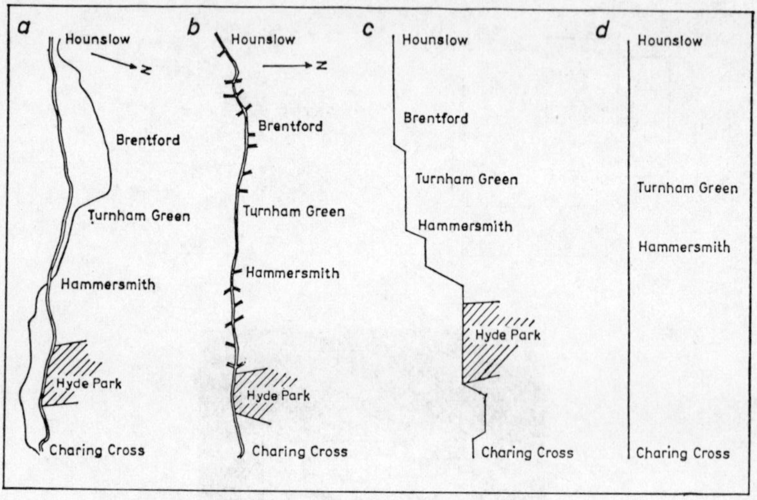


**16.20** The routes between London, Charing Cross, and Hounslow, Middlesex; a distance of about 10 miles. *From:*
(*a*) *The Times Atlas.*
(*b*) Bowles' Post-chaise Companion (1778).
(*c*) London Transport maps of bus and trolleybus routes (1946).
(*d*) London Transport map of the underground railway system (1946).

maps of a fairly small area such as the British Isles, with comparatively small East–West extent and longer North–South extent sacrifice very little if drawn on a Transverse Mercator projection. Compass courses may be followed, and both area and shape are scarcely distorted near the central meridian.

By relaxing the standards of faithfulness in one property, whilst retaining the others true, one moves a little further from reality but gains from the emphasis of particular properties. Zenithal equidistant projections are often used to show distance and orientation from the centre of the map to any other point. Such maps are commonly used to show air routes from important centres and the distances to and relative positions of places from some centre, like Wellington, New Zealand, which frequently suffers from a marginal position on conventional world maps. However these maps have the major drawback of distorting both the shape and area at the margin of the map antipodal to the centre of the projection. Areal exaggeration increases away from the centre. Other kinds of map may retain areas true to scale, in order to present distributions as honestly as possible. World equal area projections are in common use for this purpose. In these, the shapes of the land masses are generally not greatly distorted (see Fig. 16.18). Another class of equal area maps shows land area true to scale, but relaxes standards of faithfulness in distance and orientation by presenting countries as rectangles in roughly their correct positions relative to each other (Kirk, 1964, p. 12) (see Fig. 16.19C). Other maps are constructed in such a way that only distance is preserved as true to scale. One of the oldest is the Peutinger table (Bagrow, 1964, p. 143) where Roman roads are shown as straight lines with places marked at the correct distances. The coaching itineraries so popular in the eighteenth and nineteenth centuries (see Fig. 16.20B) also made use of the idea. Orientation in these maps is more often than not sacrificed to distance and simplicity and area is irrelevant to their purpose.

## Diagrammatic maps

A final class of map may be distinguished in which none of the three basic properties is faithful to reality. Only relative position may remain to remind us that the map is a representation of reality. These are often called cartograms but it is difficult to determine a dividing line between them and maps *sensu stricto*. Such maps usually have a very special purpose. The bus and underground railway maps of London Transport (1946, see Fig. 16.20C and D) are diagrammatic representations of the networks of services and merely show the different routes and interchange points. Distances and directions vary across these maps. Yet another group of diagrammatic maps show, not land area and topographic features, but population or volume of trade (Grotewold, 1961). The areas depicted on such maps are to scale in terms of some special property. The shapes of areas may be recognizable as in the case of Grotewold's maps or Wotyinsky's maps (1953). Recently Hollingsworth (1964) has constructed a map showing parliamentary constituencies in the United Kingdom so that their areas are roughly proportional to the electorate they contain, whilst retaining the topological properties of the real situation. Each constituency is contiguous on the map to those which border it in reality. This constraint (see Fig. 16.19B) results in a degree of distortion of shape and orientation which may well be intolerable to most map readers. An adaptation of the idea is at present being designed by the Ministry of Housing and Local Government in London (see Fig. 16.19E).

Although the areas of local authorities are proportional to the population enumerated within them, their positions relative to each other are nearer to reality because the constraint that contiguous areas should actually touch has been relaxed. Instead the areas float, as it were, in a black matrix, so that the outline of the coast is more recognizable than it is on Hollingsworth's map. The choice of two shapes of symbols – squares and diamonds – also gives the design a certain flexibility helping it to correspond more closely to reality.

## The map as an end product

As an end product, a representation of reality, a map may be judged either by its artistic qualities – the fineness of line, harmony of colour and lettering and balanced layout or design, or by its usefulness. Whether a work of art or of science, such a map furnishes a new view of reality, albeit a subjective one. It also gives the map reader a view of reality not perceptible to the man on the ground (or even to one in an orbiting satellite). This it achieves by compressing and codifying reality to a comprehensible document that can be decoded and employed as a tool by which to navigate or otherwise inform the reader about the real world.

Maps such as these are representations of mental or conceptual models of spatially arranged features. Skilling (1964) reminds us that one retains mental models of places, such that, in spite of their being highly imperfect, we know that if we go in a certain direction we shall pass certain landmarks. Lynch's *The Image of the City* (1960) contains several maps of American cities drawn from such mental images of landmarks. Whether such mental models of places come from memories of travel, or from reading the map, it is important to remember that they are not infallible. In the same way the map as a model is not infallible. Nevertheless it can safely be used for prediction, for interpretation, if its properties are known and understood.

At the same time, not every feature of the model corresponds to some characteristic of its subject matter. Some features are irrelevant, such as the pink colour used for the British Empire (Kaplan, 1964, p. 285). This does not mean that the British Empire is pink because the map depicts it as such. Indeed Tom Sawyer was able to correct Huckleberry Finn when he thought their balloon ought not to be over Indiana because the ground was not pink (Twain, 1894, p. 35). Of course the states were not the same colour out of doors as they were on the map. The map did not tell lies, it used those colours for a purpose. 'It ain't to deceive you – it's to keep you from deceiving yourself' (Twain, 1894, p. 40).

Maps as representational models of the real world do need revision from time to time. The medieval T in O maps with Jerusalem at the centre of the world disc was made by craftsmen who 'made no serious attempt' to show the world as it actually was (Raisz, 1948, p. 14). As geographical knowledge grew and the cartographer became emancipated from the influence of the medieval church, this model was revised. Today changes in features on the earth's surface are brought about mainly as a result of man's activities. With the advent of rapid methods of mapping made possible by aerial photography, stereoplotters and the orthophotoscope, there is little need to accept the outdated models of the past. Revision can now proceed as fast as maps can be made and financial considerations permit. The natural landscape, within the limits set out above, can be mapped satisfactorily and needs little revision thereafter; the representation of the cultural landscape is perhaps in progressively greater need of change, as man replaces outworn structures and builds new. More frequent revision is required on the larger scale representations (Imhof, 1951, p. 131), which is partly a reflection of their complexity. It is for this reason that topographic maps, for which revision is so important, carry so little detail of an ephemeral nature. Thus, land use maps at scales larger than 1:250,000 would be likely to appear out of date far sooner than those of scales smaller than 1:1 million.

## TESTING THE MODEL

Thus far we have discussed the reduction of complex reality into a codified, two dimensional form which has well-known properties. Some attention has already been paid to the purpose and utilization of the models we call maps, but this is largely incidental to the theme of representation. The distinction between thematic and general or topographic maps has already been drawn. This is paralleled by essential differences in the functions of such maps. 'Thematic maps present not only facts, but ideas, hypotheses, and the results of analysis and synthesis' (Miller and Voskuil, 1964, p. 14). Moles (1964, p. 13) clearly distinguishes between maps of two kinds. The first represent an accumulation of a reserve of information, which can be consulted rather than read. The second has the specific aim of presenting a message derived from such a reservoir of facts in as comprehensible form as possible, to the detriment if necessary, of detail, precision and accuracy. Moles observes that the average human being can effectively assimilate a message of this kind at a rate of no more than 10 bits a second. On the other hand Roberts (1962) gives an estimate that the standard, multi-coloured, large-scale topographic map of a part of the United States contains between 100 and 200 million bits of information. To assimilate all this would take well over a year of the normal map reader's time. Electronic computers, with large storage capacities may well be able to handle such stores of information by new scanning methods, but 'the map, unlike a computer memory, has the distinct advantage of permitting direct human use of its vast store of information' (Roberts, 1962, p.13). But the memory of the computer is more reliable than ours. Thus inspection of a detailed map by the human eye can yield *some immediate* return, usually in the form of a hunch or speculation about some pattern or relationship. 'La carte n'est plus un reservoir d'information à l'usage des spécialistes, elle est un outil pour faire émerger des formes et des idées dans un large public' (Moles, 1964, p. 13). This applies not only to the thematic map, but also to the less specialized, topographic map. But we must not forget that 'like carpenters' tools, maps should not be misused. More should not be expected of them than they can perform' (Wright, 1942, pp. 543–544). It is by understanding their limitations that they may be regarded as useful tools for research. By employing them in this way, the investigator is testing them against reality. In the simplest case the trained field observer goes on to the ground with a topographic or geological map. In field work the 'primary task is to make significant additions to the map. There is no better method of training in field observation: it is the sovereign cure for the attitude of the beginner to think that it is all on the map anyway; and this is to indulge in the mere incurious verification of what the map-maker has seen fit to portray' (Wooldridge and East, 1958, p. 165). But the observer does not add anything and everything, indiscriminately. To start with it may be a class of features, such as field boundary types or some aspect of current land use. The scientist, well aware of this, does not completely trust this product of human artifice to inform him of the nature of reality. We have already seen that the map maker may have seen fit to present an extremely imperfect picture of reality. The interpolation of human interpretation between reality and the scientist reminded Norbert Wiener of Einstein's dictum: 'Der Herr Gott ist raffiniert, aber boshaft Er ist nicht' (God may be artful, but He is not malicious). Nature plays fair and does not deliberately frustrate the scientist. (Wiener, 1954, p. 188).

So it may be thought that the researcher would do better to use raw data for his investigation. Having formulated a working hypothesis about the explanation of some interesting patterns, perhaps even from map evidence, the data for constructing the model can be collected directly from the real world. Indeed such a course of action avoids the interference from other sources such as maps. There are instances where maps do provide important evidence for research. In many cases long and laborious field surveys have been recorded permanently in map form. Geological, soil, vegetation and land use surveys when mapped are all documents in their own right and cannot be superseded by statistics or descriptions. It is unlikely that geological and soil surveys would be undertaken twice for the same area – at least at the same level of detail, unless for some special purpose when probably only a limited field of inquiry would be involved. Vegetation and land use surveys record relatively ephemeral features. The expense of repeating such surveys often means that they are done but once. Maps of such surveys can therefore be properly regarded as initial data with which to begin an investigation of some problem within an areal context. They are as good as raw data derived directly from observation of reality if such raw data cannot feasibly be collected.

## The special role of isarithmic maps in generalizing spatial patterns

Derived maps, not based in the first place on field observations but on statistics or values calculated from sets or series treated by modern quantitative map analysis. That isarithmic maps are not often compiled and used, even as representations, let alone as an input for a geographical analysis, is certainly thanks to the tedious methods of constructing them manually. The alternative, choropleth method of depicting raw or derived values has simplicity on its side. Robinson (1961) has suggested moreover that we accept the use of isarithms for relief maps (and also for climatic distributions) by long-established custom, but isarithmic maps for cultural distributions are avoided because they do not stress the values at particular places and they somehow transform essentially discrete data into a spatial continuum. If on the other hand it is accepted that the 'portrayal of variations from place to place is the basic aim of a geographical volume', that is some quantitative variable, the general configuration of the data is all important. Its representation by a statistical surface whose gradients may be visually assessed is most appropriate. 'When a space-series array is to be plotted on a map, only the isarithmic technique provides a graphic display of the assumption of continuous gradient, and it also smooths the data to some extent, depending upon the choice of isarithmic interval' (Robinson, 1961, p. 54). Thus the irritating and distracting vagaries of the choropleth map may be partially suppressed. The local variations that are quite likely to be emphasized in choropleth maps are probably more obvious in maps of nominally scaled data. The intricacy of land use maps even with as few categories as two – the presence of arable land and its absence for example – is well known. It is not necessary to remind generations of geographers that the task of interpreting the incredible complexity of any topographic map by eye is formidable even after years of practice. Tobler (1966) has developed techniques for use in computer mapping to generalize isarithmic patterns and reverse the process, or roughen the smoothed pattern ending with the original pattern.

No great a gap exists in geographical method than that which lies between such maps and their generalized, objective interpretation. Analytical techniques for this purpose are only now being developed, have still to be tested in a variety of conditions and are known only to a minority of investigators. The explanation of spatial patterns requires first of all the identification of pattern. The complexity of the detailed map may at first seem to reflect a random arrangement of phenomena. A search for order in this imagined chaos is therefore the first step in a scientific investigation. By some chance observation, hunch or speculation the germ of a theory may be established. Next a working model or hypothesis is developed and this tested, by experiment or comparison with the real world. Maps play a significant part in these processes. They can furnish data initially, they can suggest the hypotheses and they can be employed as tools to make and test models. In many instances they are themselves models incorporating only selected aspects of the real world reflecting some hypothetical situation.

## Trend surface mapping

In a spatial context the development of methods of identifying, delineating what are called surfaces, trend-surfaces or response surfaces has greatly assisted the recognition of the role of maps in the explanation of geographical patterns. Chorley and Haggett (1965, p. 47) in a recent review of these methods state the problems facing the map interpreter in these terms: 'Because geographical problems. . . are characterized by areal sampling restrictions, by a multiplicity of variables, and by the interaction and simultaneous variation of most of the variables, we cannot be certain how much of the information transmitted by a map may be regarded as a "signal" and how much as random variations or "noise".' By constructing maps of the trend or message separately from maps of noise, the unexplained random variations, a more satisfactory approach to pattern interpretation may be attempted. The explanations thus given are expressed in terms of statistical relationships or as process-response models where variations in form correspond to the strength and balance of the supposed controlling factors (Chorley and Haggett, 1965, p. 48). In the same review a distinction is made between selective and objective methods of constructing trend surfaces. Since there is clearly a selective and subjective element in both methods, a distinction will here be made between mainly graphical techniques and the more rigorous mathematical techniques.

## Graphical trend surface analysis

Graphical methods were for long the only ones applied to the problem of sifting the regional from the local component in spatial patterns. Graphical generalization of contours as for example used by Wooldridge (1927) to serve as a surface upon which outliers of Tertiary age could be plotted. The compilation of generalized contours suffers from variations principally due to the different ways in which individual operators interpret the ground rules. Graphical methods also include the techniques of constructing parallel and orthogonally intersecting profiles, which are themselves generalized from the available data (Krumbein, 1956). Robinson (1961) also shows that the method adopted by Tanaka Kitiro (1932), which yielded a planimetrically correct and visually pleasing impression of plastic relief, could be applied to isarithmic maps of cultural phenomena. By employing a series of uniformly spaced planes inclined at 45° the traces of which were linked to the isarithmic pattern, the general configuration of variations in population density is obtained. The advantage of this relatively objective method is that it employs an imagery familiar to those who understand the shaded relief map and it also smoothes out the local irregularities. Robinson (1961, p. 57) suggests that 'there appears to be a relatively large amount of high-frequency variation in cultural data, as compared to the purely "physical" ' possible because of the large number of factors affecting man's behaviour and a whole series of covariant factors which are less well understood in the human sciences. Furthermore isarithmic maps of cultural features do not behave in the same way as the land surface, nor are the operations of well-known causal factors readily appreciated. 'A fairly flat surface may dip gently over large areas, and, if it undulates at all, the pattern of wandering isarithms and their relation to the surrounding forms of the statistical surface may easily mask its basic flatness' (p. 58). It is therefore all the more important for geographers attempting to understand the patterns of human activity to appreciate the possibilities involved in the mapping of trends and residuals or anomalies.

## More rigorous trend surface analysis

In the mathematical methods of trend surface mapping analysis before mapping is altogether more complicated. It has been tremendously aided, indeed made feasible, by the introduction of high-speed electronic computers (Krumbein and Graybill, 1966, p. 321). Methods performed objectively by the computer enable the investigator to divide each map observation into two or more parts: 'large scale' or regional trends from one edge of the map to the other and 'small scale' or local effects. By a reiterative process the original observations may be broken down into a series of trends of progressively smaller extent. The local effects, which are apparently non-systematic fluctuations, can be identified and plotted at each stage. Their pattern and nature may well suggest the interpretation of the next higher order trend surface or new variable to be built into the explanatory model being constructed. Both linear and higher order trend surfaces and residuals (the computed differences between observed values and trend values at particular points) are mapped by isarithms. It is natural at this stage to ask how well the trend fits the reality presented by observed values. Merriam and Harbaugh (1964) when fitting successively higher-order surfaces to the elevation of geological formations measured the goodness of fit in terms of the percentage reduction in the sum of squares. The surfaces are so constructed as to make the sum of the squared residual values as small as possible. In this way objectivity is achieved and a measure of correspondence with reality is obtained. Similar criteria are used in the regression models described by Chorley and Haggett (1965). In such cases the first trend is a simple linear relationship, such as between voting behaviour and proportion of the population that is rural (Thomas, 1960; see Fig. 16.21) or between cash grain farming and flat land (Hidore, 1963). The mapping of residuals from regression showed up areas where the regression under- or over-estimated the strength of voting or the amount of cash grain farming. A closer look at these areas of anomaly suggested additional variables. In the first instance, the distance from the home town of the candidate raised the level of explanation from about a third to nearly 50 per cent. In the second instance, soil quality was considered to be of importance where the regression had underestimated

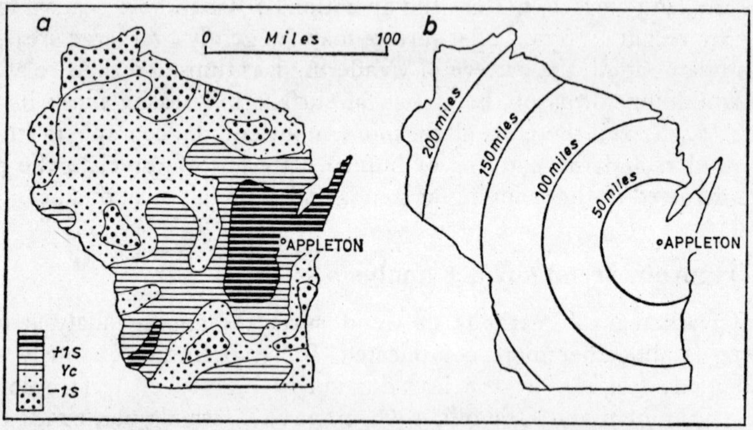


**16.21** Testing the model: developing from the first hypothesis of reasons for voting behaviour, a new hypothesis from map of residuals.

(*a*) Residuals from regression $(Y_{cn} - Y_n)/S_{Y_c}$ of per cent of total vote cast for Senator McCarthy in Wisconsin on the percentage of total population that is rural. (1S 1 standard error of estimate, $Y_c$ as predicted by the regression, −1S −1 standard error of estimate). In general the regression underestimated support for the senator nearer his hometown, Appleton.
(*b*) The basis for the second hypothesis – distance from Appleton (*Source: Thomas, 1960*).

the amount of cash grain farming. In both cases, the map of residuals had real value as a means of suggesting further lines of inquiry, thus restarting the analysis. As Robinson (1961, p. 57) pointed out the different sizes of areas and hence the spacing of control points upon which isarithmic maps are based are another source of noise. Chorley and Haggett (1965, p. 61) take this argument a stage further by observing that what may be taken for noise in coarse-grained analysis, may be partly explicable in a finer grained analysis, where the mesh of control points is closer together. On the whole however it is advisable to use the finest possible mesh right from the beginning. Not to use all the available data would in fact be inefficient because trend-surface mapping itself seeks order in complexity.

## CONCLUSION

Once the experimental iterations with progressively more complex models can explain no significant further variation the investigator is entitled to assume that his model, comprising a trend map with a multivariate message and a map of completely inexplicable residuals, fits the real world well enough. At this stage the cycle of research is temporarily completed because a new view of the real world is the outcome. Maps used in this way are not merely ornaments, or even portraits, they are vital tools of research. To recognize this is to return to maps their rightful place 'as pre-eminently the geographer's tool both in investigation of his problems and the presentation of his results' (Wooldridge and East, 1958, p. 64). Though placed on a level with verbal description, symbolic logic and mathematics, maps still have their peculiar disadvantages and intrinsically useful properties. 'It is too often forgotten that geographical studies are not descriptions of the real world, but rather perceptions passed through the double filter of the author's mind and his available tools of argument and representation' (Curry, 1962, p. 21). By recognizing maps as models of the real world and by employing them as conceptual models in order better to understand the real world, their central importance in geographical methodology is assured. Because the making of maps belongs properly to the profession of cartography, the geographer cannot afford to make his models without reference to cartographic practice. Many of these models will however, be made without reference to him. It is because of this that 'no-one claiming the title of geographer, however humbly, is entitled to be ignorant of how maps are made' (Wooldridge and East, 1958, p. 70).

## REFERENCES

BAGROW, L., [1964], *History of cartography*, (London), 312 pp.

BALCHIN, W. G. V., [1952], Cartographic portrayal by geographers; *Indian Geographical Society (Madras)*, Silver Jubilee Volume, 141–147.

BAMBROUGH, R., [1964], Principia metaphysica; *Philosophy*, 39, 97–109.

*Bartholomew's Atlas of the World's Commerce*, [1907], ed. by J. G. Bartholomew, (Edinburgh), 176 plates.

BERRY, B. J. L., [1960], An inductive approach to the regionalization of economic development; In Ginsburg, N., (Ed.), *Essays on Geography and Economic Development*, University of Chicago, Department of Geography, Research Papers, No. 62.

BERRY, B. J. L., [1961], Basic patterns of economic development; In Ginsburg, N., (Ed.), *Atlas of Economic Development Chicago*, (University of Chicago Press), 110–119.

BLACK, M., [1962], *Models and Metaphors*, (Ithaca, New York), 267 pp.

BLUMENSTOCK, D. I., [1953], The reliability factor in the drawing of isarithms; *Annals of the Association of American Geographers*, 43 (4), 289–304.

BOARD, C., [1962], *The Border Region, Natural environment and land use in the Eastern Cape*; (Oxford University Press, Cape Town), 238 pp. and volume of maps.

BUNGE, W., [1962], *Theoretical Geography*; Lund Studies in Geography Series C, No. 1, 208 pp.

CARROLL, L., (pseud. Dodgson, C. L.), [1893], *Sylvie and Bruno concluded*; (Macmillan, London).

CHORLEY, R. J., [1964], Geography and analogue theory; *Annals of the Association of American Geographers*, 54 (1), 127–137.

CHORLEY, R. J. and HAGGETT, P., [1965], Trend surface mapping in geographical research; *Transactions of the Institute of British Geographers*, 37, 47–67.

CLARKE, J. I., [1959], Statistical map reading; *Geography*, 44 (2), 96–104.

COLE, J. P., [1965], *A geography of World Affairs*, (London), 348 pp.

CONZEN, M. R. G., [1960], Alnwick, a Northumberland study in town-plan analysis; *Transactions of the Institute of British Geographers*, 27, 127 pp.

COPPOCK, J. T., [1960], The parish as a geographical-statistical unit; *Tijdschrift voor sociale en economische geografie*, 51 (12), 317–326.

CROXTON, F. E. and STEIN, H., [1932], Graphic comparison by bars, squares, circles and cubes; *Journal of the American Statistical Association*, 27, 54–60.

CURRY, L., [1962], Climatic change as a random series; *Annals of the Association of American Geographers*, 52, 21–31.

CZEKALSKI, J., [1933], Mapa izarytmiczna a obraz rzeczywisty (próba analizy metody) (The isarithmic map, its method and degree of precision); *Wiadomości Służby Geograficznej*, 7 (3), 202–234, (Warsaw).

DAHLBERG, R. E., [1962], Evolution of interrupted map projections; *International Yearbook of Cartography*, 2, 36–54.

DAVIS, L. B., [1958], Design criteria for today's aeronautical charts; *Surveying and Mapping*, 18, 49–57.

DE DAINVILLE, Fr. S. J., [1964], *Le langage des géographes*, (Paris), 384 pp.

*Dictionary of National Biography*, [1921 edn.], volume 5, (Oxford University Press).

DUMKE, G. S., [1963], *The Boom of the Eighties in Southern California*, (San Marino, Calif.), 313 pp.

ECKERT, MAX, [1908], On the nature of maps and map logic; *Bulletin of the American Geographical Society*, 40, 344–351.

ECKMAN, G. and JUNGE, K., [1960], *Psychophysical relations in the perception of visual length, area and volume with special regard to interpretation of certain map symbols*; Paper prepared for the meeting of the International Geographical congress, Commission on a World Population Map, Stockholm, 22 pp., (Mimeographed).

EKMAN, G., LINDMAN, R. and WILLIAM-OLSSON, W., [1961], A psychophysical study of cartographic symbols; *Perceptual and Motor Skills*, 13, 355–368.

FORGOTSON, J. M. Jr., [1960], Review and classification of quantitative mapping techniques; *Bulletin of the American Association of Petroleum Geologists*, 44 (1), 83–100.

FOX, J. R., [1956], *Land-use survey general principles and a New Zealand example*, (Auckland, New Zealand), 46 pp.

GOMBRICH, E. H. J., [1962], *Art and Illusion*, (London), 388 pp.

GROTEWOLD, A., [1961], Some aspects of the geography of international trade; *Economic Geography*, 37, 309–319.

*Michelin Guide*, [1963], Paris, (Pneu Michelin Services de Tourisme), 997 pp.

HAGGETT, P., [1964], Regional and local components in the distribution of forested areas in South-east Brazil: a multi-variate approach; *Geographical Journal*, 130 (3), 365–380.

HARRIS, L. J., [1959], Hill shading for relief depiction in topographical maps; *Chartered Surveyor*, 91 (9), 515–520.

HARRISON, M. J., MEAD, W. R. and PANNETT, D. J., [1965], A Midland ridge-and-furrow map; *Geographical Journal*, 131 (3), 366–369.

HARRISON, R. E., [1959], Art and common sense in cartography; *Surveying and Mapping*, 19 (1), 27–38.

HARLEY, J. B., [1962], The one-inch to the mile maps of England and Wales; *Amateur Historian*, 5 (5), 130–140.

HIDORE, J. J., [1963], The relationships between cash-grain farming and land forms; *Economic Geography*, 39, 84–89.

HOLLINGSWORTH, T. H., [1964], *The Times*, 19 October 1964; (Map on p. 18).

IMBRIE, J. and PURDY, E. G., [1962], Classification of modern Bahamian carbonate sediments; In *Memoir* No. 1, *American Association of Petroleum Geologists*, Classification of carbonate rocks, a symposium, 253–272.

IMHOF, E., [1951], *Terrain et Carte*, (Zurich), 261 pp.

IMHOF, E., [1962], Die Anordnung der Namen in der Karte; *International Yearbook of Cartography*, 2, 93–129.

JACKSON, B. G., BARNARD, C. S. and STURROCK, F., [1963], *The pattern of farming in the Eastern counties*; Farm Economics Branch, School of Agriculture, Cambridge University, Occasional Papers, No. 8, 60 pp.

JENKS, G. F. and KNOS, D. S., [1961], The use of shading patterns in graded series; *Annals of the Association of American Geographers*, 51 (3), 316–334.

JENKS, G. F., [1963], Generalization in statistical mapping; *Annals of the Association of American Geographers*, 53 (1), 15–26.

JOHNSON, F. C. and KLARE, G. R., [1961], General models of communication research, a survey of the development of a decade; *Journal of Communication*, 11 (1), 13–26 and 45.

JOHNSTON, D. A., [1902], *Ordnance Survey maps of the United Kingdom, a description of their scales, characteristics, etc.*, 17 pp.

KANSKY, K. J., [1963], *Structure of transportation networks*; University of Chicago, Department of Geography Research Papers, No. 84, 155 pp.

KAPLAN, A., [1964], *The conduct of enquiry: methodology for behavioural science*, (San Francisco), 428 pp.

KEATES, J. S., [1962], The perception of colour in cartography; *Proceedings of the Cartographic Symposium, Edinburgh*, 19–28.

KING, L. J. and HENSHALL, J., [1966], Some structural characteristics of peasant agriculture in Barbados; *Economic Geography*, 42, 74–84.

KINGSBURY, R. L., [1964], The world of little maps; *Journal of Geography*, 63, 355–366.

KIRK, D., [1946], *Europe's population in the interwar years*; (Princeton, New Jersey), (for the League of Nations), 307 pp.

KITIRO, T., [1932], The orthographical relief method of representing hill features; *Geographical Journal*, 79, 213–219.

KRUMBEIN, W. C., [1956], Regional and local components in facies mapping; *Bulletin of the American Society of Petroleum Geologists*, 40, 2163–2194.

KRUMBEIN, W. C., [1957], Comparison of percentage and ratio data in facies mapping; *Journal of Sedimentary Petrology*, 27 (3), 293–297.

KRUMBEIN, W. C. and GRAYBILL, F. A., [1965], *An introduction to statistical models in geology*, (New York), 475 pp.

LANGDALE-BROWN, I. and SPOONER, R. J., [1963], *Land use prospects of Northern Bechuanaland*; Department of Technical Co-operation Directorate of Overseas Surveys, Development Study No. 1, 40 pp.

LEWIS, W. V., [1959], The karstlands of Jamaica: cockpits or rounded hills?; *Geographical Journal*, 125, 289.

LINTON, D. L., [1948], The ideal geological map; *Advancement of Science*, 5 (18), 141–149.

LUNDQUIST, G., [1959], Generalization – a preliminary survey of an important subject; *Nachrichten aus dem Karten – und Vermessungswesen*, 2 (3), 46–51.

LUNDQUIST, G., [1963], Generalization of communication net-works; Bulletin No. 4, International Cartographic Association, *Sonderdruck* from *Nachrichten aus dem Karten – und Vermessungswesen*, 5 (5), 35–42.

LYNCH, K., [1960], *The image of the city*, (Cambridge, Mass.), 194 pp.

LYONS, H. G., [1914], Relief in cartography; *Geographical Journal*, 43, 233–248 and 395–407.

LYONS, H. G., [1928], The sailing charts of the Marshall Islanders; *Geographical Journal*, 72, 325–328.

MACKAY, J. R., [1949], Dotting the dot map: an analysis of dot size, number and visual tone density; *Surveying and Mapping*, 9 (1), 3–10.

MACKAY, J. R., [1953], The alternative choice in isopleth interpolation; *Professional Geographer*, 5 (4), 2–4.

MACKAY, J. R., [1954], Geographic Cartography; *Canadian Geographer*, 4, 1–14.

MACKAY, J. R., [1963], Isopleth class intervals: a consideration in their selection; *Canadian Geographer*, 7 (1), 42–45.

MALING, D. H., [1963], Some quantitative ideas about cartographic generalization; Bulletin No. 4, International Cartographic Association, *Sonderdruck* from *Nachrichten aus dem Karten – und Vermessungswesen*, 5 (5), 6–22.

MARSCHNER, F. J., [1943], Maps and a mapping program for the United States; *Annals of the Association of American Geographers*, 33, 199–219.

MARSCHNER, F. J., [1944], Structural properties of medium- and small-scale maps; *Annals of the Association of American Geographers*, 34, 1–46.

MEAD, W. R., [1954], Ridge-and-furrow in Buckinghamshire; *Geographical Journal*, 120 (1), 34–42.

MERRIAM, D. F. and HARBAUGH, J. W., [1964], Trend-surface analysis of regional and residual components of geologic structure in Kansas; *Kansas State Geological Survey, Special Distribution Publication* No. 11, (Lawrence, Kansas), 27 pp.

MERRIAM, M., [1965], The conversion of aerial photography to symbolised maps; *Cartographic Journal*, 2, 9–14.

MILLER, O. M. and VOSKUIL, R. J., [1964], Thematic-map generalization; *Geographical Review*, 54, 13–19.

MOLES, A. A., [1964], Théorie de l'information et message cartographique; *Sciences et Enseignement des Sciences*, (Paris), 5 (32), 11–16.

MONMONIER, M. S., [1965], The production of shaded maps on the digital computer; *Professional Geographer*, 17 (5), 13–14.

MOTT, P. G., [1964], Topographical model making using air photographs; *Cartographic Journal*, 1 (2), 29–32.

PILLEWIZER, W. and TOPFER, F., [1964], Das Auswahlgesetz: Ein Mittel zur Kartographischen Generalisierung; *Kartographische Nachrichten*, 14 (4), 117–121.

PORTER, P. W., [1958], Putting the isopleth in its place; *Proceedings of the Minnesota Academy of Science*, 25/26, 373–384.

PORTER, P. W., [1964], *A bibliography of statistical cartography*, (Minneapolis), 66 pp.

PUMPELLY, J. W., [1964], Cartographic treatments in the production of orthophotomaps; *Surveying and Mapping*, 24 (4), 567–571.

RAISZ, E., [1948], *General cartography*, (New York), 354 pp.

REEVES, E. A., [1929], in Withycombe (1929).

RIDDIFORD, C. E., [1952], On the lettering of maps; *Professional Geographer*, 4 (5), 7–10.

ROBERTS, J. A., [1962], The topographic map in a world of computers; *Professional Geographer*, 14 (6), 12–13.

ROBINSON, A. H., [1952], *The Look of maps*, (Madison, Wisconsin), 105 pp.

ROBINSON, A. H., [1960], *Elements of cartography*; 2nd edition, (New York), 343 pp.

ROBINSON, A. H., [1961], The cartographic representation of the statistical surface; *International Yearbook of Cartography*, 1, 53–63.

ROBINSON, A. H., LINDBERG, J. H. and BRINKMAN, L., [1961], A correlation and regression analysis of population densities in the Great Plains; *Annals of the Association of American Geographers*, 51, 211–222.

SCHMIDT-FALKENBURG, H., [1962], Grundlinien einer Theorie der Kartographie; *Nachrichten aus dem Karten- und Vermessungs-wesen*, 1 (22), 5–37.

SCHULZ, H. C., [1939], Elizabethan map of Wotton Underwood, Buckinghamshire; *Huntington Library Quarterly*, 3, 395–407.

SCHULZ, H. C., [1954], A Shakespeare haunt in Bucks?; *Shakespeare Quarterly*, 5 (2), 177–178.

SINNHUBER, K. A., [1964], The representation of disputed political boundaries in general atlases; *Cartographic Journal*, 1 (2), 20–28.

SKILLING, H., [1964], An operational view; *American Scientist*, 52, 388A–396A.

STAMP, L. D., [1948], *The land of Britain its use and misuse*, (London), 507 pp.

STAMP, L. D., [1961], *A glossary of geographical terms*, (London), 539 pp.

STAROSTIN, I. I. and YANIKOV, G. V., [1959], *Osnovy Topografii i Kartografii* (Principles of Topography and Cartography), (Moscow).

SWEETING, M. M., [1958], The karstlands of Jamaica; *Geographical Journal*, 124, 184–199.

SWEETING, M. M., [1959], Reply to W. V. Lewis; *Geographical Journal*, 125, 289–291.

THOMAS, E. N., [1960], *Maps of residuals from regression: their characteristics and uses in geographic research*; Department of Geography, State University of Iowa, Publication No. 2, (Iowa City), 60 pp.

TOBLER, W. R., [1965], Automation in the preparation of thematic maps; *Cartographic Journal*, 2 (1), 32–38.

TOBLER, W. R., [1966], Numerical Map Generalization; In *Discussion Paper* No. 8, *Michigan Inter-University Community of Mathematical Geographers*, 24 pp.

'*Tomlinson Report*', [1956], *Summary of the Report of the Commission for the Socio-Economic Development of the Bantu Areas within the Union of South Africa*, (Government Printer, Pretoria), U.G. 61/1955, 213 + 64 pp.

TÖPFER, F. and PILLEWIZER, W., [1966], The principles of selection, *The Cartographic Journal*, 3 (1), 10–16, translated and with explanatory notes by D. H. Maling.

TOULMIN, S. E., [1953], *The philosophy of science*, (London), 176 pp.

TRIESMAN, A., [1966], Our limited attention; *Advancement of Science*, 22 (104), 600–611.

TWAIN, M., (pseud. Clemens, S. L.), [1894], *Tom Sawyer Abroad*, (London), 208 pp.

*The University Atlas*, [1946], Goodall, G. and Darby, H. C., (Eds.), (George Philip, London), 96 Plates.

UHORCZAK, FR., [1930], Metoda izarytmiczna w mapach statystycnych (The isarithmic method applied to statistical maps); *Polski Przeglad Kartograficzny*, 4, 95–129.

VON HUHN, R., [1927], Further studies in the graphic use of circles and bars: I A discussion of the Eells experiment; *Journal of the American Statistical Association*, 22, 31–39.

WEAVER, J. C., [1954], Crop-combination regions in the Middle West; *Geographical Review*, 44 (2), 175–200.

WEAVER, J. C., HOAG, L. P. and FENTON, B. L., [1956], Livestock units and combination regions in the Middle West; *Economic Geography*, 32 (3), 237–259.

WEINER, N., [1954], *Thr human use of human beings: cybernetics and society*, (Garden City, New York), 199 pp.

WILKINSON, H. R., [1951], *Maps and Politics: A Review of the ethnographic cartography of Macedonia*, (Liverpool), 366 pp.

WILLIAMS, R. L., [1956], *Statistical symbols for maps: their design and relative values*; (New Haven, Connecticut), (Yale University Map Laboratory), 115 pp.

WILSON, F. R., [1965], Collection of traffic data by aerial photographs; *Traffic Engineering and Control*, 7 (4), 258–261.

WINTERBOTHAM, H. ST. J. L., [1929], in Withycombe (1929).

WITHYCOMBE, J. G., [1929], Lettering on maps; *Geographical Journal*, 73, 429–446.

WRIGHT, J. K., [1942], Map makers are human: Comments on the subjective in maps; *Geographical Review*, 32, 527–544.

WOOLDRIDGE, S. W., [1927], The Pliocene history of the London Basin; *Proceedings of the Geologists' Association of London*, 38, 49–132.

WOOLDRIDGE, S. W. and EAST, W. G., [1958], *The spirit and purpose of Geography*; 2nd edition, (London), 186 pp.

WOYTINSKY, W. S. and WOYTINSYK, E. S., [1953], *World population and production trends and outlook*; (New York), (20th Century Fund), 1268 pp.

CHAPTER SEVENTEEN

# Hardware Models in Geography

M. A. MORGAN

## INTRODUCTION

It is curious that geographers who are often regarded by others if not by themselves as practical men should have paid relatively little attention to models or physical constructs either in teaching or research. Exceptions are found mainly in the field of physical geography and in particular geomorphology, but even here geologists and hydrologists have been much more productive. The comparative neglect of hardware models in geography may indicate that they have been considered inappropriate or that we have been a little tardy in realizing their value. It may be that a widespread and traditional preoccupation with the unique has diverted our attention from the problems and rewards of trying to explain the more general. The absence until recently of any rigorous mathematical or symbolic formalization in human geography and to a lesser degree in physical geography has not encouraged a search for understanding by means of analogues.

Models are of the greatest value in teaching, particularly if the students are involved in their design, construction and operation. One recalls the old saying, 'Hear and forget, see and remember, do and understand'. The educational advantages that attend students finding out for themselves are increasingly being recognized amongst science teachers. The process of making, doing, observing and measuring develops the student's intuitive capacity, and relationships and principles are often perceived long before he is able to understand a complete and rigorous explanation of what he is observing or creating.

The physical constructs however cannot and should not be isolated from conceptual models. To design a physical model we often without realizing it depend on some conceptual model; conversely observations made from a construct may lead to some mathematical or verbal formalization, in other words to a conceptual model. So the two types of models are closely related.

In the research field especially in human geography the scope for physical models seems relatively restricted. This is largely because many of the models being developed in human geography can be derived and tested with the use of a computer, which is in many respects an infinitely more subtle and flexible tool than other hardware models. In the latter too great a complexity is disastrous because the model becomes tedious to construct and is operationally difficult. The essence of a good model is simplicity and inevitably this limits the circumstances under which its use is appropriate.

This chapter has three objectives. Firstly to present a broad selection of models from a variety of disciplines, models that appear to have some bearing on the sort of topics that interest geographers; secondly to indicate in a very tentative way a few new models that could be useful and thirdly, if mainly by implication, to suggest that this is still a new field and in many respects one well worth cultivating.

## STATIC MODELS

All hardware models or constructs represent some degree of abstraction of reality. The model with which most people are familiar, the three-dimensional relief model, is an abstraction. It cannot with complete fidelity represent every detail of a part of the earth's surface. Not only do we readily accept the inevitable distortions and generalizations, we frequently accept avoidable distortions probably without realizing it. Most relief models have an unnecessarily great degree of vertical exaggeration which is to be deprecated if only because it needlessly reduces the effectiveness with which surface texture can be represented. Scale factors are thoroughly examined by Imhof (1965). Techniques and materials available for relief model construction have been enhanced with the development, for example, of thermoplastics and resins (U. S. Army Map Service, 1950; Spooner, 1953). The use of computers programmed to feed taped instructions from three-dimensional co-ordinates to cutting machines which carve the final relief models is a recent and promising development (Noma and Misulia, 1959) which should do much to make a wide range of relief models more generally available.

It is not always appreciated that many other spatial variables besides relief lend themselves to modelling. The whole concept of statistical gradients and surfaces, whether physical or socio-economic in character, can very often be made more comprehensible in teaching if a model is used. Generally whenever three variables are being considered a model offers the most obvious mode of illustration. Application of models to climatological data has been discussed by Conrad and Pollak (1950) (Fig. 17.1). The difficult problem of describing the structure of a cyclone was tackled by Bjerknes (Shaw, 1934) using a wire model and Shaw (1934) made glass models of the distribution of upper air temperature. Chorley and Morgan (1962) represented two geologically similar areas by models so proportioned that each embodied to scale characteristic features associated with drainage networks and basin shapes, obtained by morphometric analysis (Fig. 17.2). Neither model represented

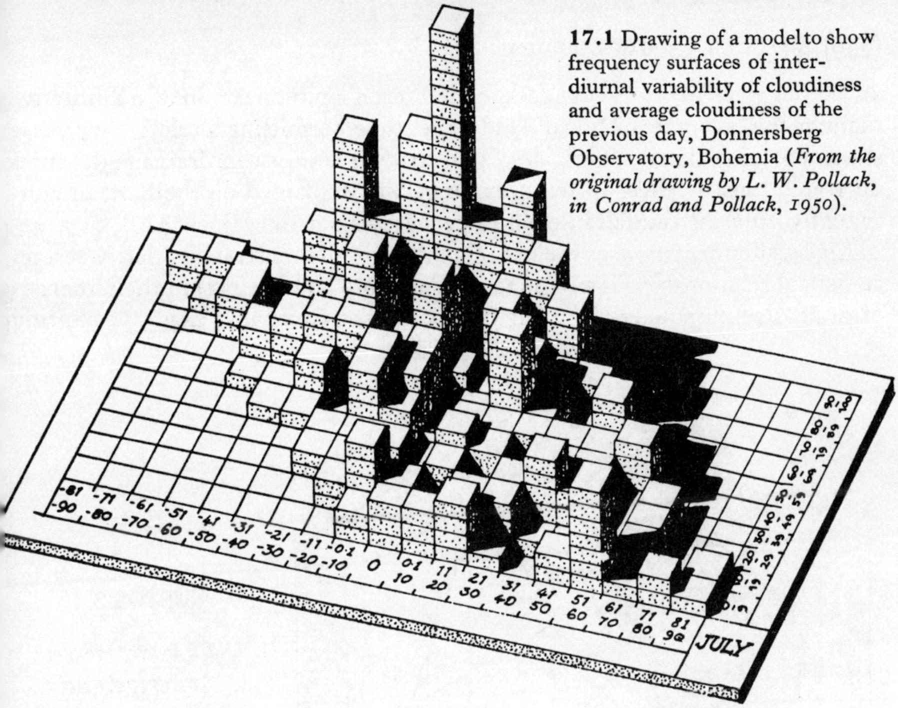


**17.1** Drawing of a model to show frequency surfaces of inter-diurnal variability of cloudiness and average cloudiness of the previous day, Donnersberg Observatory, Bohemia (*From the original drawing by L. W. Pollack, in Conrad and Pollack, 1950*).

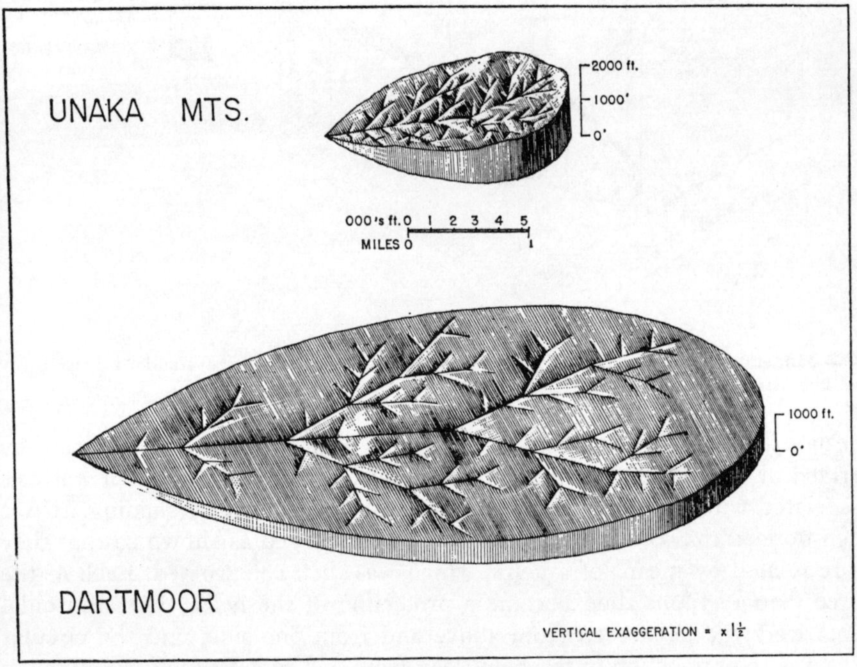


**17.2** Drawing of relief models illustrating idealized drainage basins (*Source: Chorley and Morgan, 1962*).

an actual part of a real landscape, but each epitomized in a quantitative manner the 'average' or 'ideal' landscape, thus permitting a relatively precise statement of regional differences. The striking contrast in drainage densities on which most of these differences hinge was explained on the basis of consistently differing rainfall intensities probably since the Miocene.

An excellent example of the use of a model to demonstrate or clarify certain statistical relationships is provided by Folk and Ward (1957) in their analysis of grain size parameters in a bar on the Brazos River in Texas. By plotting

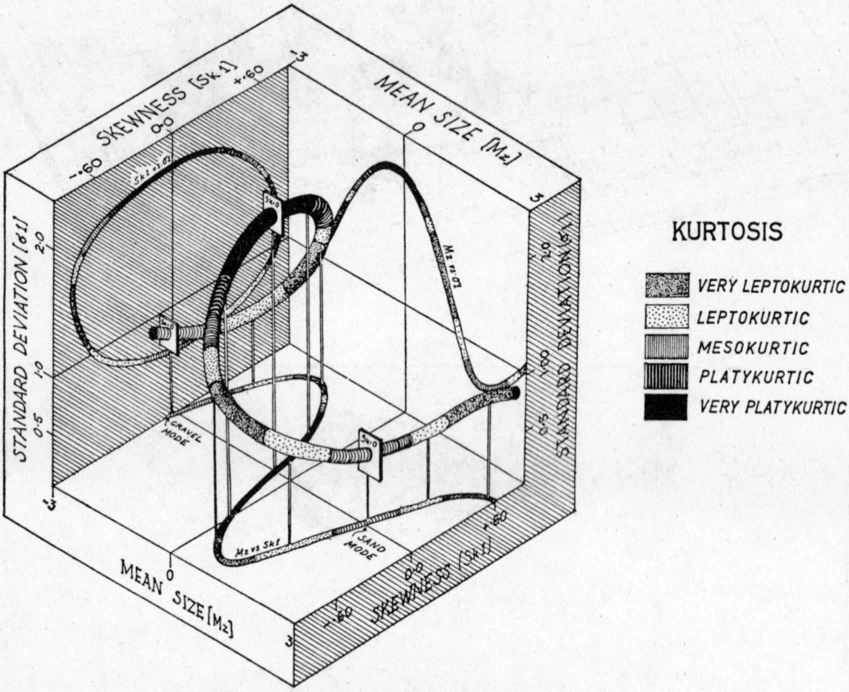


**17.3** Model demonstrating a relationship between skewness, mean size, standard deviation and kurtosis (*Source: Folk and Ward, 1957*).

mean size, standard deviation and skewness against each other in turn they arrived at the three graphs shown in Figure 17.3. The degree of kurtosis associated with different parts of each graph was shown by shading. It was then noticed that the three graphs could be arranged as shown so that they were related by means of a helix, which was then constructed. Each of the three planar graphs then became a projection of the helix. The sinusoidal plots were the helix seen from above and from one side, and the circular plot was the projection of the helix seen from one end. Regular variations in kurtosis along the helix are faithfully reproduced as it is projected on each

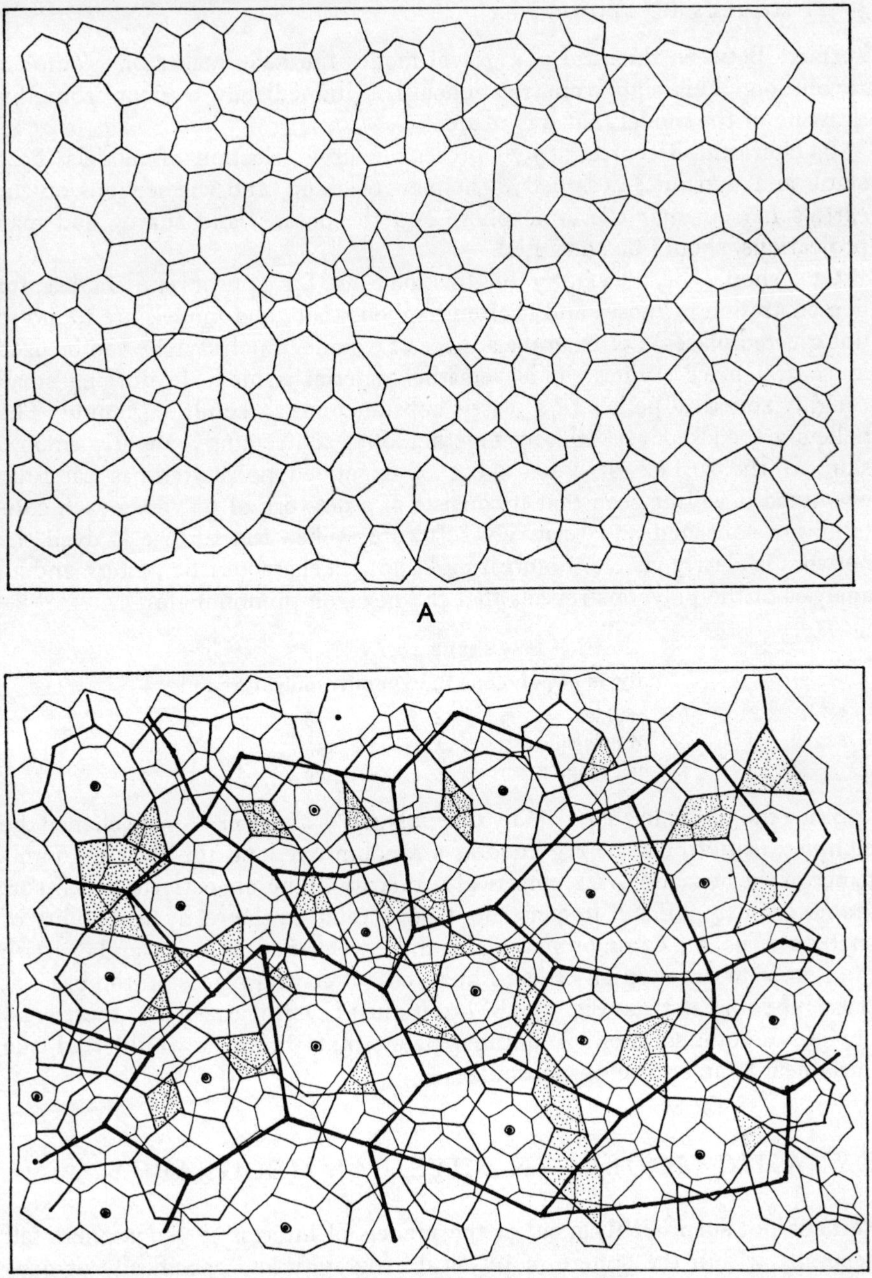

A

B

**17.4** A Random network of polygons generated within expanded polystyrene.
B A sort of central place hierarchy generated from this random network. Stippled areas indicate parts of the network that cannot be incorporated within the hierarchy without considerable distortion.

surface. Between the sand and gravel modes the helix makes one complete revolution. This significant relationship is immediately evident from this drawing of the model that was made.

Balchin and Richards (1952) present a large selection of models, both static and dynamic, designed to help in teaching, and the sections on the earth's movements; the atmosphere and the oceans, and survey and map projections, should be consulted.

One should be constantly on the look-out for appropriate models for demonstration purposes and it often happens that good models are found in unexpected places. For example a piece of expanded polystyrene can be used in central place studies. This versatile material is made by forcing small roughly spherical pellets of a plastic substance into a confining mould. The pellets are packed very closely together but each retains a slightly smooth skin. If the surface of a flat piece of expanded polystyrene is carefully examined it will be seen that it consists of a network of polygons each containing a squashed and relatively soft core. When the surface is dyed the network is clearly visible. Figure 17.4A shows a characteristic pattern and an analysis of the polygons reveals that the hexagon predominates.

TABLE 17.1

**Analysis of polygons in expanded polystyrene**

| No. of sides | 3 | 4 | 5 | 6 | 7 | 8 |
|---|---|---|---|---|---|---|
| No. of polygons | 7 | 12 | 33 | 153 | 15 | 1 |
| % total, (approx.) | 3 | 5 | 15 | 69 | 7 | 1 |

Since the ultimate shape of any one polygon is a matter of pure chance the complete pattern can be regarded as a random one, and the clear predominance of the hexagon as an 'empirical regularity'. Random patterns of this sort are notoriously difficult to construct from scratch so the ready availability of such patterns in a common material is to be welcomed. It is a useful exercise to attempt to draw up an ordered hierarchy of central places on the basis of these random patterns (Fig. 17.4B) but it must be emphasized that there is no obvious unique solution. A number of attempts however would lead one fairly near to an optimum solution.

## DYNAMIC MODELS IN PHYSICAL GEOGRAPHY

Perhaps not surprisingly most of the models of interest to geographers fall generally within the sphere of physical geography and specifically of geomorphology. There are good historical reasons for this emphasis. It is natural to try to create miniature replicas of parts of the earth's surface the more conveniently to study structure and process, and the growth of large scale engineering projects has stimulated work in this field since the latter half of the last century. Many models have been made by civil engineers and others to solve unique problems of great practical importance; others have been designed to explore more general or universal relationships and each to a degree has benefited from the other. Models have been used at all levels of inquiry – from the behaviour of individual grains of sand to the gross features of planetary wind circulation – to illustrate or test some hypothesis, and their success is ultimately judged by the extent to which they confirm observations about the real world.

## Models of physical processes in geomorphology and structural geology

The effect of temperature changes on a homogeneous material such as clay makes a simple and instructive experiment. A shallow metal tray is filled to an even depth with a thin layer of wet clay or a thin mixture of kaolin and water. The tray is then placed in a gentle source of heat. The pattern of shrinkage cracks which develop can be photographed at various stages and the characteristics of the pattern evaluated. An interesting variation might involve freezing the material in an ice-box and then allowing it to regain room temperature. Other variants could include using non-homogeneous materials in a controlled fashion.

Van Burkalow (1945) evolved an exhaustive series of simple and repeatable experiments to evaluate the factors determining subaerial angles of repose of loose material which are of great interest to geomorphologists. With fragments less than one inch in diameter the method used to obtain a true angle of repose was to place two straight boards edge to edge and to drop a conical pile of material over the junction. When one board was moved the material left on the other slumped until the angle of repose was obtained. To produce slopes that were convex or concave in plan the boards were cut to fit together along semicircles of appropriate dimensions. For fragments between one and three inches in diameter the angle of repose was found by dropping them into a relatively tall glass sided box with a movable interior partition which allowed the width of the box to be maintained at six times the diameter of the fragments being tested. When the box was filled one side was removed and the fragments assumed a true angle of rest. In either case the angle of repose was measured with a clinometer. For uniformity artificial materials were used ranging from chilled lead shot to three inch diameter wooden blocks of various shapes. The series of experiments was designed to demonstrate general relationships rather than to solve specific geomorphological problems and in fact succeeded in formulating certain precise conclusions concerning the effect of different types of material and conditions on the angle of repose of natural slopes. The geomorphologist is frequently heavily dependent upon being able to interpret correctly the significance of transported material.

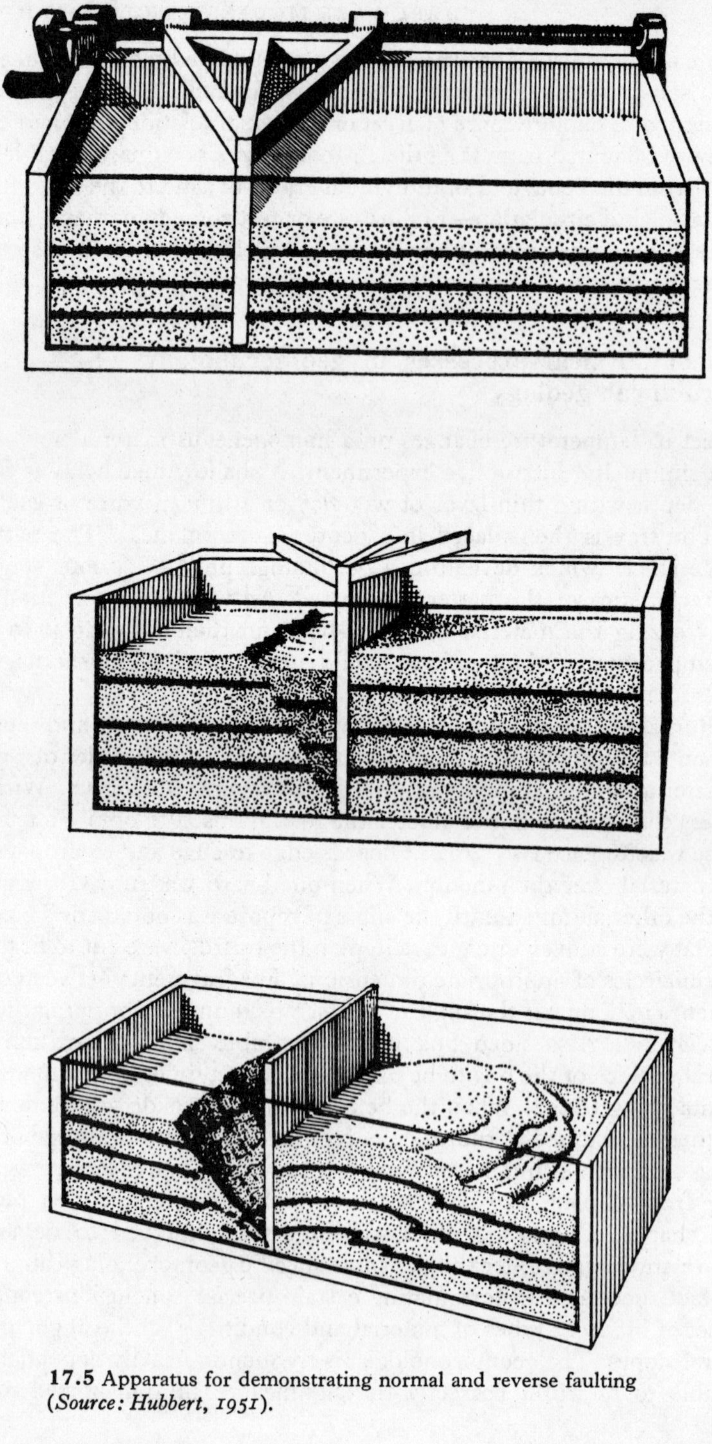

**17.5** Apparatus for demonstrating normal and reverse faulting (*Source: Hubbert, 1951*).

Sands, gravels and pebbles, for example, contain in their shapes often the only clues to the history of their movement and it is consequently extremely important to be able to attribute certain shape characteristics to different modes of transport. While experimental work in the field is obviously indispensible the problem of sorting out complex variables is likely to remain a stumbling block and under these circumstances controlled laboratory experiments are suggested. Daubrée (1879) was the first of many to undertake experimental work in this field. Kuenen (1956 and 1960) has reported a series of investigations into abrasion by current and eolian action under carefully controlled conditions. The equipment used is too substantial to be easily duplicated outside a properly equipped laboratory, but his results have been valuable in clarifying a number of relationships.

While the behaviour of loose or poorly consolidated material can often be studied in the laboratory using the materials themselves it is not possible to reproduce accurately the gross behaviour of other materials such as rock or ice unless we use some other material to represent them, unless in fact we use an analogue. In terms of structural geology there are a number of physical analogues that appear satisfactory. Hubbert (1951) developed a simple model (Fig. 17.5) to demonstrate faulting. The dark layers in the model are made from any suitable contrast material with the same properties as the dry sand with which the box is filled. The middle drawing shows the development of a normal fault following the forcible shift of the dividing partition. Further movement of the divider produced reverse faulting (lower drawing) whose characteristics were found to accord well with field observation. Billings (1946) reports an experiment by Cloos in which soft mud was used successfully and it should be instructive to repeat these experiments with a variety of materials of differing competence and composition. On a much larger scale efforts have been made to simulate the effects of mountain-building, linking the observations that mountain ranges have roots penetrating into the lower part of the earth's crust with the hypothesis that sub-crustal convection currents are responsible for mountain-building. Griggs (1939) used a layer of oil loaded with sand lying upon a tank filled with water glass to simulate the surface and sub-crustal layers. Two contra-rotating drums in the water glass produced the equivalent to convection currents and deformed the oil and sand layer so that it had roots and protruberances. Kenn and Wooldridge (1965) constructed a basically similar apparatus (Fig. 17.6) in which a layer of engine lubricating oil floated upon a quantity of glycerol. Two Perspex drums in the glycerol were made to rotate in opposite directions by means of a crown-geared reversible electric motor. The lower sketch shows the patterns produced when the drums were run steadily at 81 r.p.m. As Wooldridge observed, when the drums were rotating inwards the pattern of a broad gentle uplift and extended roots recalls the classic features of epirogenetic movement. When the drums, and by analogy the sub-crustal convection

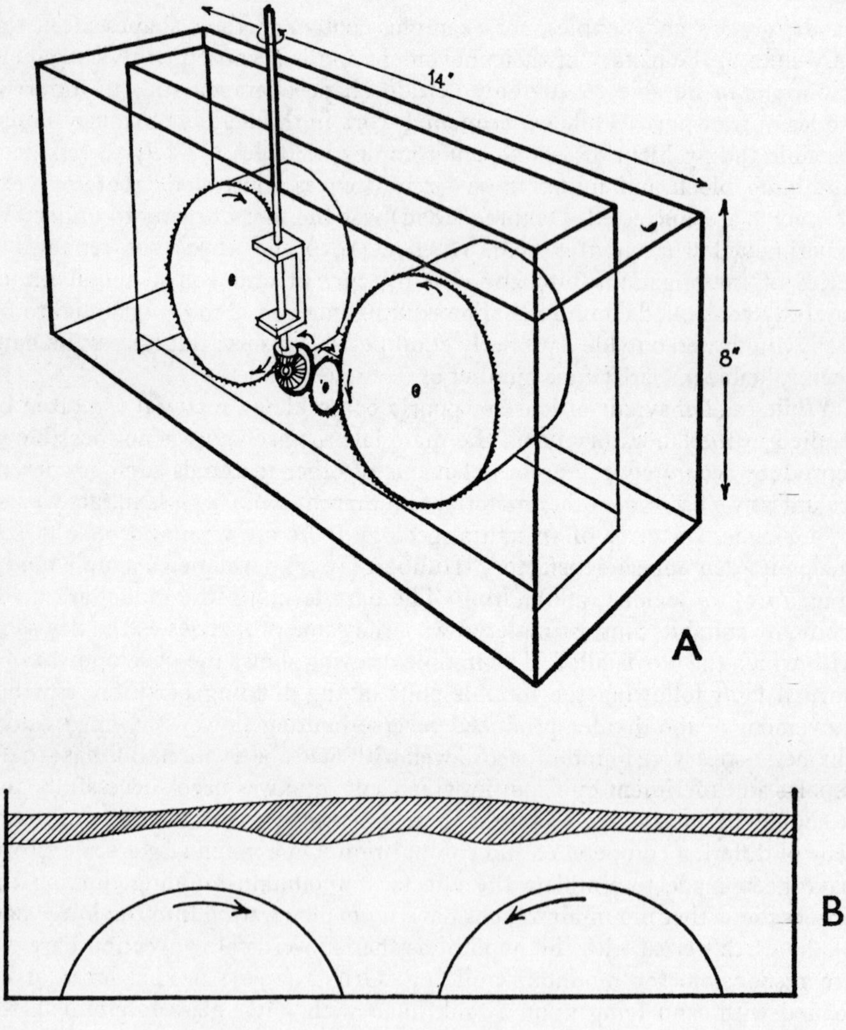


**17.6** A Construction details of the mountain-building simulator.

B and C The deformation of the oil film (shaded) when the drums are rotated (*Source: Kenn, 1965*).

currents, are moving outwards the model may be thought of as demonstrating a process rather similar to continental drift.

Reiner (1959) shows how some problems of deformation and fluid motion can be represented by appropriate mechanical analogues and this has obvious bearing on the process of deformation of rocks.

The problem of simulating the movement of ice in glaciers has been dealt with by Lewis and Miller (1955). A material was needed that would exhibit the same plastic behaviour on a small scale as that of the large mass of ice in the real world that it was representing. It had necessarily to be weaker than real ice, but still capable of yielding to stress and deformation by a degree of faulting and fracturing. Eventually Lewis discovered an ideal material in the form of a mixture of two parts of kaolin (china clay) and one part of water. A relief model of a valley was made in Plaster of Paris, about 150 cms. long and 30 cms. wide, and the kaolin mixture was placed in the upper half of the valley. Plates 17.1–3 show a model roughly similar to that used by Lewis and Miller. The patterns of faults and crevasses in the model can be seen clearly. This type of model is excellent for qualitative purposes. Ideally one should arrange for a steady supply of the kaolin mixture in the cirque area, but this rather spoils the development of the bergschrund. It is usually sufficient to add quantities of kaolin as the névé becomes impoverished. If the kaolin shows a reluctance to move the valley can be greased and it may be necessary to tilt the model to as much as 20° from the horizontal. During construction a glass panel can be inserted to give a view into the cirque area. The 'glacier' can be provided with material (in the form of powdered coal, cork, etc.) to transport. Matchsticks can be placed in the surface in order to calculate speed of movement (Plate 17.2) and the general behaviour of the model will be found to accord well with field observations. Lewis and Miller deal at length with the interpretation of features developed on and in their model glacier and also discuss the possibilities of making quantitative operational models.

Chemical or electrochemical analogues have not been widely used but Muskat (1949) and Botset (1946) whose work is summarized in Karplus (1958) developed a method to assist oil engineers increase oil recovery from long-worked fields. In older fields water is sometimes injected into centrally placed wells so that the water front moves outwards and increases the oil pressure in marginal wells. It is obviously important to know the likely progress of the water front and to know the most advantageous point at which to inject water into the field. The analogue model consists of a conducting layer cut to the shape of the oil field and made of a 1 per cent agar gelatin solution with a small quantity of zinc ammonium chloride. The wells are made of plastic tubes inserted into the gelatin sheet. Injection wells are filled with a mixture of agar and a 0·1 molal deep blue copper ammonium chloride solution and the production wells contain the same material as the model field.

The injection wells are made electrically positive and supplied with potentials up to 1,000 volts and the producing wells are all made negative. When the power supply is connected a potential gradient exists between the injection and producing wells. With a flow of current blue copper ammonium ions stain the field as the 'water front' moves outwards. It is possible that the ion diffusion method may have some application in diffusion studies in human geography but the implications remain to be worked out.

## Hydrological models

Models of streams, rivers and estuaries have been used for a long time, originally for demonstration purposes, but increasingly today to solve a wide range of civil engineering problems. Many of us have cause to be grateful to the late W. V. Lewis for our introduction to this peculiarly pleasurable aspect of experimental geography. Streams and rivers in miniature can be observed or created on the beach, in the gutter or garden on a rainy day. Spoil heaps of all kinds, if reasonably homogeneous can be seen to have miniature drainage patterns etched upon their surfaces. A simple river can be created upon a bed of sand in a waterproof trough and the processes of erosion and deposition can be followed. A river system can be originated on a sloping wedge of sand by gently watering the surface with an overhead sprinkler. Certain streams can be given an advantage by lowering their base level, shortening their courses or increasing their discharge. A layered basis of sand and marl can be used instead of pure sand in order to introduce an element of lithological complexity. Simple models of this sort are ideally suited to introductory courses in geomorphology in schools; they make few demands in terms of apparatus and they encourage the practice of accurate observation and recording.

When model rivers are used to solve specific problems the experimental techniques and the type of constructs have necessarily to be more sophisticated. Just how much more refined relates very much to the nature of the problem. Lewis (1944) used a stream trough four metres long and 50 cms. wide to experiment on the influence of load on gradient in the hope of throwing light on certain features associated with flood plains, in particular river terraces formed without change of sea-level. The behaviour of knickpoints in non-cohesive material has been experimentally studied by Brush and Wolman (1960) using a flume (trough) 52 feet long and 4 feet wide. Complex stratification and cross-stratification in streams and deltas (Nevin and Trainer, 1927; McKee, 1957) and meandering (Friedkin, 1945) have been studied with the use of physical models.

Civil engineering today is slowly moving away from the often inspired empiricism of its Victorian past. Vast capital outlays cannot be hazarded by inadequate information, not only concerning the structures themselves but also their effect on the site. New harbours, barrages, flood control and irrigation schemes, if only by virtue of their great size are likely to create local environmental changes. A very great deal of experimental work has been done in relation to these sorts of problems by the construction of scale models in which alternative proposals can be evaluated and their respective merits accurately assessed. One difficulty however arises from the fact that in small scale models the change of scale affects the relationships between certain properties of the model and the real world in different ways. If, for example, we make a model of the Severn Estuary at a scale of 1:10,000 the geometrical and topographical relationships can be preserved fairly easily. When we add the water though we find that an actual depth of water of say about 20 feet is represented in our model by a layer of water less than 1/40th inch thick. Not only will surface tension assume intractable proportions but it would be vir-

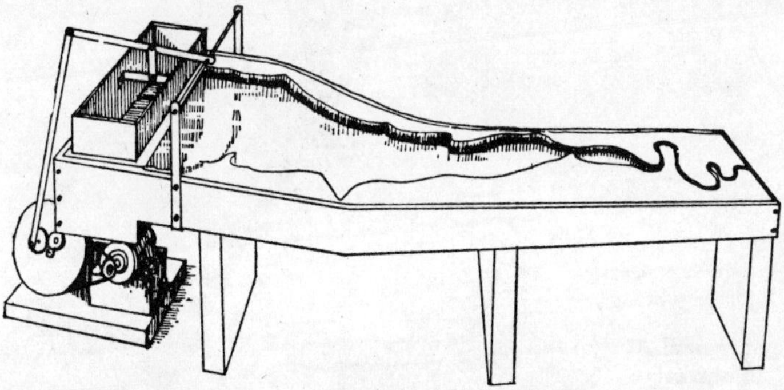

**17.7** Gibson's small model of the Severn Estuary (*Source: Allen, 1947*).

tually impossible accurately to simulate tidal range and currents. Equally the sand along the bed of the river could not be reproduced to scale and even were it possible a suitable material would mostly float on the surface of the water. A coarser material in the model would represent large boulders in the real world.

In order to overcome scale difficulties it has been found possible to distort the scale of one attribute in the model in order to preserve that of another. Models of rivers and estuaries for example often have the vertical scale greatly exaggerated in order to reproduce more or less correctly the effects of turbulence in the water. Figure 17.7 shows a sketch of a model of the Severn estuary made by Professor Gibson (discussed in Allen, 1947). It has a horizontal scale of 1:40,000 and a vertical scale of 1:366. The tide in this model is produced by calculated periodic displacement of water in the reservoir by means of a float. More complicated sequences of tides are usually generated by complex linked systems of cams and plungers.

For detailed experimental work empirical distortion of one attribute to preserve another is inadequate. Most models of this sort have been proportioned using the principles of dimensional analysis. Langhaar (1951) and Duncan (1953) are the principal authorities on this subject. Normally the first requirement of a model is that it shall be geometrically similar, at least in the two principal dimensions, to the prototype; the distance between any two points in the model must bear a constant ratio to the distance between the

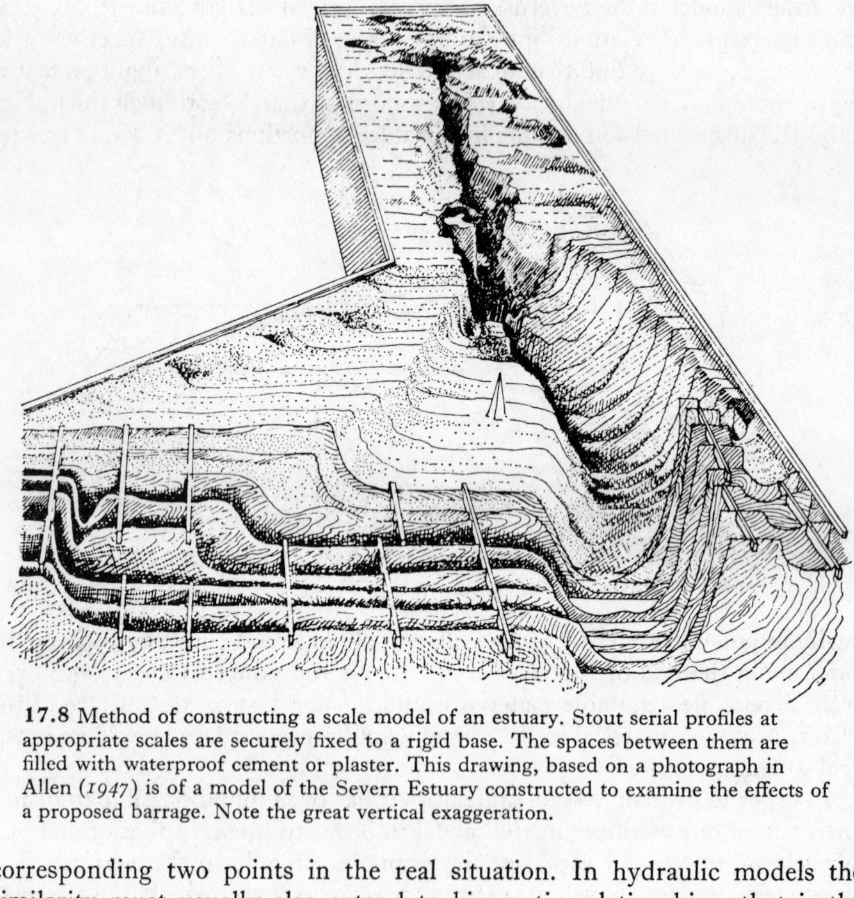

**17.8** Method of constructing a scale model of an estuary. Stout serial profiles at appropriate scales are securely fixed to a rigid base. The spaces between them are filled with waterproof cement or plaster. This drawing, based on a photograph in Allen (*1947*) is of a model of the Severn Estuary constructed to examine the effects of a proposed barrage. Note the great vertical exaggeration.

corresponding two points in the real situation. In hydraulic models the similarity must usually also extend to kinematic relationships, that is the velocities and paths and patterns of movement in the water in the model should have a one-to-one correspondence with what happens in the prototype. The details of dimensional analysis and its practical implications are too complicated to be dealt with here but Pankhurst (1964) gives an excellent introduction to the subject as a whole and Allen (1947) is an outstandingly interesting and valuable source for all aspects of hydraulic models. In essence though dimensional analysis and the application of distortions calculated with the use of dimensionless numbers permits models to be made so as to preserve correctly a number of properties: to the extent that the models are made to be good analogues one can be certain that what happens in the model would happen under similar circumstances in the real world. Nevertheless it is a wise precaution constantly to test one's model against the real world because each situation contains elements of uniqueness. The most usual

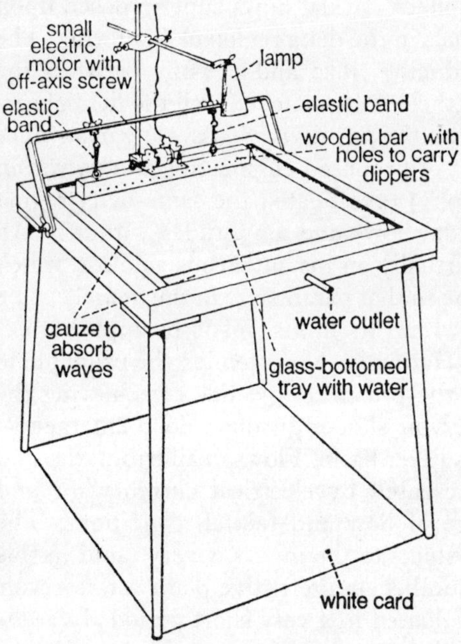


**17.9** A ripple tank. The ripples are produced by vibrations of the wooden bar which in turn are induced by mounting a small off-axis screw or other weight on the shaft of the motor. The bar dipper will produce plane waves. Circular waves are made by inserting a short rod in one of the holes on the bar dipper (*Source: 'How and Why?' BBC, 1965*).

method of construction is suggested by Figure 17.8 which shows a basis of serial cross-profiles (with appropriate vertical exaggeration). The intervening areas are filled with cement or plaster of Paris (if the latter it must of course be waterproofed). The cross profiles define in this case the bedrock surface. If the model is to contain loose material over its bed than a second set of templates is made so that loose material put on to the bedrock can be accurately arranged in terms of its thickness.

Waves and coastal phenomena associated with wave action can be studied at different levels of sophistication. Simple basic experiments can be done with an easily made ripple tank, much used in physics teaching (Fig. 17.9) but in addition some sort of wave tank or flume is needed if, for example, constructive and destructive action of waves is being studied. Here, unfortunately, there is no substitute for really rugged construction and accurately designed wave generating equipment. One tends to forget the great weight of relatively small quantities of water. Stanton *et al.* (1932) and more recently Saville (1950), Johnson and Rice (1952) and McKee (1960) have accounts of work with model wave tanks which should be consulted as much from the point of view of operation and experimental design as of conclusions reached.

Until recently most hydraulic models were scale models, at least in important respects. In the past decade, however, certain electrical analogue models have been developed which have the advantage of greater speed of construction and operation and also of substantially lower cost. Glover *et al.* (1953) used such an analogue to study the effect of tidal flows superimposed upon the steady flows within a series of canals in the delta region of California. The effect of changes induced by introducing links and cut-offs between the canals was simulated electrically and the results matched well the field observations. This model technique opens up the possibility of assessing in advance of actual construction work the effect of proposed engineering works, within the system. Einstein and Harder (1961) investigated the same delta region mainly from the point of view of tidal flows and amplitudes within it. An electrical analogue of course gives virtually an instantaneous solution which can be displayed on a cathode ray tube so that parameters in the model can be adjusted rapidly and the analogue, within its limits, made to duplicate exactly the behaviour of the prototype. Harder (1963) extended the principle to simulate flood control systems with the practical objective of achieving the best phasing of the operation of reservoir sluices during a flood emergency. Harder's model related to the Kansas River basin. Flows in all tributaries and along the main channels were represented by electrical current flow and adjustable to represent a wide range of flow and rainfall conditions. The speed of response of the electrical systems to the inputs is very rapid in this case, only 0·03 seconds, so a large number of alternative plans for reservoir operation in an emergency can be evaluated in a very short period. In addition to its short term benefits the long range advantages are the comparative ease with which more permanent solutions to perennial flooding problems can be evaluated.

An unusual analogue based on heat conductance was developed by Appleby (1956). In a drainage network, whether it consists of rivers or urban sewage systems, the value of the run-off function is most difficult to calculate since it is a function both of variable input – variable in terms of time and quantity – and losses through evaporation and absorption of various kinds. Appleby's analogue is based on the fact that stream network flow and heat flow behave in the same manner and can be represented by the same differential equations. His apparatus allows a series of trial values for the run-off function to be rapidly tested against the form of an actual hydrograph. When similarity is achieved then the actual values of the run-off function can be recorded from the analogue.

## Models in meteorology, climate and oceanography

Electrical analogues are beginning to be employed in meteorological research. Wallington (1961) has studied the behaviour of lee waves and Tyldesley (1965) proposed an electrical analogue for solving atmospheric diffusion equations. Non-electrical analogues have been used in evaporation and condensation studies (Turner, 1965) and tornado simulation (Turner and Lilly, 1963).

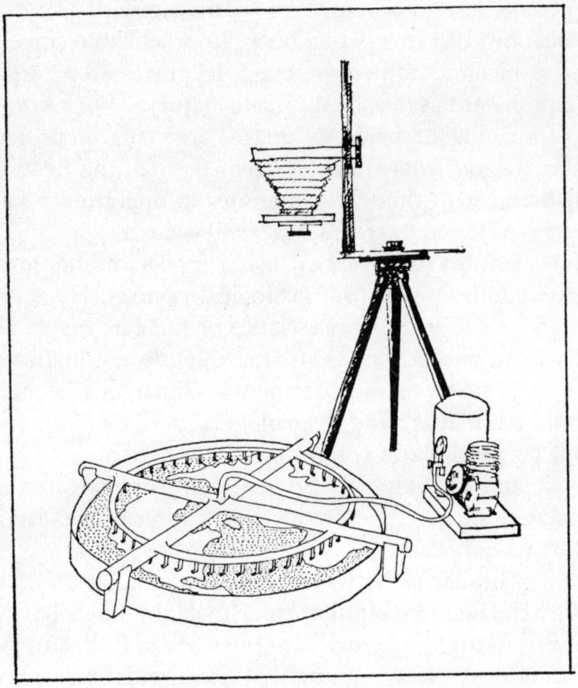

**17.10** Lasareff's (1929) dishpan model for the oceanic circulation.

The problem of explaining gross planetary circulations of both wind and waters has led to an interest in simulation by model building. Lasareff (1929), then the Director of the Moscow Geophysical Institute, constructed a series of models to show the relationship between oceanic currents and trade winds which had been suggested much earlier but never experimentally verified. His apparatus (Fig. 17.10) consisted of a circular dish (later a rectangular dish was used) in which the continental masses were represented by raised plaster blocks. Wind was represented by peripheral air currents directed through adjustable nozzles over the surface of the dish which was filled with water. Powdered aluminium sprinkled on the surface revealed the pattern of water movement. The currents were recorded by time lapse methods using a vertically mounted camera. The map of the North Atlantic (Fig. 17.11A) shows not only gross features of the resulting circulation but also a number of significant secondary circulations and fits well with what is known of the real situation. It was found that small island arcs and archipelagos are of great significance in affecting circulation patterns in their immediate vicinity and occasionally over unexpectedly wide areas. Lasareff dealt in detail with the problem of choosing a suitable projection upon which to model a hemisphere. For true kinematic similarity one should use a gnomonic projection, but this is obviously impracticable since a hemisphere cannot be shown on a gnomonic projection. In practice a projection similar to the zenithal equidistant is reasonably satisfactory. (Von Arx, 1962, p. 303 gives details of spacing of parallels in the best sort of projection for this purpose.) Thermal gradients were established by fitting heating coils to the edge of the dish and with the wind generator in operation a satisfactory disposition of warm and cold currents was achieved.

Lasareff then went on to show that using models it should be possible to recreate climatic conditions of past geological periods. He assumed that the gross thermal and rotational characteristics of the earth have remained substantially unchanged and argued that consequently the main variable would have been the disposition of the continents. Given this he made models to illustrate oceanic circulation in past geological periods (Fig. 17.11B) and by extension gross palaeoclimatic conditions.

Von Arx (1957 and 1962) has developed more sophisticated models taking into account such additional factors as the rotation of the earth and certain effects caused by its curvature.

The underlying similarity between planetary circulation of water and air is emphasized by the fact that similar experimental models have been evolved to study them. Starr (1956) in a review of the work of Fultz (1956) has pointed to the defects in classical views of atmospheric circulation – specifically their failure adequately to explain the source of energy for the mid-latitude westerlies. A mathematical description of the complex dynamics of the atmosphere is extremely difficult and its interpretation would be beyond the capacity of many people interested in it. There is a strong suggestion that a combination of thermal gradients and rotational effects will explain the westerlies and Fultz has designed a model which experimentally supports this view. The apparatus consists of a flat-bottomed dishpan representing a hemisphere, filled to a depth of one inch with water. Peripheral heating and polar cooling produces a simple overturning flow pattern. When the pan is rotated from two to six revolutions a minute the simple initial circulation breaks up dramatically into eddies and vortices like the familiar sequence of cyclones and anticyclones of the mid-latitudes. From equator and pole eddies move towards the middle latitudes where westerly currents develop. The

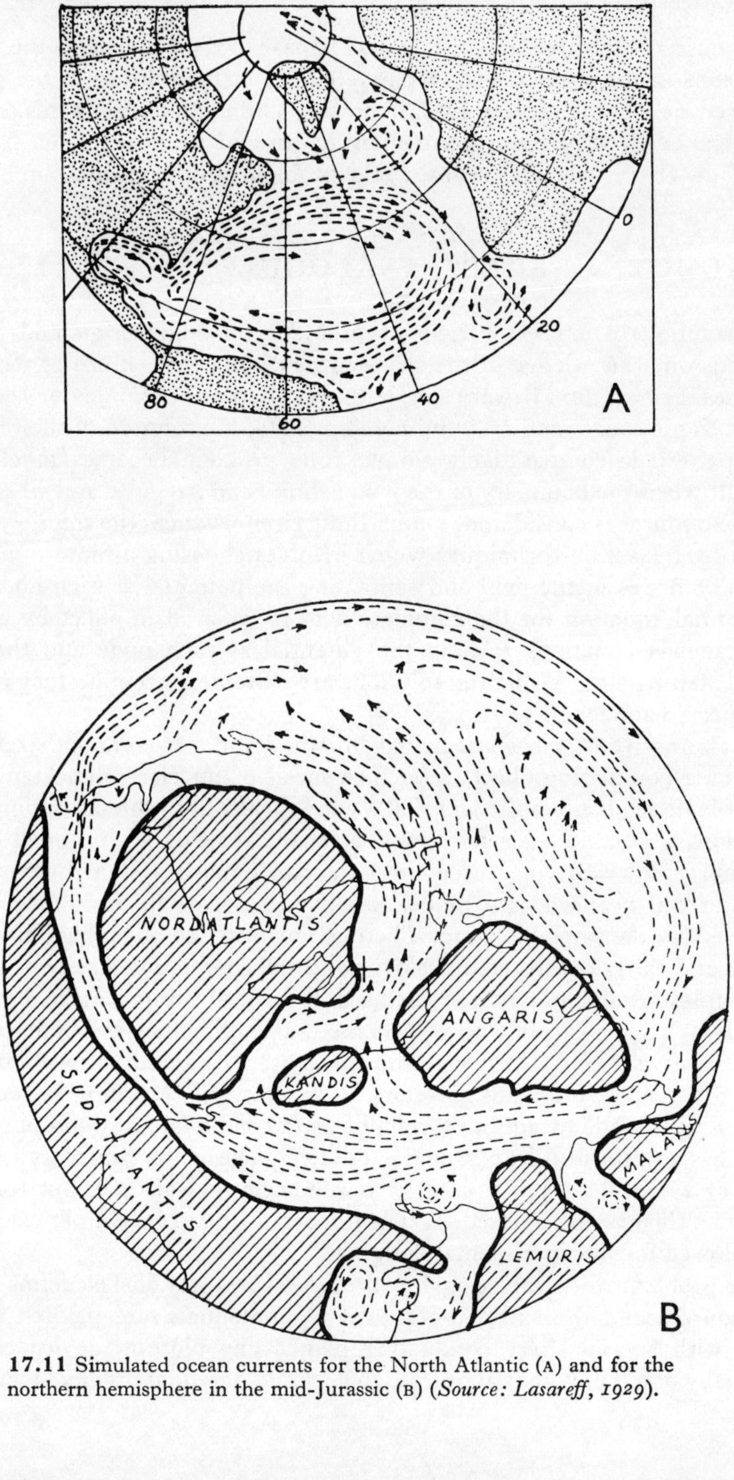


**17.11** Simulated ocean currents for the North Atlantic (A) and for the northern hemisphere in the mid-Jurassic (B) (*Source: Lasareff, 1929*).

water currents in three dimensions also appear to reproduce well the vertical variations in the atmosphere. Separate eddies at the bottom of the pan are replaced nearer the surface by fast flowing undulating jet streams. A more technical account of the experimental design and results is given by Riehl and Fultz (1957) who also deal with the problem of dimensional similarity.

## CONDUCTING SHEET ANALOGUES

The solution to a large variety of problems in engineering and physics depends on a knowledge of the shape of fields and the pattern of magnetic, electrical or heat flux. In very simple cases, where for example one is dealing with a single source and an infinitely large field, the solution of the resulting field pattern is comparatively simple. The problem becomes much more difficult when the boundary of the field is finite and irregular and when more than one source is considered. Under these circumstances the solution can be found by relaxation techniques which involves choosing a finite number of points or nodes in the field and estimating the potential at each node. The differential equation for the complete field is replaced in effect by a set of simultaneous equations relating the potential at each node and these are solved step by step. While the solutions are extremely accurate they are also complicated and lengthy.

An alternative is to use an analogue method and this commonly takes the form of a conducting sheet, resistance mesh or an electrolytic tank. The principle depends upon the fact that a wide variety of problems in fluid flow, heat flow, gravitation, electrical and mechanical fields are founded on the equation: Flow=driving force×acceptance. Thus a system obeying this equation can be used as a model to solve problems in any other system governed by the same equation. There is obviously an advantage in using a simple and easily constructed analogue and the conducting sheet analogue is the simplest available. Kirchoff used a thin sheet of copper for field plotting as early as 1845 (Karplus, 1958) but there are difficulties in obtaining uniform resistivity in thin sheets. The most used conducting sheet today is Teledeltos paper, originally developed for electrical recording purposes in 1948. This is made by adding carbon black (a conductor) to wood pulp. The paper has a uniform thickness and is relatively cheap; its resistivity is of the order of 2,000 ohms per square. It is somewhat anisotropic, but for most problems this can be ignored. If it is important it can be simply calculated and allowed for in experimental design.

The problem to be solved is drawn out on the sheet of Teledeltos paper. The sources and the sinks, the known equipotentials, are painted on the paper with special silver conducting paint. The plotting device consists essentially of a d.c. potential source, an accurate potential divider, a galvanometer and a probe. The output terminals of the plotter are connected, one to the source and one to the sink. When the plotter is switched on the electrical field is established over the paper. The potential divider has a scale with 100 divisions so each division is equivalent to 0·01 per cent of the potential across the terminals. The actual value in volts of this potential is unimportant for most purposes. The potential divider is set at a desired reading, say 50, and the probe moved over the surface of the paper until there is zero deflection on the galvanometer. At this probe position the potential is exactly 50 per cent of the input potential at the source. Several of these points with the same value are located and joined to form the 50 per cent equipotential line. When a sufficient number of equipotential lines have been drawn it is a relatively simple problem to sketch in the flux lines since they flow always at right angles to the equipotentials. In drawing the flux lines it is normal to space them so that they form sets of curvilinear squares each with the characteristic that mean width is equal to mean length. It is often a help to draw

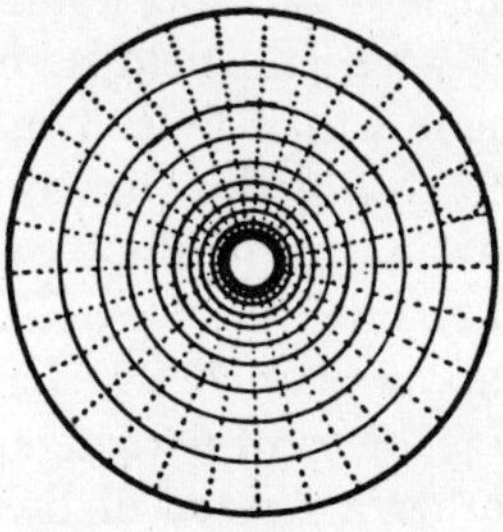

A

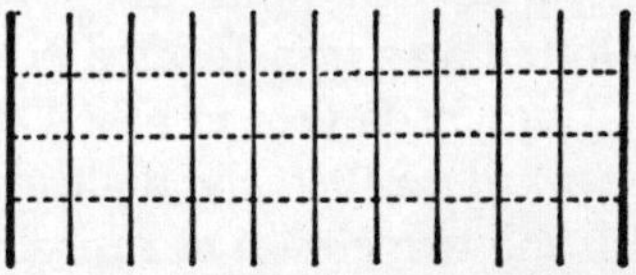

B

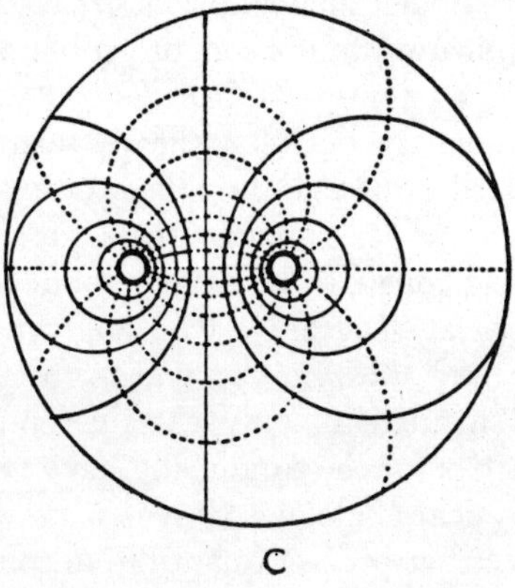

C

**17.12** Characteristic field plots.

A Central source and finite circular boundary.

B Two parallel straight electrodes.

C Two sources in an infinite field.

Equipotential lines are shown by continuous lines and streamlines by dotted lines orthoganal (at right angles) to the equipotential lines. The dotted circle in A indicates one way of drawing curvilinear squares based on the plotted equipotentials.

circles whose diameter is equal to the distance between adjacent equipotential lines as a basis for inscribing the flux lines (see Fig. 17.12A). Characteristic simple plots are shown in Figures 17.12A and B. In each case boundary conditions are simple and the boundaries are finite. It should be noted that the spacing of the equipotential lines in the two examples is different. The lines about a point source are bunched closer together near the source but between two parallel plates the equipotentials are equidistant. From a point source $X$ located in an infinitely large two-dimensional plane field the potential at a point $Y$ at distance $r$ from $X$ is proportional to log $\frac{1}{r}$ (in a three-dimensional field it is proportional to $\frac{1}{r}$). With a finite boundary, however one needs to apply Laplace's equation in two dimensions if the equipotentials are to be calculated. The field plotter, however, gives the graphic solution in a fraction of the time needed to calculate it.

It is sometimes necessary to plot fields in which the boundary or sink is at infinity. One solution is to place all the sources as near as possible to the centre of a large piece of resistance paper – around the edge of which the boundary is painted – in the hope that it is sufficiently remote not to affect the plot in the central area. For more precise work where the field is known to be symmetrical about a point the plot can be made with a finite boundary and then replotted on a transformed grid. Alternatively a method has recently been suggested by Olsen (1963) based on an adaptation of the double layer electrolytic tank described by Boothroyd, Cherry and Makar (1949). This involves cutting two circles of resistance paper about 20 inches in diameter and placing them back to back with an intervening piece of cartridge paper for insulation. The edges in contact with each other are freed of the insulated backing and finally clamped together with insulated clips. The source or sources are painted on one side and the centre of the opposite side is made a sink. The upper sheet is then able to behave as if it were part of an infinite domain. Figure 17.12C shows a characteristic plot made under these conditions.

While it is beyond question that this particular analogue is of great value in physical and mechanical sciences, its application to the sort of problems geographers concern themselves with may need some justification.

In physical terms there is an exact analogy between the pattern of an electrical field and the pattern of heat distribution and fluid flow for example, because the different physical phenomena are described by the same mathematical formalism and quantitative conclusions can be drawn about one by studying the other. In the field of physical geography true analogues are certainly feasible. But in human geography, despite the search for formal order, there is as yet no mathematical formalism, only certain 'empirical regularities' (Isard, 1956). These regularities for the time being have to serve as our models if we are to use these physical analogues, but it is as well to recognize their limitations, and to realize that precise quantitative solutions are not generally possible. On the other hand if we accept the limitations we can apply field plotting techniques in a qualitative sense to a large range of problems with considerable benefit.

We may consider possibilities first on uniform, and then on non-uniform fields. The first published account of the use of field-plotting techniques in geography is that by J. R. McKay (1965) who used an analogue field plotter to simulate the macroscopic pattern of glacier flow for large ice sheets of the Wiconsin glaciation. The experimental design is illustrated in Figure 17.13. The equipotential lines are not shown but the arrows indicate flux lines. The

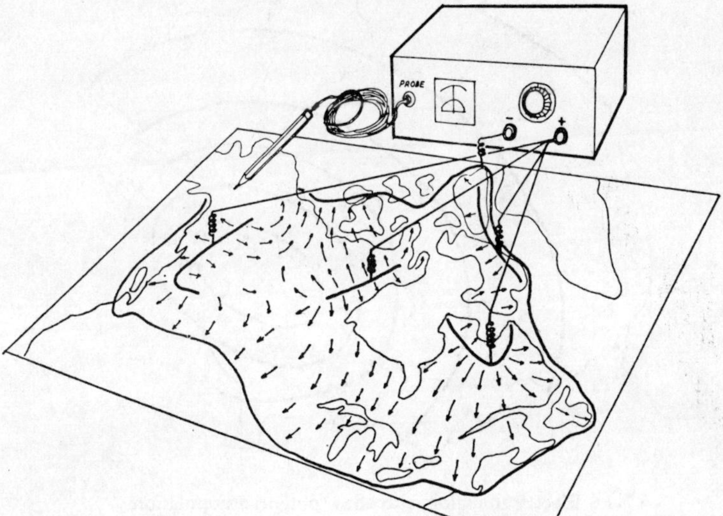


**17.13** Sketch of apparatus used by MacKay (*1965*). Direction of flux lines, and (by analogy) lines of ice flow, are indicated by arrows.

sources are painted electrodes in their correct position and the boundary of the ice sheet is the painted line at zero potential. The results of the experiment were sufficiently promising (in the sense that the pattern of flux lines fitted the observed direction of ice flow under certain simplified conditions) to suggest that it would be worth developing a more refined model, which would enable different conditions to be simulated.

There are grounds for believing that the concept of 'population potential' as developed by Stewart and Warntz (1958) represents a suitable field for analogue simulation. If we equate mass with electrical potential the resulting equipotential lines from a field plot should give results comparable with those arrived at by the different method used by Stewart and Warntz. Inevitably problems of scale limit the accuracy of the final result but the method suggested should be useful both in teaching and in giving first approximations of population potentials in areas for which data is sparse or in which a crude picture is acceptable. Plate 17.4 shows the apparatus used – though the arrangement of the sources relates to another experiment. In this case a 250-volt D.C. supply was used and was fed to the resistance sheet via a number of resistors whose value was calculated to represent roughly the populations of the largest cities in Canada and USA. Very large resistances

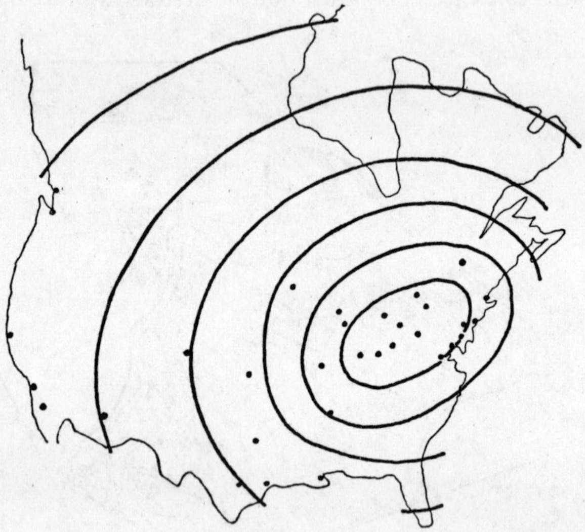

**17.14** Electrical analogue to show potential population in North America. This sketch has been included merely to illustrate the order of result that may be expected using this method and is not in any sense in this form capable of yielding a precise result. The dots represent the major cities and each dot is supplied with an electrical current proportional to the population to which it refers. The resulting equipotential lines are then plotted. With careful experimental design the voltages along each equipotential line can be converted into population potential figures.

were employed to give a range of current inputs from 1 milliamp for New York to 0·06 ma. for cities with a population of from 4–500,000. The current was supplied to the resistance paper by needles correctly located and the equipotential lines were plotted (Fig. 17.14). With this apparatus actual voltages can be allocated to each equipotential line and these can be converted into population potential figures. More accurate work along these lines is currently being undertaken and it does appear to hold some promise.

There is a possibility that a similar technique could be used to simulate the territories of central places, but there are operational difficulties in applying a large number of sources with only a peripheral sink. Work is now in progress on a system with small settlements acting as sources and central places as controlled sinks. Figure 17.15 shows field plots in which the experimental design has been made progressively more complicated and could be thought of as crudely simulating the movement of a frontier of settlement from the right of each diagram to the left. In Figure 17.15B, a line of conducting paint ($x$) has been placed in the direction of movement of the frontier and acts as a

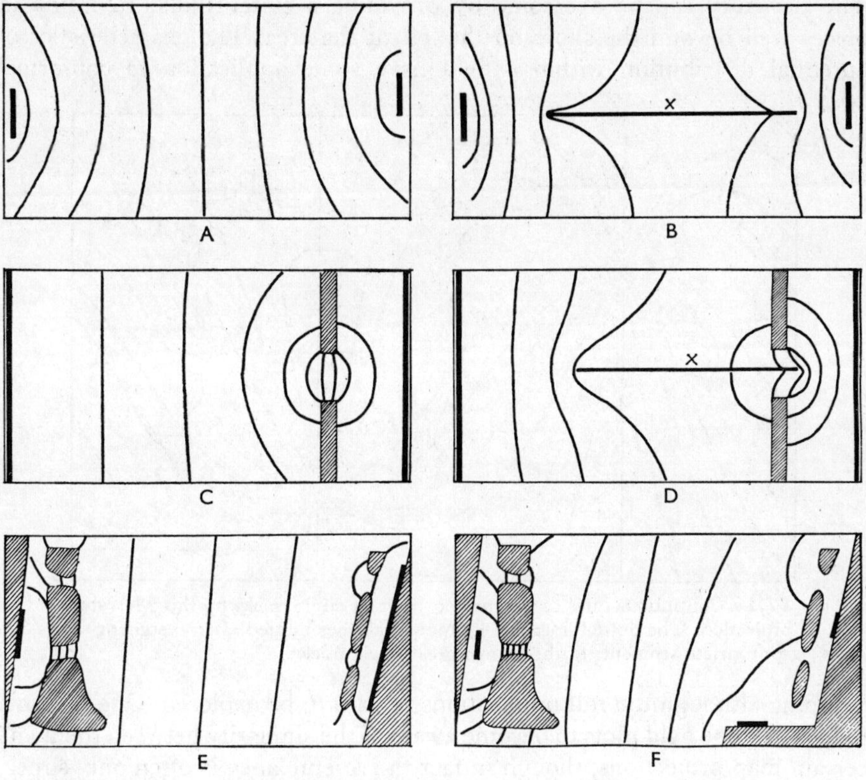


**17.15** A selection of equipotential line plots. The silver paint electrodes are indicated by thick lines. 'X' represents a line of silver paint applied to the surface but not connected to the power supply. The shaded areas show where the paper has been cut away.

line of greater ease of movement. Figure 17.15C introduces a barrier with one gap (the shaded area represents the barrier which is actually cut out of the paper). Figure 17.15D places an easy route inland for some distance from the gap. The last two figures introduce two barriers with several gaps of varying width and in the final one another source of potential colonizers is added on the seaboard. The basis of the analogy is that a frontier of settlement represents a balance between pressures behind pushing it forward and attractions and difficulties in front. If these were the only factors and if they could be quantified then a simple model could be constructed. Unfortunately, if equipotential lines are equated with frontier positions the experimental conditions are such that no matter how hard or easy it is for the current to move across any part of the paper the potential difference between the source on the right and the sink on the left is always constant; so there will always be an equal number of equipotential lines between source and sink and therefore in the model the frontier reaches the final point simultaneously from all points. The difficulty can be overcome by providing a suitably distorted sink in places well beyond the shore on the left of the area. The characteristics of potential distribution within a field have some application to conformal

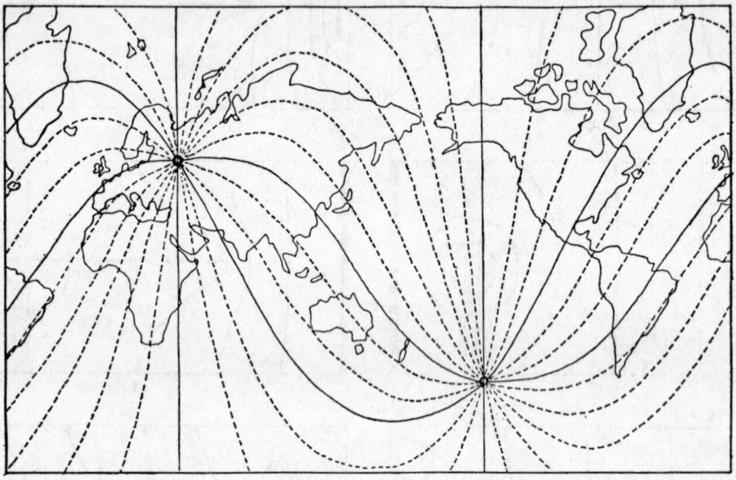

**17.16** Continuous lines represent known great circle tracks on this Mercator projection. The dotted lines are equipotential lines plotted after energizing appropriate segments of the known great circle tracks.

mapping although the full implications remain to be explored. One has only to look at some field plots to become aware of the similarity between them and certain map projections, though in fact the resemblance is often only superficial. Warntz (1965), in a typically stimulating paper, has drawn attention to the geometry of surfaces and paths on conformal map projections. In one map he plots two complete great circle courses from London on a Mercator projection (equatorial case) and shows how the great circle tracks are orthogonal (i.e. at right angles) to the iso-distance lines based on London when plotted on the same graticule. Figure 17.16 develops an idea suggested by this map and uses the field plotting analogue to plot a series of great circles based on Moscow on Mercator's projection. Any great circle can be defined on a globe using a piece of taut string. Two are determined in this fashion and transferred to the Mercator graticule marked on Teledeltos paper. In fact since the meridian upon which Moscow and its antipodean point stand are also great circles only one need be plotted from the globe. Each segment of the great circles so drawn on the paper is painted with silver paint (taking care that lines at junctions are not linked up with paint) and adjacent parts of the tracks are energized in turn. The equipotential lines are drawn in and represent other great circles emanating from Moscow. If checked against a globe they will be found to be correct.

Since the Teledeltos paper can easily be cut with scissors it is amusing and instructive to play with different shapes. One can illustrate certain basic features of rivers, for example, with pieces of resistance paper. If we accept as a working hypothesis that there is at any point on a river's course, a relationship between the volume, and the width and the height above base level we can symbolize or represent a river by a triangular wedge of paper. The scale for the width is conveniently many times greater than the scale of the

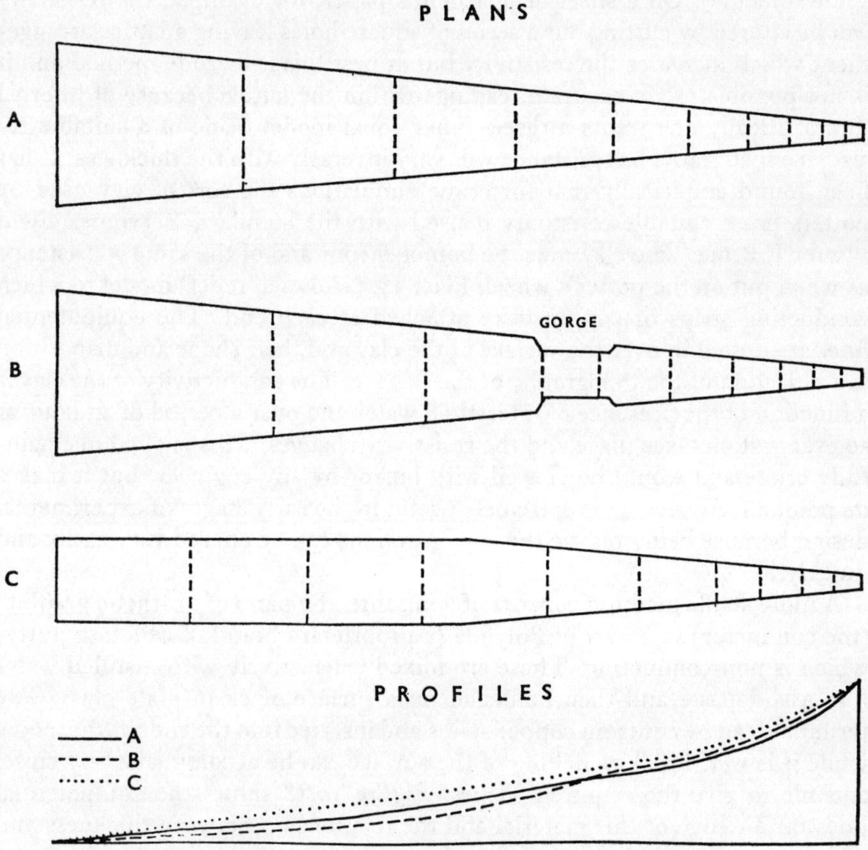


**17.17** Simulation of stream characteristics by means of Teledeltos paper strips.

length. If we apply a potential at the source of the river and arrange a sink at the point where the river either enters the sea or joins another river of the same order we can plot the equipotential lines. If we equate these with height we can then derive a profile which accords fairly well with observation (Fig. 17.17). If we introduce a restriction at some point in the course by narrowing the width of our paper river we find the profile has a characteristic irregularity at that point. It is useful as an exercise not only to vary the width of the model river but to introduce tributaries, vary the flow at different points, and even perhaps attempt to simulate some features of an actual river system.

While most of the discipline in which analogue field plotting has developed are interested primarily in uniform fields, geography more often than not deals with non-uniform surfaces. The irregularity may be topographic, socio-economic or statistical for example. There are certain advantages in being able to simulate non-uniform fields but practical problems present some difficulty. On a sheet of Teledeltos paper, for example, the resistivity can be altered by cutting out a series of square holes leaving a lattice arrangement which increases the resistivity but at best this is a crude method and it is not possible to get accurate readings within the lattice because of internal fluxes. Ideally one wants a three-dimensional model made of a suitable resistance material. The resistance will vary inversely with the thickness. It has been found empirically that for crude simulations the sort of clay used by potters has a suitable resistivity if used with the standard Servomex Field Plotter F.P.144. The clay must be homogeneous and of the same consistency as when put on the potter's wheel. Plate 17.5 shows a 'relief' model to which conducting strips of copper were attached at each end. The equipotential lines are dotted in over the surface of the clay and their shape and disposition is a reflection of the topography of the surface. The conductivity of the clay is a function of the presence of interstitial water and over a period of an hour or so evaporation takes place and the resistivity changes. This is of course painfully crude and would be viewed with horror by any engineer, but it makes its point effectively and is extremely useful in the early stages of experimental design because being plastic the configurations can be altered very easily and quickly.

A more stable medium consists of a mixture of 2 parts of synthetic graphite (the conductor) to 1 part of Polyfilla (a proprietary brand of cellulose putty) which is non-conducting. These are mixed exhaustively with distilled water into a stiff paste and then modelled on a surface of clean plate glass. The terminals can be cut from copper strips and inserted into the ends of the model while it is wet. Final modelling of the surface can be accomplished by chisel and file to give the required shapes. Figure 17.18 shows the equipotential lines on a wedge of this material and the relationship between thickness and resistance can easily be seen. In a colonization or settlement model it is possible to shape the relief so that it expresses the ease or difficulty with which areas can be traversed or settled. Where the surface is thick the going is good, the relief is low, the soil is fertile, there is an absence of hostile natives, etc., etc. Merely assessing terrain potential is from a teaching point of view a valuable part of the exercise. When the potential is applied at the point of initial settlement its value is adjusted to the capacity for expansion at that time and place and the equipotential lines can then be held to reveal the rate and direction in which the frontier might be expected to move over time. The disadvantage of a single sink can be overcome by burying different sinks at various points towards the far edge of the model.

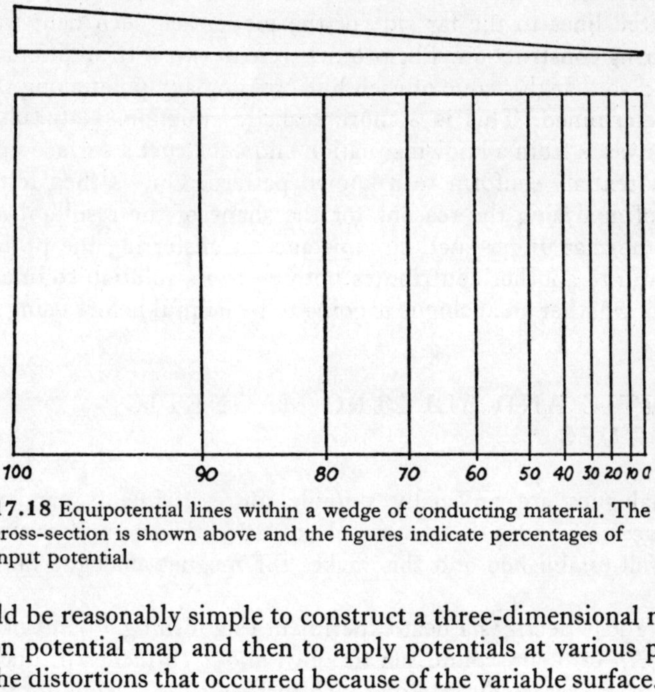


**17.18** Equipotential lines within a wedge of conducting material. The cross-section is shown above and the figures indicate percentages of input potential.

It would be reasonably simple to construct a three-dimensional model of a population potential map and then to apply potentials at various points and observe the distortions that occurred because of the variable surface. A simple concentric pattern about a point would be deformed if at the point of application the surface were not uniform. Plate 17.6 shows a cardboard model of a sloping statistical surface. The rings upon it would be circular if the surface were horizontal. The distortion of the pattern is caused by the slope of the surface. There is an analogy here with Stewart and Warntz's observation that Federal Reserve Banks are displaced from the geographical centre of the area they serve towards the zone of highest population potential. This is possibly one of the most promising applications of the method and could be extended into the study of deformations over many different types of surfaces.

It is not practicable to model a continuous surface; reasonable results can be obtained by sectioning the desired surface into a number of discrete profiles. The profiles are then cut out of Teledeltos paper and the equipotential lines drawn on each in turn. They can then be made into a model and lines of equal value linked by elastic threads. Plate 17.7B shows a simple model in which its assumed movement is controlled by relief. The sections of course represent the reverse of the actual profile, i.e. areas of low resistance are high, and of high resistance are low. Plate 17.7A shows the profiles in 'topographic' terms, the mountain ranges appearing as high points, the passes and plains as low points. In each case the equipotential lines have been placed in the same relationship. Plate 17.7C shows a model in which the disposition of the equipotential lines to the far side of the model has been deliberately controlled during construction. The profiles were drawn with additional strips on the far end so that the value of each line crossing any point along the far end was predetermined. This is a more realistic situation. Naturally one can work backwards from a known situation and construct a surface upon which the equipotentials conform to a known pattern. One is then left with the problem of analysing the reasons for the shape of the resultant surface. It may well be that under such circumstances transferring the problem from one medium to another contributes nothing to its solution so one must obviously ask whether an analogue is going to be helpful before using it.

## MAGNETIC AND ELECTRO-MAGNETIC ANALOGUES

These analogues are especially suitable for experiments concerned with central places. The idea that central places 'attract' custom from surrounding areas is well established and this makes the magnet analogue fairly easy to accept.

Bunge (1964) decribes a neat experiment establishing a point of some importance. An ordinary round plastic bowl about 18 inches in diameter was half filled with water. Twenty one-inch diameter corks were dumped all together and at random into the water. Each cork had a bar magnet forced through the centre with the same pole towards the top and a small lead weight at the bottom to make sure that the corks all floated the same way up. The corks represented central places and the surface of the water the uniform surface upon which the central places are disposed. The corks rapidly achieved a position of equilibrium in relation to each other based on mutual repulsion and the constraints imposed by the size of the bowl. After one hundred random tosses it was found that the corks in the middle of the bowl arranged themselves 43 times in the form of a nearly perfect hexagon with a central point, 25 times as a nearly perfect hexagon without a central point and

22 times as a nearly perfect pentagon with a central point. The remaining ten throws produced a variety of other patterns.

The simplicity of the analogue should not be allowed to blind one to its significance. While it provides some confirmation of one's intuition that there will be a unique solution (the hexagonal arrangement results in 68 per cent of the throws, cf. Table 17.1) it also demonstrates that alternative solutions to this equilibrium problem are possible. One might therefore use the experiment as an introduction to stochastic models.

Extra point is given to the use of magnetic analogues by the similarity between Reilly's Law of Retail gravitation (1931) and Coulomb's Law of Force, which is valid for practically all fields of force including magnetic and gravitational fields. Reilly's Law states that 'two cities attract retail trade from any intermediate city or town in the vicinity of the breaking point (between their two spheres of dominance) approximately in direct proportion to the populations of the two cities, and in inverse proportion to the square of the distance from these two cities to the intermediate town'. Coulomb showed in 1785 that 'the force between concentrated (point) changes varies directly with the product of the individual changes, and inversely with the square of the distance between them'. Of course Reilly's law lacks the universal validity in economic terms that Coulomb's possesses in the physical sphere, but for many purposes it can be regarded as an acceptable statement of a tendency. The inverse square law, therefore, can be expressed in magnetic terms and so a magnetic analogue is suitable for certain experiments in central place studies. In all cases where magnets are used to represent central places it is important that the polarity of the magnets should be the same at the surface over which they are disposed. Like poles repel and the flux lines between like poles form a saddle – a narrow zone in which the flux lines turn abruptly through a right angle.

Perhaps the most convenient experimental arrangement is for the magnets to be arranged with like poles upwards underneath a sheet of glass upon which a sheet of tracing paper is laid. The patterns of the field can be shown by scattering iron filings over the paper and gently tapping it (Plate 17.8). In practice friction between the iron filings and the paper limits the area over which the patterns can be made to appear, and should a collection of iron filings develop more over one magnet than the others its strength will alter in relation to the rest. A more efficient method involves use of a simple field-plotting compass. Moving from one magnet to another the needle changes direction sharply at the point of balance between them and so the individual fields can be plotted easily. The needle is more sensitive than iron filings.

Figures 17.19A–C show the results from plotting fields of equal strength permanent magnets evenly and randomly spaced. To simulate the effect of centres of different importance it is better to use electro-magnets.

In a simple electro-magnetic simulator solenoids are made with copper wire, the number of turns being made proportional to a particular property of the central places represented. This may be population, number of retail outlets, population employed in service trades or any other attribute

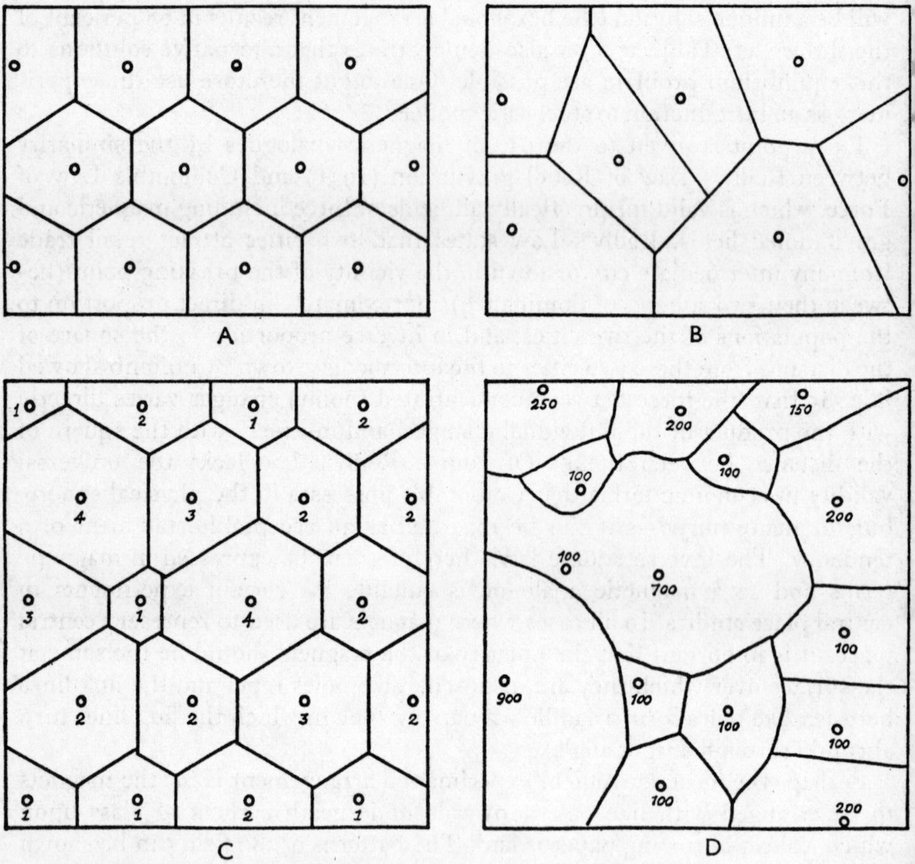


**17.19** Plots of magnetic fields.

A Equal strength magnets, regular spacing.
B Equal strength magnets, irregular spacing.
C Varying strength magnets, regular spacing.
D Varying strength magnets, irregular spacing.
The numbers in C and D refer to relative strengths of the magnets used.

thought appropriate. Soft iron cores are fitted into the bobbins and the coils energized by a suitable D.C. source. The electro-magnets are placed under a sheet of glass in their proper positions and the fields outlined with a plotting compass.

Plate 17.9 shows a more sophisticated general purpose model in which the central places are represented by a series of electro-magnets to each of which is linked a rheostat by which the current entering each coil can be controlled. With coils that all have an equal number of turns variations in magnetic strength depend simply on the amount of current passing through. Rheostats can be calibrated in terms of population and the magnets whose strength they control can be disposed at will under a sheet of glass over which field plots can be made. This type of analogue from which spatial patterns can be derived has a number of applications. Many central place studies can perhaps be criticized for placing too much emphasis on the functional attributes of the centres themselves and too little on the actual areas served. Analogues of this sort can be used to simulate actual patterns of spatial organization. The inputs for each centre can be adjusted by degrees until the model situation conforms to the real situation. Then the inputs needed in the model can be related to certain attributes or combination of attributes in the real situation. The problem of creating non-uniform surfaces is always vexing in analogues of this sort and it may be that it is best to confine first experiments to real situations where a degree of uniformity is present.

## BLOTTING PAPER ANALOGUES

This analogue is based on a simplified version of the technique of paper chromatography. It is a matter of common observation that a drop of water spilled on a piece of blotting paper will spread out and leave a roughly circular damp patch. If the water is supplied to the surface of a horizontal piece of blotting paper by means of a small wick the damp area will spread out until a

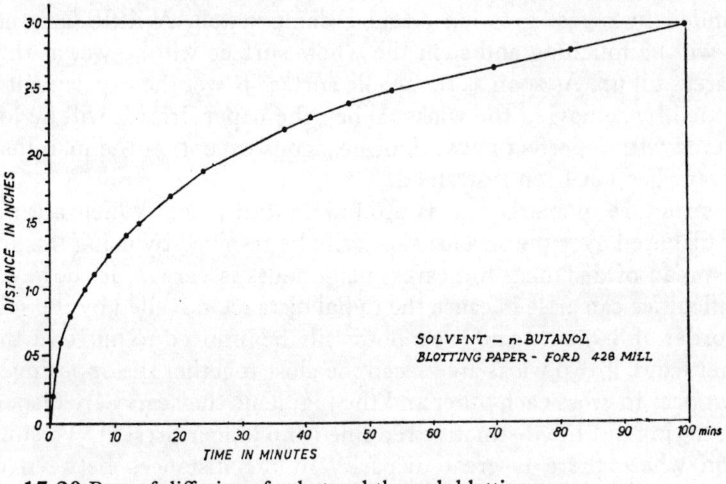


**17.20** Rate of diffusion of n-butanol through blotting paper.

balance between supply and evaporation prevents further movement. In paper chromatography a compound substance is placed at the point where the wick touches the surface and as the solvent front advances outwards it carries with it at different speeds the constituents of the original compound depositing each as a ring of distinctive colour. To achieve the best separation of particles and the smoothest progression of the front requires a solvent composed of 45 parts of N-butanol, 25 parts of pyridine and 40 parts of distilled water. For many purposes, however, valuable results can be achieved by using only N-butanol as the solvent and ordinary pink blotting paper. As the solvent front advances it carries with it some of the red pigment in the blotting paper and when the process is finished and the paper dried the limits of the solvent front are marked by a thin red line. The solvent front advances very rapidly at first but soon moves more and more sluggishly. A series of experiments at normal room temperatures of about 68° F and average humidity reveal the pattern illustrated in Figure 17.20. The analogue is particularly helpful in the field of central place and diffusion studies. The circular area encompassed by the solvent front from any one point may be likened to the area of influence of a central place or market, and the analogue can easily be made to demonstrate very elegantly the reason for the hexagonal market area of a Christaller central place system (Plate 17.10). The pink blotting paper is pierced by a network of small holes, arranged at the apexes of equilateral triangles so that each hole is about two inches from its neighbour. Small wicks made of tightly rolled blotting paper are inserted tightly in each hole and rested in a dish filled with butanol. It is important that all wicks should be the same length, be made of the same amount of paper and rolled with equal tightness. It is best to fit the wicks first and then present all of them simultaneously to the butanol. Within a minute each wick will be surrounded by a fast growing circular damp patch. A little later adjacent circles will be touching and soon the whole surface will be wet as the interstitial areas fill up. As soon as the whole surface is wet the experiment should be stopped by removing the wicks. When the paper dries it will be found to be covered with a perfect network of hexagons except on the margins where the solvent has not been restrained.

The structure of market areas around central places which are not uniformly disposed over the surface can easily be resolved by using this method. If the spread of distances to nearest neighbours is very wide, however, certain difficulties can arise because the radial distance travelled by the solvent is of the order of 8–9 cms. and it cannot easily be induced to move further. On the other hand, if two wicks are placed too close together the opposing solvent fronts appear to cross each other and the pigments they carry are dispersed so that on drying out the distinctive red line is no longer present. To simulate a situation where there is great disparity in the distances between centres therefore, it is best to break the area up into smaller sections and simulate each on an appropriate scale. The result can then be compounded on one diagram. In certain cases the problem can be solved by introducing a dye into the solvent. The dye is moved more sluggishly across the paper and where the red material is absent the boundary between different centres can be determined from the disposition of the added dye.

Plate 17.11 shows a simulation of the territories of the centres of S.W. England. The model was run originally on a rectangular sheet of paper and the coastline was cut out when it had dried out. It will be apparent that since all wicks deliver the same amount of solvent the assumption is that all centres have potentially the same influence. Moreover no constraints are applied on the uniformly absorptive surface of the paper. So the illustration shows the pattern of the territories under unrealistic conditions, i.e. as it would be if all centres wielded the same influence. However, it is interesting that the irregular size and disposition of the territories arises from very simple premises. While uniform surfaces are conceptually very valuable geographers have also to be concerned with non-uniform conditions. It is not easy to introduce constraints into this particular model. However, negative areas can be created by cutting out pieces of blotting paper. A particular problem is impeding the passage of the solvent: were it not so powerful it might be possible to spray or paint on lacquers or paste on extra pieces of paper to impede the passage of the front. A promising technique of introducing variable resistance into the surface, borrowed from bio-chemistry, involves replacing blotting paper by a plastic mixture of cellulose powder and water which can be sculpted at will, and through which butanol will diffuse at different rates. Yet such merit as this analogue possesses depends on its simplicity and it is probably not worth forcing sophistication upon it.

A further series of experiments could be done to illustrate the hierarchical concepts of Christaller. By starting off with a uniform distribution of settlements as in Plate 17.10 one derives the first set of hexagons. These can then be drawn in on a fresh piece of paper at a smaller scale and wicks placed in the next lowest grade of centres (the number and arrangment depending on the factor used) from these another set of hexagons will arise defining the market areas of these more important centres. The process can be repeated several times to give the required ranks in the hierarchy, the final product being a whole landscape with all elements represented.

Another method involves using a dye composed of several different colours. Green ink, for example, will, under the action of the solvent, split into blue yellow, one colour always moving further from the wick than the other. By starting the model with centres as widely spaced as possible one will get, for example, a set of blue hexagons each containing a yellow circle inside of considerably smaller radius. One can then add new sources of supply (smaller centres) between the yellow rings eventually creating a network of yellow hexagons bearing a measurable relationship with the larger blue ones.

Experiment will produce compounds composed of a suitable proportion of mobile and reluctant colours to meet most requirements.

The analogue can also be used to represent certain diffusion features. For example Figure 17.21 represents an urban area growing near a river which can be regarded as a semi-permeable barrier. The permeability is effected by means of a blotting paper 'bridge'. If the centre of the city is represented by a wick near the bridge the bridge will permit a limited extension of the growing city on the opposite bank. There is a close parallel between the degree of movement to the paper bridge and a real bridge. The opposite bank seems to be controlled both by the width of the 'river' and the breadth of the bridge but no quantitative measurements have been done to establish relative importance. Stages through which the pattern passes in acquiring its final shape can be recorded in two ways. One can use time-lapse photography to record the various stages (bearing in mind the relation between time and speed of movement of the solvent front noted in Fig. 17.20); or one can run the model through to its final stage, dry it, replace the old wick with a new one (to provide extra dye) and run it again for a shorter time. If this process is repeated carefully a series of parallel curves is produced (Fig. 17.21F). Their distance apart is a function of relative ease or difficulty of movement. In fact, as Yuill (1964) has pointed out, it is a characteristic of river towns that they are usually better developed on one bank than another and this 'affords an excellent starting place for a solid study of city shapes and the response of the growth mechanisms to barriers'. Yuill's paper is a computer simulation study of the role of barriers in structuring human activity patterns in space, but at a very crude level the blotting paper analogue is valuable in that the method of operation is easily grasped by the mathematically innocent.

## LOCATION MODELS

It must be stated at the outset that contemporary location theory has become far too complex in terms of the problems it sets itself and the techniques it uses for physical models to make any worthwhile contribution to advanced studies. But in teaching elementary location theory models have their uses. In the simplified form in which it is usually presented Weber's theory seems to offer few problems and yet the average student if asked to work out a simple example very often finds himself unable to structure the problem correctly. The two models presented here are designed to enhance understanding of the principles involved by permitting simple mechanical or graphical solutions of elementary location problems.

The simplest analogue is based on systems of pulleys and weights, an early example being quoted by Friedrich (1929). It is cheap to make and easy to operate, giving an immediate solution. Assume the problem is to find the

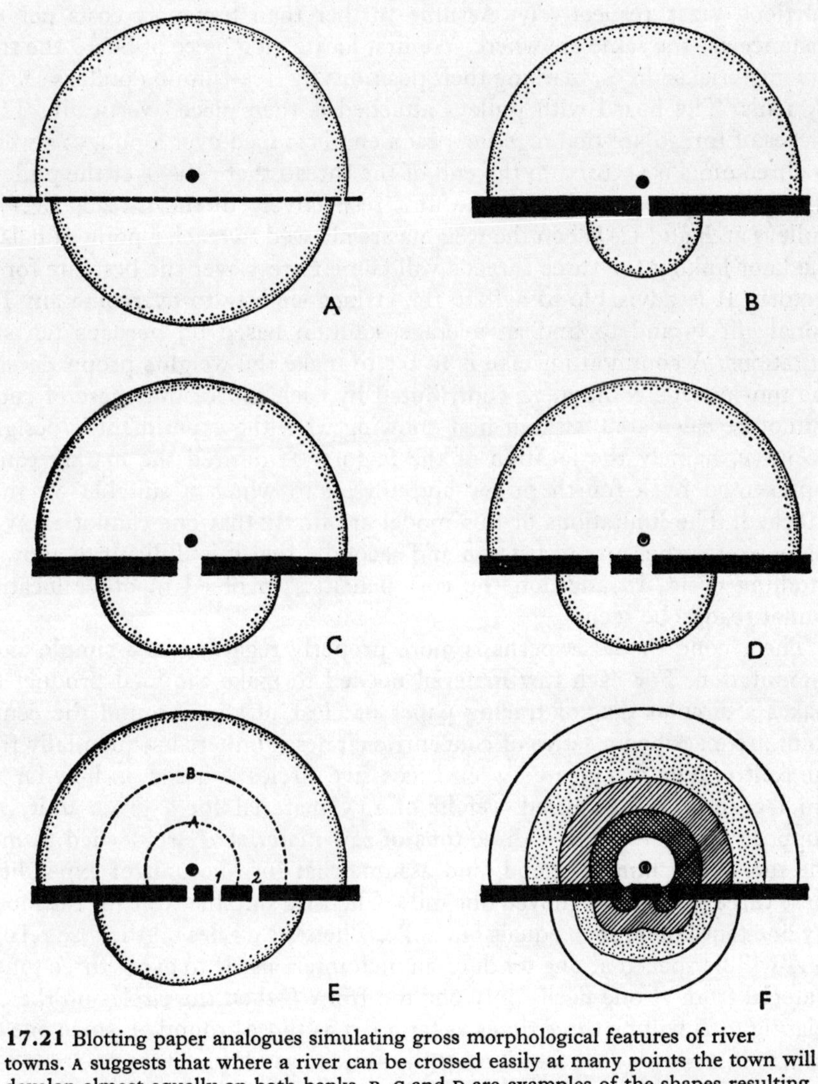


**17.21** Blotting paper analogues simulating gross morphological features of river towns. A suggests that where a river can be crossed easily at many points the town will develop almost equally on both banks. B, C and D are examples of the shapes resulting from various restrictions to easy access to the opposite bank. E is a rudimentary growth model, in that when the solvent front reached the line A bridge 1 was added. Bridge 2 was added when the solvent front reached the line B. This process leads to a marked asymmetry in the shape of the part of the town on the far side of the river. F is a plot of the various stages of growth.

best location for a factory using three raw materials, *A*, *B* and *C* in the proportions 3:2:1 respectively. Assume further than transport costs per unit distance are the same anywhere. We first locate on a piece of board the three raw material sources, marking their positions by free-running pulleys secured by nails. The board with pulleys attached is then placed vertically. Three pieces of thread are tied together, each end is looped over a pulley; a weight of three units is secured to the end of the thread that runs over the pulley at *A*, and weights of two and one unit respectively to the threads over the pulleys at *B* and *C*. When the weights are allowed to reach a point of balance the knot linking the three threads will come to rest over the best site for the factory. It is advisable to agitate the strings slightly to overcome any frictional effects and to find an average solution based on perhaps ten such agitations. A common mistake is to try to make the weights proportional to the ton-mileage component contributed by each source but these of course cannot be calculated without first knowing what the experiment is designed to prove, namely the location of the factory. If desired the market can be represented by a fourth pulley and thread to which a suitable weight is attached. The limitations of this model are firstly that one cannot allow for many sources because of friction and secondly that it is difficult to allow for handling costs. In addition the cost penalties involved in other locations cannot readily be seen.

The second model is perhaps more properly regarded as a simple aid to computation. For each raw material needed to make the final product one makes a circular disc of tracing paper or clear plastic. Around the central point are inscribed a series of concentric circles numbered sequentially from the centre. The distance between successive circles is based on how far one can transport the required weight of raw material for a given unit cost. Suppose for example that three tons of raw material *A* are needed to make one ton of the finished good, and assume that for one unit of expenditure these three tons can be moved one mile. Choose a suitable scale for the model, say one tenth of an inch equals one mile. Then the circles on the disc relating to *A* will be spaced at one tenth of an inch intervals. If to every three tons of material from *A* one needs only one ton from *B* then the circles on the disc relating to *B* will be three times as far apart as those around *A*, since one can assume in a simple case that it costs the same to send three tons one mile as to send one ton three miles. By the same argument one and a half tons of material *C* can be moved two tenths of an inch on the model for one unit cost.

If we now mark on a piece of paper the three raw material sources, *A*, *B* and *C* in their correct scale relationships we can pin to each the appropriate transparent disc. Covering them all with another piece of tracing paper we choose one of the many points where three circles of different origins intersect and add up the values ascribed to them. The total can be recorded as a 'spot height'. When enough of these points have been plotted lines of equal transport costs (isodapanes) can be drawn. The property of the isodapanes is such that the cost of assembling the raw materials in the proportions selected is the same at all points along the line of any one isodapane. The lowest value isodapane, often represented by a point, is the least cost location for the constraints imposed by the model design.

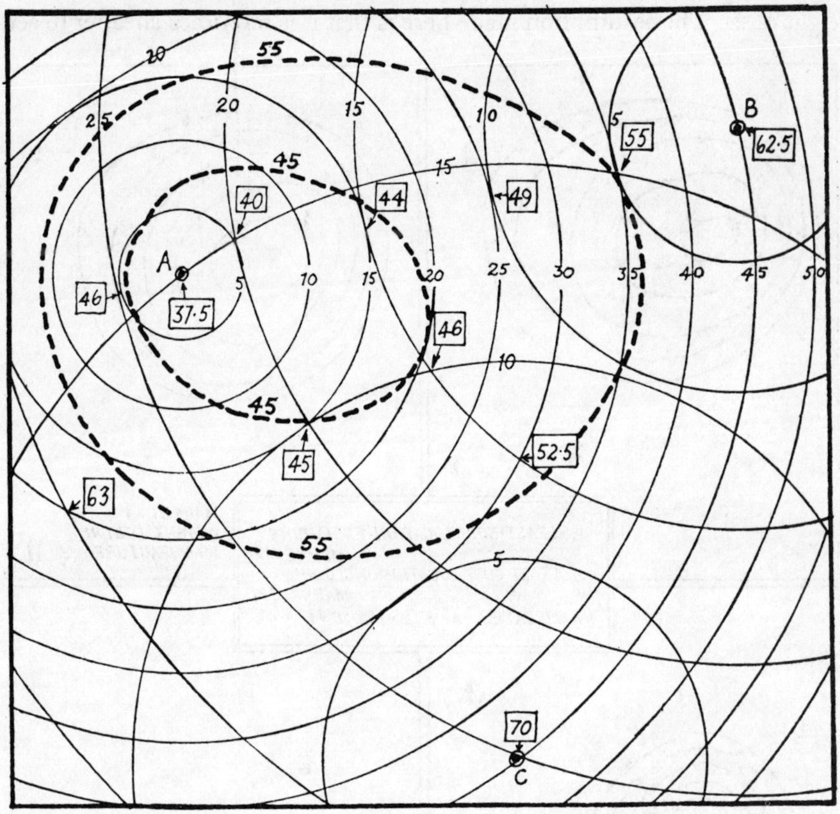


**17.22** Graphical method for deriving isodapanes in simple problems. Isolines at appropriate intervals surround the three raw material sources, A, B and C. The boxed figures record the sum of the values of three sets of isolines at particular points. Given sufficient of these points the isodapanes can then be drawn. They are shown here by the thick dashed lines.

If more than three material sources are involved they need not be solved simultaneously: it is often much less confusing to deal with them in small groups and to combine the partial solutions in a final plot. Figure 17.22 should make the general procedure clear. Handling costs can be incorporated readily by converting them into distances the material could be moved and weighting the values of the relevant circles by an appropriate amount. The market can be allowed for (provided that it is regarded as existing at a single point). Figures 17.23 (II–IV) show the effect of loss and gain of weight during production both on the optimum location of the plant and on the pattern of the isodapanes. It is very easy to structure the models so that they demonstrate the basic principles of location.

Figure 17.24 shows how it is possible to build in the effect of differential freight rates. The assumption made here is that it is ten times cheaper to send

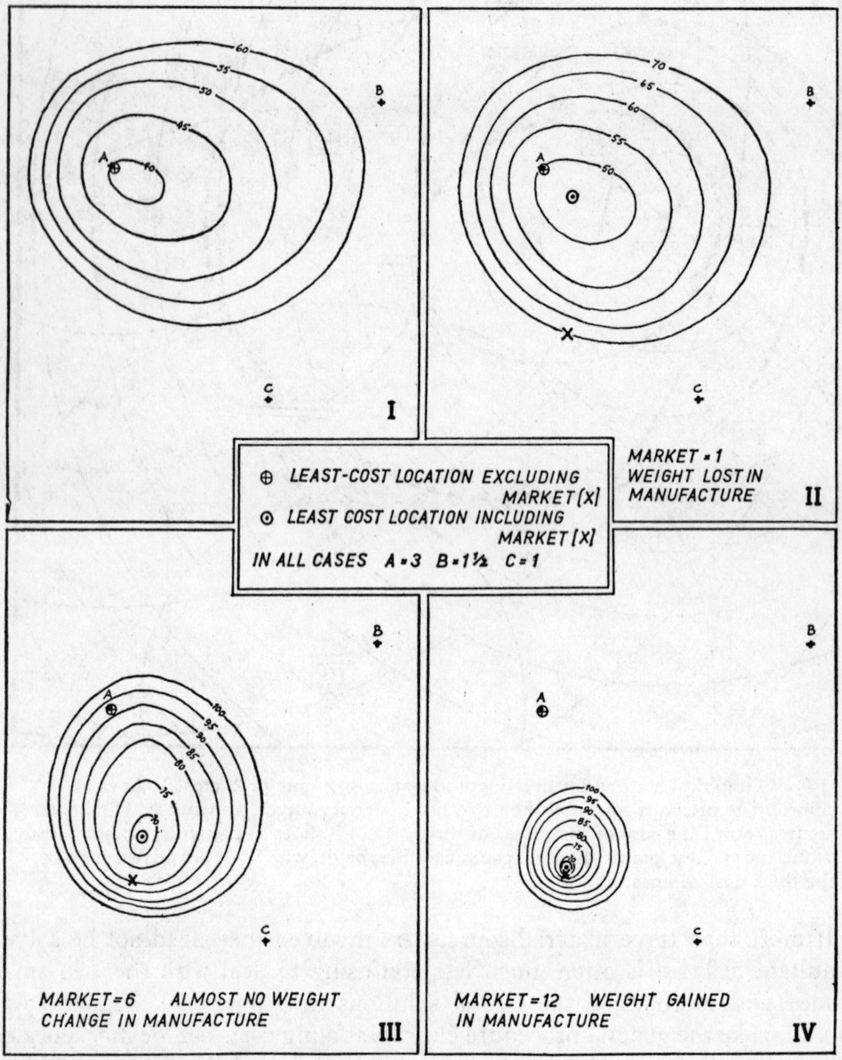


**17.23** The effect of loss and gain of weight during production on the optimum location of the plant and the pattern of the isodapanes.

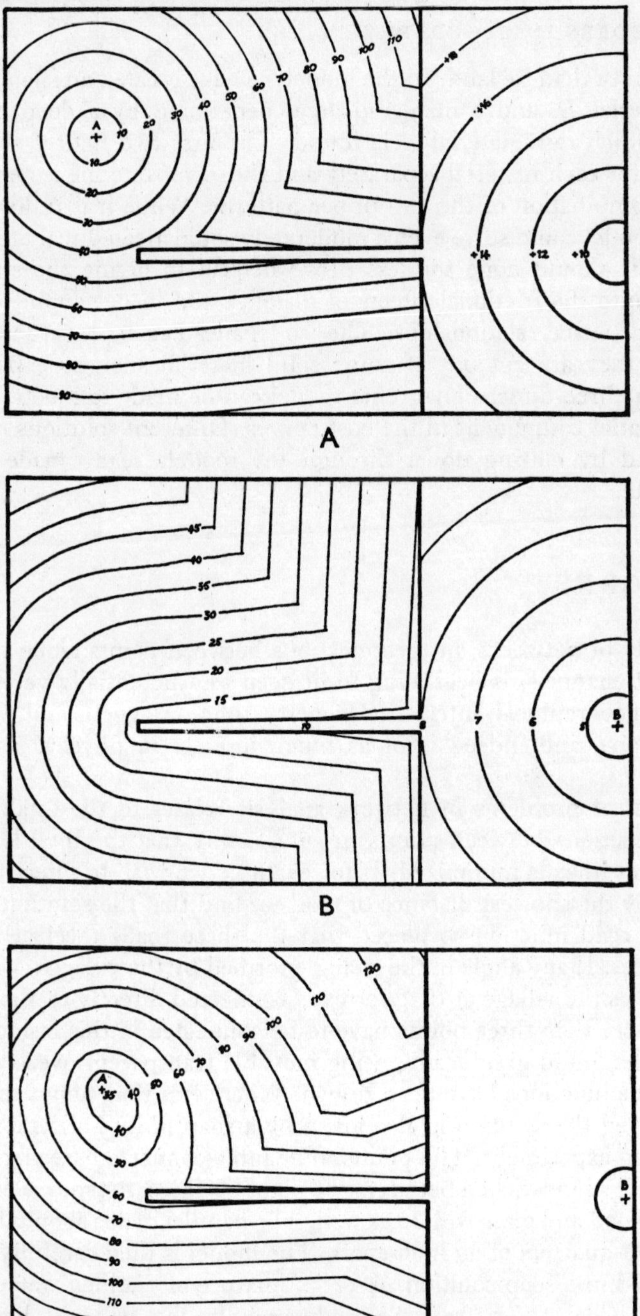


**17.24** Model involving differential freight rates.

goods by sea than by land. In the model we have created an island containing raw material *B*, and a mainland area, penetrated by a deep estuary, and within which raw material *A* is found. The first two figures show the isodapanes for each material separately and the lower one the isodapanes based on the combination of the two upper patterns. While it is unlikely that this type of model could solve even a moderately sophisticated real life problem of location it should none the less prove helpful in giving the student some insights into the frictional effects of distance and in developing an intuitive feeling for space relationships. The isodapanes can be regarded as contour lines; if they are cut out of some solid material such as expanded polystyrene a three-dimensional relief model can be made to illustrate the effect of the spatial component in the cost curves. Different solutions can be compared and by cutting down through the models ready made graphs are obtainable.

## NETWORKS

The study of networks, interconnections between points along more or less restricted channels, is beginning to be seen as a potentially very productive field of geographical interest (Haggett, 1965; Haggett and Chorley, In preparation), and indeed is of as much interest in physical as in human geography.

One set of problems in network analysis relates to the establishment of interconnections between points in such a way that the total length of the constituent links is minimized. If for example we wish to interconnect three villages by the shortest distance of road we find that the solution is to create a central road junction so placed that the three roads meet at an angle of 120 degrees. If any angle in the triangle formed by the villages is greater than 120 degrees the village at that vertex is connected directly to the other two. Where more than three points have to be connected in this fashion the solution can be found graphically, using movable transparent overlays on which 120 degree junctions have been drawn. Where less than about fifteen points are involved the solution is also given by a soap film. This is a particularly fascinating experiment. The points to be inter-connected are arranged in the form of pegs sandwiched between two plane surfaces. Perspex is excellent but glass or wood and glass will do as well. The parallel plates should be not more than three-quarters of an inch apart. The model is then carefully and slowly submerged in a soap solution. A 50:50 mixture of distilled water and liquid detergent behaves very well and produces a film that will with luck last up to twenty-four hours. When the model is slowly and carefully withdrawn soap films are left perpendicular to the parallel plates and they gradually contract until the minimum length links are established. All the junctions of the film are at 120 degrees. Plate 17.12 shows clearly the sort of configuration of the soap film in a typical situation. The film can be touched with a wet finger without danger of breaking. The model works because a soap film always adopts a shape which minimizes its area for the constraints imposed upon it. (Miehle, 1958; Courant and Robbins, 1941).

If too large a number of points are included a unique solution is not given by the soap film. However this disadvantage can be turned to good account. If we make a model with say about twenty-five or thirty randomly disposed pegs in it, we can dip it a number of times in the soap solution and record the pattern of interconnections established on each occasion. The minimum solution can be found by inspection, and its length compared with other solutions. If a compound drawing is made showing the frequency with which individual pairs of points are linked limited experience suggests that certain general regularities will be found, and such a procedure could form an introduction to stochastic models.

If a unique solution to a large number of points is needed it can be obtained by discretizing the network into a number of smaller parts and solving each of these in turn. The process of breaking up a large network for analysis may involve simply taking various sections in turn, but in the case of a road network for example it may be worth while first to minimize the major road net for the whole area, then the minor roads in the interstices or sectors of the major net and so on down to the local roads. These progressive changes of scale can be reconciled in a final composite map of the whole system.

Where many points are involved possibly a better solution is that proposed by Miehle (1958). Fixed points are represented to scale by pegs fixed in a board, each peg carrying a small pulley. Movable pegs, also with pulleys represent mobile junctions. A thread is looped about each peg in turn and around an appropriate number of movable pegs, the thread being finally brought back to its starting position (Plate 17.13). The loose end is then tightened. If there were no friction, and if the interconnections had been wisely chosen the first time, the system would move into equilibrium and the total length of thread used would be the least possible. In practice friction has to be overcome by easing and teasing the thread so that it is equally taut throughout the model. Faced with a large number of points and junctions one is at first bewildered by the complexity of the situation but experience helps one to decide on appropriate interconnections. Local and difficult minimizations can be solved using soap film models at first. J. A. Silk (1965) constructed a mechanical link-length minimization of the road system of Monmouthshire. The model was constructed on a one-inch O.S. map and was found to accord well with the actual road pattern.

The advantage of the mechanical minimizer is that integral weights can be assigned to individual links (by multiplying the number of linking loops of thread), constraints can be applied to the movable points and arbitrary selections of interconnections is possible, advantages not available in soap film models. Miehle in fact developed the mechanical method in order to find the answer to a problem involving the best location for a given number of control centres serving a fixed disposition of missile sites. It is not difficult to see that the method could as well be applied in central place studies, enabling one for example to judge the efficiency of the location of existing central places under certain predetermined constraining conditions. Brink and Cani (1957) in fact used an electrical analogue to determine the location of service points to serve a number of customers with a known demand pattern so as to minimize transport costs, but a link-length minimizer should be capable of giving approximately the same answer.

Networks suggest the use of electrical analogues since after all an electrical circuit is a form of a network in itself. Movement of groundwater in a three-dimensional aquifer has been simulated by means of an electrical resistance analogue (Robinove, 1962; Skibitzke, 1960), described in Chapter 5.

The application of electrical analogues to problems in spatial economics is relatively new and was elegantly illustrated by Enke (1951). He dealt with the difficult problem of determining the equilibrium prices and commodity

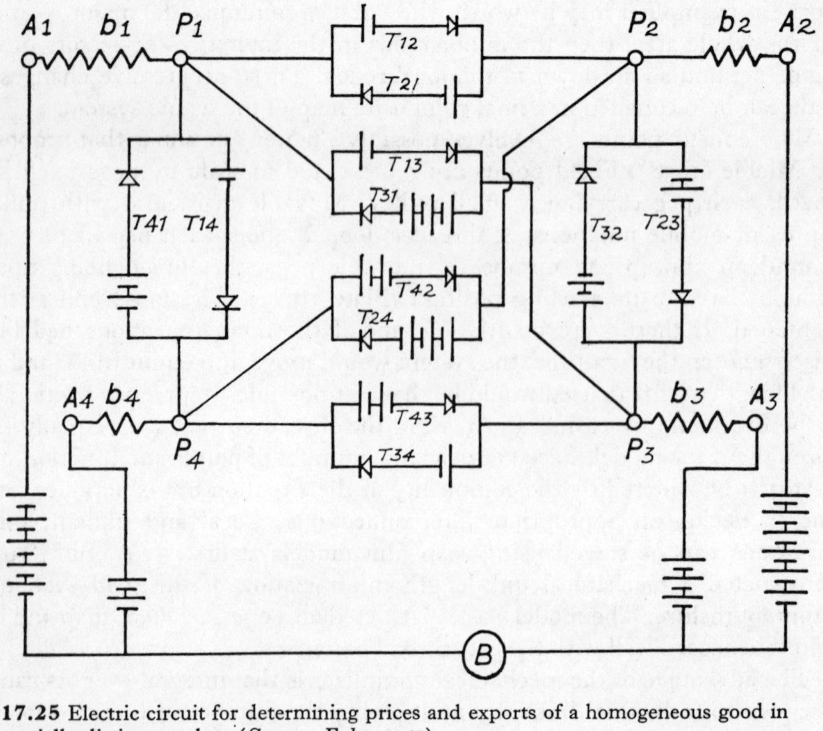


**17.25** Electric circuit for determining prices and exports of a homogeneous good in spatially distinct markets (*Source: Enke, 1951*).

flows that will result when a number of interdependent trading units (four in this case) are in a position to buy or sell a particular good and where freight rates between potential trading units are significant. Each trading unit may be a seller at a relatively high price and a buyer at a low price. Transport costs are never so high as to prevent one trading unit from participating in the process of exchange. In the electrical analogue commodity flows are represented by current flow (measured in amperes) and prices by voltages. The circuit used by Enke is shown in Figure 17.25. For a detailed explanation of its design reference should be made to the original article. The analogue had particular point in 1951 because the only other way at that time of solving a problem of such complexity was through laborious iterative methods. Samuelson (1952) subsequently showed how equilibrium problems can be solved using computers, a technique made possible by the development of linear programming.

## REFERENCES


ALLEN, J., [1947], *Scale models in hydraulic engineering*, (London).

APPLEBY, F. V., [1956], Run-off dynamics – a heat conduction analogue of storage flow in channel networks; *International Association of Hydrology*, Publication No. 38, (Rome, 1956), 3, 338–348.

VON ARX, W. S., [1957], *Physics and Chemistry of the Earth*; Vol. 2 (ed. by Ahrens et al.), (London), Ch. 1, 1–29.

VON, ARX, W. S., [1962], *Introduction to physical oceanography*, (Cambridge, Mass.).

BALCHIN, W. G. V. and RICHARDS, A. W., [1952], *Practical and experimental geography*, (London).

BILLINGS, M. P., [1946], *Structural geology*, (New York).

BOOTHROYD, A. R., CHERRY, E. C. and MAKAR, R., [1949], An electrolytic tank for the measurement of steady state response, transient response, and allied properties of networks; *Proceedings of the Institute of Electrical Engineers*, 96 (Pt. 1), 163–177.

BOTSET, H. G., [1946], The electrolytic model and study of recovery problems; *Transactions of the American Institute of Mechanical Engineers*, 165, 15–25.

BRINK, E. J. and DE CANI, J. S., [1957], An analogue solution of the generalised transportation problem with specific application to marketing location; *Proceedings of the First International Conference on Operational Research*, (Oxford), (English Universities Press).

BRUSH, L. M. and WOLMAN, M. G., [1960], Knickpoint behaviour in a non-cohesive material – a laboratory study; *Bulletin of the Geological Society of America*, 71, 59–74.

BUNGE, W., [1962], Theoretical geography; *Lund studies in geography, Series C, General and mathematical geography*.

BUNGE, W., [1964], Patterns of Location; *Michigan Inter-University Community of Mathematical Geographers, Discussion Paper No.* 3.

BURKALOW, A. VAN, [1945], Angle of repose and angle of sliding friction – an experimental study; *Bulletin of the Geological Society of America*, 56, 669–708. (Contains a large bibliography.)

CHORLEY, R. J. and MORGAN, M. A., [1962], A comparison of morphometric features, Unaka mountains, Tennessee and North Carolina and Dartmoor, England; *Bulletin of the Geological Society of America*, 73, 17–34.

CONRAD, V. and POLLACK, L. W., [1950], *Methods in climatology*, (Harvard), Ch. 3.

COURANT, R. and ROBBINS, H., [1941], *What is mathematics?*; (New York), Ch. 7.

DAUBREE, A., [1879], *Études synthetiques de géologie experimentale*; 2 Vols., (Paris).

DUNCAN, W. J., [1953], *Physical similarity and dimensional analysis*, (London).

EINSTEIN, H. A. and HARDER, J. A., [1961], Electric analog model of a tidal estuary; *Transactions of the American Society of Civil Engineers*, 126 (4), 855.

ENKE, S., [1951], Equilibrium among spatially separated markets – solution by electric analogue; *Econometrica*, 19, 40–47.

FOLK, R. L. and WARD, W. C., [1957], Brazos River bar; a study in the significance of grain size parameters; *Journal of Sedimentary Petrology*, 27 (1), 3–26.

FRIEDKIN, J. F., [1945], A laboratory study of the meandering of alluvial rivers; *U.S. Waterways Experimental station*, (Vicksburg, Miss.).

FRIEDRICH, C. J., [1929], *Alfred Weber's theory of the location of industries*, (Chicago).

FULTZ, D., [1956], Fluid models in geophysics; *Proceedings of the first symposium on geophysical models, Baltimore* 1953, (Washington).

GLOVER, R. E., HERBERT, D. J. and DAUM, C. R., [1953], Application to a hydraulic problem; *Transactions of the American Society of Civil Engineers*, 118, 1010.

GRIGGS, D. T., [1939], A theory of mountain building; *American Journal of Science*, 237, 611–650.

HAGGETT, P., [1965], *Locational analysis in human geography*, (London), Ch. 3.

HAGGETT, P. and CHORLEY, R. J., [in preparation], *Network models in geography, an integrated approach*, (London).

HARDER, J. A., [1963], Analog models for flood control systems; *Transactions of the American Society of Civil Engineers*, 128 (1), 993.

HOTELLING, H., [1921], *A mathematical theory of migration*; Unpublished master's thesis, University of Washington.

HUBBERT, M. K., [1951], Mechanical basis for certain geologic structures; *Bulletin of the Geological Society of America*, 62, 355–372.

IMHOF, E., [1965], *Kartographische Gelandedarstellung*, (Berlin).

ISARD, W., [1956], *Location and the space economy*, (New York).

JOHNSON, J. W. and RICE, E. K., [1952], A laboratory investigation of wind generated waves; *Transactions of the American Geophysical Union*, 33 (6), 845–854.

KARPLUS, W. J., [1958], *Analog simulation*, (New York).

KENN, M. J., [1965], Experiments simulating the effects of mountain building initiated by the late Professor S. W. Wooldridge; *Proceedings of the Geologists' Association*, 76 (1), 21–27.

KUENAN, P. H., [1956], Experimental abrasion of pebbles; 2, Rolling by current; *Journal of Geology*, 64, 336–368.

KUENAN, P. H., [1960], Experimental abrasion; 4, Eolian action; *Journal of Geology*, 68, 427–449.

LANGHAAR, H. L., [1951], *Dimensional analysis and theory of models*, (New York).

LASAREFF, P., [1929], Sur une méthode permettant de démonstrer la dépendance des courants océaniques des vents alizés et sur le rôle des courants océaniques dans le changement du climat aux époques géologiques; *Gerlands Beitrage zur geophysik*, (Leipzig), Bd. 21, 215–233.

LEWIS, W. V., [1944], Stream trough experiments and terrace formation; *Geological Magazine*, 81 (6), 240–252.

LEWIS, W. V. and MILLER, M. M., [1955], Kaolin model glaciers; *Journal of Glaciology*, 2, 533–538.

MACKAY, J. R., [1965], Glacier flow and analogue simulation; *Geographical Bulletin*, 7 (1), 1–6.

MCKEE, E. D., [1957], Experiments on the production of stratification and cross-stratification; *Journal of Sedimentary Petrology*, 27 (2), 129–134.

MCKEE, E. D., [1960], Laboratory experiments on form and structure of off-shore bars and beaches; *Bulletin of the American Association of Petroleum Geologists*, 44, 1253.

MIEHLE, W., [1958], Link-length minimisation in networks; *Operations Research*, 6, 232–243.

MUSKAT, M., [1949], *Physical principles of oil production*, (New York).

NEVIN, C. M. and TRAINER, D. W., [1927], Laboratory study in delta building; *Bulletin of the Geological Society of America*, 38, 451–458.

NOMA, A. A. and MISULIA, M. G., [1959], Programming topographic maps for automatic terrain model construction; *Surveying and Mapping*, 19, 355–366.

OLSEN, G. H., [1962], Field Plotting; *Wireless World*, 68 (2), 58. (The author has drawn attention to errors in figures 8 and 11.)

OLSEN, G. H., [1963], *Servomex field plotter manual*, (Crowborough, England).

PANKHURST, R. C., [1964], *Dimensional analysis and scale factors*, (London).

REILLY, W. J., [1931], *The law of retail gravitation*, (New York).

REINER, M., [1959], The flow of matter; *Scientific American*, 201 (6), 122–138.

RIEHL, H. and FULTZ, D., [1957], Jet streams and long waves in a steadily rotating-dishpan experiment: structure and circulation; *Quarterly Journal of the Royal Meteorological Society*, 83, 215–231.

ROBINOVE, C. J., [1962], Ground water studies and analog models; *U.S. Geological Survey, Circular* 468, (Washington), (Reprinted 1965).

SAMUELSON, P. A., [1952], Spatial price equilibrium and linear programming; *American Economic Review*, 42, 283–303.

SAVILLE, T., [1950], Model study of sand transport along an infinitely long straight beach; *Transactions of the American Geophysical Union*, 31 (4), 555–565.

SHAW, N., [1934], *Manual of Meteorology*, (Cambridge), 1, 100 and 2, 396.

SILK, J. A., [1965], *Road network of Monmouthsire*; University of Cambridge, Department of Geography, B.A. dissertation.

SKIBITZKE, H. E., [1960], International association of scientific hydrology, Publication No. 52.

SPOONER, C. S., [1953], Modernisation of terrain model production; *Geographical Review*, 43, 60–68.

STANTON, T., MARSHALL, D. and HOUGHTON, R., [1932], The growth of waves on water due to the action of the wind; *Proceedings of the Royal Society, Series A*, 137, 283–293.

STARR, V. P., [1956], The general circulation of the atmosphere; *Scientific American*, 195 (6), 40–45.

STEWART, J. Q. and WARNTZ, W., [1958], Macrogeography and social science; *Geographical Review*, 48, 167–184.

TOBLER, W. R., [1959], Automation and cartography; *Geographical Review*, 44, 536–544.

TURNER, J. S., [1963], Model experiments relating to thermals with increasing buoyancy; *Quarterly Journal of the Royal Meteorological Society*, 89, 62–74.

TURNER, J. S., [1965], Laboratory models of evaporation and condensation; *Weather*, 20, 124–128.

TURNER, J. S. and LILLY, D. K., [1963], The carbonated water tornado vortex; *Journal of Atmospheric Science*, 20, 468–471.

TYLDESLEY, B., [1965], The solution of atmospheric diffusion equations by electrical analogue methods; *Meteorological Office, Scientific Paper* No. 22, (London, HMSO).

U.S. ARMY MAP SERVICE, [1950], Refinements on production of molded relief maps and aerial photographs; *Army Map Service Bulletin* No. 29, (Washington).

WALLINGTON, C. E., [1961], An electrical analogue method of studying the behaviour of lee waves; *Weather*, 16, 222–228.

WARNTZ, W., [1959], *Towards a geography of price*, (Philadelphia).

WARNTZ, W., [1965], A note on surfaces and paths and applications to geographical problems; *Michigan Inter-University Community of Mathematical Geographers, Discussion Paper* No. 6.

WILLOUGHBY, E. O., [1946], Some applications of field plotting; *Journal of the Institute of Electrical Engineers*, 93 (3), (Communications) 275–293.

YUILL, R. S., [1965], A simulation study of barrier effects in spatial diffusion problems; *Technical Report* No. 1, *Spatial diffusion Study, Department of Geography, Northwestern University*, (Evanston, Ill.).

CHAPTER EIGHTEEN

# Models of Geographical Teaching

S. G. HARRIES

## INTRODUCTION

In a sense models have always been used in teaching for they are inseparable from communication. Only recently, however, has much systematic attention been given to the use and limits of the various models used in teaching geography. A consideration of this change and its implications is best viewed against the background of developments in science as a whole. Ackerman observes that the progress of science as a whole, has at any time, depended largely on the growing points in a few fertile subjects, which provided salients for the advance of other disciplines. The strategic potential of the subject-salients varied from time to time. From 1910 to the mid-1940's physics and mathematics were the most active. Chemistry, biology, geology and the social sciences were less favourably placed for advance. What of geography, which has some affinity with both the physical and social sciences? Ackerman's answer is clear. 'We have not been on the forward salients in science, nor, until recently, have we been associated closely with those who have. The reasons are not difficult to find. During the early part of this 50-year period . . . our closest associations were with history and geology. Geological study of that period, and of the thirties, was not among the inspiring growing points of science. The history and geology connections did not correct the predisposition of our scholars . . . to the deceptive simplicity of geographic determinism' (Ackerman, 1963, p. 430). As determinism became a less acceptable model of the relations of human societies to their environments, geographers turned with more interest to the social sciences and to the alternative model in possibilism. Only later, in the fifties, was this concern with the social sciences to exert its very considerable influence on geography, but not before geography had undergone an anguished search for a unique identity. Meanwhile in biology, anthropology, genetics, psychology and economics, the applications of mathematical statistics and systems analysis were actively pursued even before the advent of the electronic computer. Only in the last decade have some geographers availed themselves of this extended range of model theory, with its productive concepts and attendant techniques. It is thus not an occasion for surprise that in geography teaching old models still persist, while newer and more sophisticated models and techniques remain untried. Not that all older models are ready for the scrap heap, or that heavy teaching loads and unimaginative examinations do anything but make change difficult and slow.

In the remainder of this essay we shall attempt an appraisal of older models, before going on to examine some of the models that take their rise in modern science. Finally we shall consider some of the models of the learning process as psychologists describe them and make some tentative indications for their application.

## MODELS IN GEOGRAPHY TEACHING

The use of the word 'model' in the sense that concerns us in this paper is relatively recent. Niels Bohr was among the first to use it in his investigations of the structure of the atom in 1913. Now its current use is widespread in physical, biological and social science, in history and in art. But although this sense of the word 'model' is new, its implied usage is old. Teaching is very much a conservative activity, and continues to use pre-scientific models.

An example may best serve us by way of introduction to the model most widely adopted by western medieval man. In the fourteenth century Guillaume Deguileville writes that the division between the realm of Nature and that of the Heavens was the orbit of the Moon. Something of this notion derives from Aristotle. He was interested in both biology and astronomy. He saw the characteristic of the world we inhabit was ceaseless change. Things did not happen uniformly or invariably, but 'on the whole' or 'for the most part'. The world of astronomy was different. So far as could be seen the heavenly bodies, permanently in their order, followed regular paths. The lower region of the Universe, that of change and irregularity, was Nature. The upper region of this two-part universe was the Sky. That variable element the weather was naturally included with inconstant Nature. Sky must therefore begin higher up at the orbit of the Moon. As C. S. Lewis (1964, p. 4) writes: 'It seemed reasonable to suppose that regions which differed in every observable respect were also made of different stuff'. Nature was made of the four elements, earth, water, fire, air. Air, nature and inconstancy then must end before the Sky began. In the Sky the substance was called aether. Here were the divine bodies. Below was the home of the changing and the perishable.

This example of medieval man's notion of the Universe well illustrates the bookish and scribe characteristics of the Middle Ages. When we speak of medieval man as living in an age of authority, we generally have in mind that of the Church. But in another, although partly overlapping sense, the Middle Ages was a time of many authorities whose thoughts were recorded in manuscripts. Every writer based himself on an authority – an auctour – an earlier writer, preferably in Latin. Medieval man was essentially taken with the practice of sorting out and classification of information, ideas and behaviour. It is said that of modern inventions he would have admired the card index. Supreme examples of his bringing masses of material to an ordered unity are the *Summa* of Aquinas and Dante's *Divine Comedy*. But supreme above all else in the period was the achievement of the medieval synthesis itself – the construction of a complex, harmonious mental model of the Universe from the scholarly materials of theology, science and history. This model resolved the contradictions of divergent writings and traditions, a process begun in the Graeco-Roman world and perfected in the Middle Ages. The medieval church taught on the authority of its accepted books. Although the Reformation divided Christendom, it in no way lessened the emphasis of the several churches on this model of teaching. What was taught was an authorized body of learning or knowledge. This conception of an essential body of knowledge as the stock in trade of the teacher, did not cease to be important as education became more and more a secular matter.

The notion of a body of required knowledge proved resistent to change. The scientific ideas of Bacon and Descartes on the study of nature were slow in effecting any significant change in the curriculum of schools and universities. Geography and other subjects outside the range of strictly physical science continued to be regarded as traditional bodies of knowledge, a feature well illustrated by textbooks of the eighteenth century, and re-enforced in several ways by social developments in the nineteenth century, when teachers for the growing elementary school system were recruited as pupil-teachers from the abler pupils in the elementary schools. For them geography provided convenient material to learn, teach and examine – a kind of literacy fodder. Above the elementary level geography gradually ceased to be a gloss on the scriptures and the classics. More attention was paid to the contemporary world largely as an encyclopaedic commentary on social, political and economic matters. In the new grammar schools, founded after the Education Act of 1902, geography became a more important element in the curriculum, a position it owed partly to a semi-scientific content with notions of causality on the nineteenth century determinist model, and partly to its association with history.

Only in the post-war expansion of university geography has there been evident an increasing concern for a more rigorous statistical and theoretical approach to teaching and investigation. Much of the stimulus for this has come from the social sciences. It has, however, been in some ways counterbalanced by contemporary stress on subject matter in examinations. Thus the conception of geography as a body of subject matter is still implicit in much teaching today and it is not easy to see what can effectively be done to lessen the addiction to the use of this model. Nevertheless there is available an alternative approach that may do something to reduce the over-dependence on the feeding-in-of-information model. An extract from the autobiography of R. G. Collingwood provides an introduction to its use.

Collingwood tells how his work in archaeology led him to revolt against the current logical theories of the time, rather like the revolt against scholastic logic which was effected through Bacon and Descartes, by their experience of scientific research. He writes: 'I began by observing that you cannot find out what a man means by simply studying his spoken or written statements, even though he has spoken or written with perfect command of language and perfectly truthful intention. In order to find out his meaning you must also know what the question was (a question in his own mind, and presumed by him to be in yours), to which the thing he has said or written was meant as an answer. It must be understood that question and answer, as I conceived them, were strictly correlative. A proposition was not an answer, or at any rate could not be the right answer, to any question which might have been answered otherwise. A highly detailed and particularized proposition must be the answer, not to a vague and generalized question, but to a question as detailed and particularized as itself. . . . People will speak of a savage as "confronted by the eternal problem of obtaining food". But what really confronts him is the problem, quite transitory like all things human, of spearing this fish, or digging up this root, or finding blackberries in this wood' (Collingwood, 1939, p. 31–32).

We may agree that information in books, in geography texts no less than others, is over-generalized, condensed and inert – inert in the sense of being divorced from the specific questions to which it was originally a response. Such knowledge, however, is successfully taught, learned, examined and evaluated. Some see this as a self-justifying activity; others as a mere assimilation model lacking the dimension of imagination and creation. To follow Collingwood's way is difficult. But a beginning can be made if the pupil is brought to see that the content of the average textbook, divorced from the original probing questions needed to enliven it, is, in effect, only part of the creative model, quite apart from any questions that have since been raised.

The practice of attacking problems by means of questions is very old and has been 'rediscovered' in various contexts. Inquiry by a continuous dialogue of question and response has a very long history. Aristotle tells us that it was Zeno of Elea who 'invented' the way of disputation by question and answer, or dialectic as it came to be called. Socrates used the method to expose some of the errors of contemporary belief. His pupil Plato developed it metaphysically in his theory of 'ideas' and in his search for the ultimate Idea of the Good. The christian church has long used the catechetical approach to the teaching of dogma. A similar method found favour with many eighteenth-century writers of geography textbooks.

Herbert Spencer thought that in the development of the child the same questions would emerge, as marked the progress of the race. He argued that, 'If there be an order in which the human race has mastered its various kinds of knowledge, there will arise in every child an aptitude to acquire these kinds of knowledge in the same order' (Spencer, 1861, p. 76).

Spencer, in favour of science in education, seemed to think it could be communicated in a ready-made form. He was persuaded of the need for a reformed subject-matter but not of the need of a new method. Dewey's theory of knowledge was developed in the light of recent advances in the biological sciences. Moreover, he accepted the evolutionary view of the continuity of organic life up to and including man. He realized that the child's environment could be directed and arranged so as to raise appropriate questions at the level of the child's experience. He therefore discarded the antithesis between empirical knowledge and the higher rational knowledge. For Dewey, knowledge was a mode of participation furthered and refined by the experimental method of science. He wrote of the experimental method, that, 'its significance is regarded as belonging to certain technical and merely physical matters', but he added with insight, 'it will doubtless take a long time to secure the perception that it holds equally as to the forming and testing of ideas in social and moral matters' (Dewey, 1916, p. 394).

As to geography, Dewey was alive to its possibilities and weaknesses. He saw its field as the 'connection of natural facts with social events and their consequences'. He was fully aware that to 'present specific geographical matter in its vital human bearings' required 'an informed and cultivated imagination'. Without that geography became 'a veritable rag-bag of intellectual odds and ends' (Dewey, 1916, p. 248).

A more recent attempt to use the question and answer model was the 'heuristic' method of teaching, which places the students as far as possible in the attitude of the discoverer – 'methods which involve the finding out instead of being merely told about things'. It was proposed mainly in connection with the physical sciences, but it had and still has obvious applications to geography as one of the earth sciences (Armstrong, 1903).

The question-and-answer model has an immense range of application and sophistication from the frontiers of science to detailed imprinting at the lowest levels of stimulus and response. In geography as in other fields the fruits of the method are proportional to the skill and subtlety of the user.

## SCIENCE AND TEACHING MODELS

Science is a concept with much relevance to the use of models in geography.

One need not dwell for long on the history of the emergence of science as part of culture. A brief look at the achievements of science in the seventeenth century will suffice. But in passing it may be remarked that, in one sense, 'science has no beginning in human history but is as old as perception, that it begins in the evolutionary scale with the capacity to generalize in perceiving an object' (Boring, 1957, p. 5). The human organism is so constructed that its perception, like that of a number of organisms, sees the same object as the same size regardless of the distance between it and the subject. Thus, in observation both science and perception look to underlying generalities, seeing in the observed object the uniformities of nature. But whatever we may wish to make of this interesting point, modern science in the sense of a social institution did not make its appearance until the Renaissance and its accompanying revival of learning. The first century of the new scientific age was rich in achievements that are fundamental to geography as it is taught today. A brief record of items is impressive: (a) In 1593, on the threshold of the startling century, Galileo had invented the thermometer. (b) In 1600 Gilbert published his work on magnetism in Latin; the first great scientific work to be published in England. (c) Kepler laid down the three laws of planetary motion in 1609–1619. (d) Galileo used his new telescope to discover Jupiter's moons. (e) In 1628 Harvey discovered the circulation of the blood – the first great biological discovery of the new age. (f) Many of Galileo's discoveries were finally published in 1638, e.g. the law of the pendulum, the law of falling bodies, acceleration, inertia and the components of motion of a projectile etc. (g) Galileo's pupil Torricelli invented the barometer in 1643. (h) Next was found 'the spring of the air' – that gases expand to fill all free space. From this came the invention of the air pump. (i) Robert Boyle's laws of gases followed in 1660; also an important date for it marked the founding of the Royal Society for the exchange and publication of scientific communications. Newton's first communication to the Royal Society was in 1672. (j) Leeuwenhoek in 1674 used the microscope to discover micro-organisms. (k) Finally, after long delays, Newton published *Principia* in 1687. He, in his work and attitudes, reveals the essence of physical science – an insight of genius, followed by mathematical deduction and experimental verification and prediction. The model of science, at least of physical science had arrived.

The eighteenth century was less productive in scientific discovery than the seventeenth. The full implications of the revolutionary work of Kepler, Galileo and Newton needed a period of assimilation and reflection. This was as true for philosophers as for scientists; although at this time the distinction between the two was not as sharp as it is today. But it was already clear that a new method of inquiry into the processes of nature had been established. In this procedure, now known as the hypothetico-deductive method, the scientist starts with an hypothesis, he deduces therefrom a consequence that can be observed, either directly or by experiment especially arranged, and is thus enabled to verify or not the predicted observation. In this particular respect then, if he is successful, the hypothesis is confirmed. This mathematical deductive method found more acceptance in England and France; in Germany biological science made more headway. But biology at this time did not lend itself to the formulation of large generalizations like the law of gravitation, from which facts could be deducted for empirical verification. The approach of biology was phenomenological, morphological and descriptive. This form of study was equally applicable to man and his activities. Immanuel Kant was able to produce, without the experimental method, but by the power and logic of his penetrating intellect, an unrivalled account of man's cognitive and moral powers. The philosophical problems that confronted him were largely set by the advance of the physical sciences we have just been speaking about. If one may put it shortly and not too erroneously, Kant thought that the freedom of the will in man was a necessity of nature. The very notion of morality was unthinkable without free will, just as much as the conception of natural science demanded the notion of necessity as a basic postulate. There arose from this the problem of the two kinds of knowledge. It has recently appeared as the problem of the two kinds of culture. On the one side we have the realm of knowledge of nature where law and necessity prevail; on the other the knowledge of man, his activities, societies and beliefs, where we act as if man had some freedom of choice. This is the dilemma that geographers find themselves in today; they have to operate on both sides of this dichotomy. This in fact is the model of the subject over which geographers presently contend. Perhaps at this stage one should point out that for Kant himself, although he was well acquainted with this division of knowledge, the subject of geography was regarded as overwhelmingly descriptive, with the exception of a small amount of mathematical geography. Indeed, as a very cursory examination of nineteenth-century geography texts will make evident, geography remained a largely descriptive study until very recently. It is therefore unhistorical to project this dichotomy back into the nineteenth century as if it were a living issue for geographers of that time. Nevertheless, we can gain a useful insight into the nature of the subject as seen by modern geographers if we take a look at the nineteenth century attempts to bridge the gap between the study of nature and that of man and his works. This will entail a brief excursion into the background development of the social sciences in the nineteenth century.

The degree to which the methods of physical science have seemed acceptable to students of society has varied from time to time and from country to country. Hegel, when he contemplated all that lies outside the realm of nature, outside those areas where scientific law prevails, asked what logic or order of growth prevailed there. His answer was that the study of history and philosophy showed that the moving spirit was *geist* or mind. Hegel, like Kant, thought there must be a deep unity between the knower and what is known, and that knowledge is impossible without that unity. He regarded this as a unity of opposites. Hegel saw all progress, whether in the history of man or in organic life, as a series of revolutionary steps. At each such step opposites are united. This method is the dialectic, a method as old as Socrates, in which we begin with a thesis – say with man as a person, who seeks to know. To this thesis the impersonal world offers an antithesis. A conflict ensues which the synthesis resolves. This method earlier writers had applied to the work of the mind; Hegel applied it to the realities of ordinary life – to the history of institutions and to the state, and its aspirations. This is an account much simplified of what Hegel wrote but will probably serve in this context. Dilthey was against the use, in the study of the cultural sciences, of methods proper to the physical sciences. He in fact sharpened the difference between the physical sciences and the cultural sciences and history. For him each of these branches of knowledge has a distinctive method of approach, physical science aimed at explanation and cultural sciences and history at interpretation. Marx, however, saw the regulative principle of society in the material interests of classes operating through the dialectic, or conflict of interests. The Utilitarians, and later in the nineteenth century, the Positivists, held that only the natural sciences, on the model of classical mechanics could give valid empirical knowledge. The man who did more than any other to bridge the theoretical gap between the extreme positions noted above, was Max Weber, the German sociologist, much of whose thought is yet to be assimilated. He put the emphasis not on given wants, either economic or psychological, but on the total cultural pattern which made the attachment to certain classes of wants meaningful. Moreover, he assumed that the components of such a gestalt or pattern of culture should be treated as independently variable. He set out to analyse certain of these relations by the comparative method, which in terms of the logic of science is perhaps the nearest empirical equivalent to the experimental method, that is accessible to the subject matter of social science (Parsons, 1965, p. 58). An illuminating comment from an historian of science comes from a recent monograph, (Kuhn, 1962, p. x), in which he writes: 'I was struck by the number and extent of overt disagreements among social scientists about the nature of legitimate scientific problems and methods. Both history and acquaintance made me doubt that practitioners of the natural sciences possess firmer or more permanent answers to such questions than their colleagues in social sciences. Yet somehow, the practice of astronomy, physics, chemistry or biology normally fails to evoke the controversies over fundamentals that today often seem endemic among, say, psychologists or sociologists (and one may add geographers). Attempting to discover the source of that difference led me to recognize the role in scientific research of what I have since called *paradigms*. These I take to be universally recognized scientific achievements that for a time provide model problems and solutions to a community of practitioners'.

These thinkers then, reflect a growing consciousness of society throughout the nineteenth century, and a consequent development in these specialist studies of society, each seeking to establish itself on more secure theoretical bases by more precise scientific methods. It seems inevitable that those aspects of geography that are linked with the cultural sciences of economics, anthropology and sociology, should in various degrees respond to this model. Experimental as well as mathematical and conceptual models have played a part in recent advances on the physical side of the subject.

If the use of models, of whatever kind, has not effectively closed the gap between the different kinds of knowledge, it has lessened geographers' pre-occupation with the unique identity of their subject, whether they base it on an integrative purpose or on a prescribed subject matter. This has come about largely by the application of research attention to concepts and techniques that transcend subject boundaries. It is significant that even in the study of history, with its regard for particularity, there is less concern with social facts as a separate type of data and more with 'social analysis as a particular set of questions which can be asked of every type of data that involves human relationships' (Hays, 1965, p. 374).

Some of these concepts have arisen in the development of modern physics and biology. Among the more important intrinsically and in their possible applications to the study of geography are: energy, feedback, information and general systems.

## Energy

This is a useful concept because it has many forms that can be transformed into each other. But though it may be transformed its quantity is invariant. The concept of energy is important in biology when studying the transformations of energy in living systems. But an 'energy' concept is used in several fields of knowledge, often metaphorically rather than in its precise physical sense. 'One speaks of psychic energies, historical energies, social energies. In these senses energy is not really measurable, nor is it directly related to physical energy. Nevertheless, like physical energy, it can be released in the form of enormous physical, mental, or social activity; and, when it is, we tend to think of it as somehow "potential" in the pre-existing situation' (Brooks, 1965, p. 76). The language of energy derived from physics has proved a very useful metaphor in the study of social, political, geographical, and psychological phenomena.

## Feedback

This is one of the fundamental ideas in modern engineering, particularly in automatic control and automation. 'In recent years the feedback concept has been extended still further to embrace the idea of "information feedback", which is important in biological and social phenomena as well as in the engineering of physical systems. The idea has been stated by Forrester in the following way: "An information feedback system exists whenever the environment leads to a decision that results in action which affects the environment and thereby influences future decisions". At first this may seem unrelated to amplifiers and control systems, but if we identify "environment" with "input" and "decision" with "output" we can readily see how the more general definition includes amplifiers and control systems as a special case. In the case of the amplifier the decision is completely and uniquely determined by the environment, but the concept of information feedback applies equally well when the decision is a discrete rather than a continuous function and when it is related to the environment only in a probabilistic sense' (Brooks, 1965, p. 77). 'Environment' and 'decision' used in this more general sense suggest further uses of the information feedback concept in biology, the social sciences and geography. The natural selection process in evolution and the process of learning may be regarded as information feedback systems. The teaching machine is designed to provide a feedback through the process of 'reinforcement' which helps the student to decide whether he has learned correctly. Information feedback systems and their properties have great possibilities in the analysis and interpretation of biological, cultural, economic and learning systems.

## General systems theory

General systems theory is an attempt at theoretical model building somewhere between the generalizations of mathematics and the specific theories of the specialized sciences. One of its advantages is that it affords a pattern of coherence and reference to much current interdisciplinary study and research. Boulding (1965, p. 200) suggests two complementary ways in which general systems theory might be structured. First, to select certain phenomena, encountered in many disciplines, and construct general theoretical models appropriate to them. He suggests the phenomena of 'population'. In various fields 'the interaction of population can be discussed in terms of competitive, complementary, or parasitic relationships among populations of different species, whether the species consist of animals, commodities, social classes or molecules'. Other phenomena of widespread occurrence suggest themselves. A second possible approach to structure in general systems theory is a hierarchy corresponding to the complexity of the 'individuals' of the various empirical fields. This more systematic approach would lead to a system of systems. Boulding identifies a number of levels at which theoretical discourse might be conducted, extending from the level of static framework to that of the human, social and transcendental.

Only recently have social scientists and geographers taken an interest in the application of these concepts to research. The question arises whether these concepts and associated techniques are to figure in undergraduate and sixth-form teaching or are they to be regarded solely as post-graduate accomplishments? It is difficult to see how they can be so restricted. The seminars now a feature of honours geography courses already entail close reading of recent specialist papers from journals. If such papers are to be properly understood and evaluated students need some theoretical and practical acquaintance with mathematical and statistical techniques. As for the schools there is, in view of the adoption of the newer mathematics, a case for the introduction of elementary statistical skills in secondary school geography. This can only come about through a drastic re-shaping of geography teaching with appropriate changes in teacher training. Indications of a new dynamic at work in the curriculum are not lacking; despite the cramping effect of examinations, there are encouraging signs of greater theoretical rigour, higher standards of achievement, and a readier acceptance of new teaching techniques. For geography, in alliance with mathematics, the data and concepts of physical and social science afford a new and productive field of study.

## LEARNING MODELS

In this section we will consider models for the learning process, a topic to which successful teaching practitioners are sometimes a little allergic. It is clear that even at the university level many teachers get by without ever raising any question about the nature of the learning process, and on what sort of models it operates. It is equally clear that they nevertheless act unconsciously as if certain models of learning were either true or at least were more acceptable than others. Perhaps the most common method by which we derive a model of learning for ourselves, and in light of which we teach others, is by trial and error. This principle is perfectly acceptable. The difficulty arises that the trial is not always a fair trial for we are seldom aware of a majority of the factors operating; equally, uncertainty prevails about errors. How can one know what is an error without a standard to judge by, and more puzzlingly, how can one get personal standards without trials and errors? This is a central and difficult problem to be worked through personally.

Before one considers models of learning, a few observations are required on two factors that even further complicate this difficult question. Learning is a process that proceeds hand in hand with *maturation*. These are the two ways in combination through which changes in students occur. They can more easily be separated conceptually than in practice. Maturation we may regard as a developmental process in which pupils manifest different characteristics, the blue-prints of which they carry in their genetic endowment. Models of learning need to be considered separately from the developmental or maturation models. Both raise in an acute form the question of the differences between the models of 'knowing that' and 'knowing how'.

The second point is that we usually consider learning models separately from cultural, and affective factors. Pupils not only learn from books and in the laboratory. They also learn about food, pop records, prejudices and other people. For this wider, and more general pattern of culture there are also a number of contending models about its nature and mode of acquirement. In the pre-scientific period these notions of culture tended to filter down as theories or models of learning. Since the seventeenth century, however, more or less systematic theories of learning have been advanced to compete with the older traditional forms. It has been estimated that it takes something between 25–75 years for a new model to be translated into classroom practice. Then it often exists alongside earlier models. It would be an interesting but professionally dangerous exercise to attempt to find out what conflicting models of learning coexist more or less peacefully within the confines of a single common room or department.

Some models of learning were current long before the twentieth century, and still have a wide vogue. I shall say something in turn about three of them; first, *Mental Discipline*; then *Natural Unfoldment* and, lastly, *Apperception*. All of these are non-experimental, were derived from a particular philosophical position, and are linked with a particular school of psychology. All were formulated by the method of introspection, much as we still construct our common-sense theories of learning. It is not then surprising that we still encounter these models in their vigorous old age. Mental discipline regards learning as the discipline of the mind, through suitable exercises in the curriculum. Behind this is a notion of the kind of knowledge best able to accomplish this end. But the idea of curriculum is less emphasized than the nature of the minds that are supposed to undergo this disciplinary process. The philosophic theory basic to it is that persons are made of mind and matter. If one holds to this mind-substance model, learning is a process of training such powers of the mind, as memory, imagination, will and thought; or, in short, a process of mental discipline. Plato thought that mental training in mathematics, particularly in geometry and philosophy, was good for those who were to conduct public affairs. The training proposed for the Guardians was of about 50 years duration. Much of this model was absorbed into Christian teaching. This interesting theme will have to be left with a bare mention.

At the Renaissance, man himself, rather than the scriptures, was taken as the measure of man's individual development. The understanding of man and his possibilities was thought to be enshrined in the classics of Greece and Rome. It was assumed that a person's direction of growth was provided and controlled from within, not by giving way to impulses, but by formulating principles which the individual set up to guide his conduct and behaviour. Learning was the harmonious development of one's inherent powers, under self-discipline, so that no faculty was under-developed. Socrates gives an authentic model in a teaching context, admittedly with smaller classes than are now customary. He sought to help students to realize what was already in their minds. In parenthesis one may remark that there is a pseudo-Socratic method to be seen today, wherein the teacher puts the idea in the student's mind with one hand and unerringly finds it with the other. Socrates thought that the environment mattered little. He did not profess to impart much information, but to draw it from the pupils by skillful questioning.

The nineteenth century was very much the century of mental discipline. Both the Arnolds, and especially Matthew, thought that Greece and Rome were the best example of the human spirit's activity. In the late nineteenth century mental discipline and the faculty theory became firmly linked and classics was regarded as the best mind-training material as well as the best repository of great truths of human experience. One of the results of this was that education was regarded solely as an art, and hence there was no point in applying the methods of science to models of learning. This attitude is readily discernible in schools and universities today.

Learning through unfoldment is usually associated with Rousseau but its influence was extended by Pestalozzi and Froebel who developed appropriate teaching methods. It stated that the natural powers of the child were good, and only needed a natural environment free from corrupting influences to allow the natural unfoldment of those powers. The duty of the teacher was to allow the child to live close to nature and to follow his natural impulses. The model of learning was one of growth or development; not the imposition of knowledge and standards. On the whole this model has been less widely accepted in England than that of mental discipline. It has perhaps more possibilities in the teaching of geography.

The last of the three older models I wish to say something about is apperception. Its origins lie in the writings of John Locke (1632–1704) and the German philosopher J. F. Herbert (1776–1841). Whereas the two earlier models assume man to be possessed of innate ideas, apperception assumes that mind is entirely a content of irreducible elements or ideas brought together by dynamic association. For Herbart the mind was an aggregate not of faculties but of ideas or mental states such as sense impressions and images. Herbart recognized in the child the stages of sense activity, memory and conceptual thinking, which he thought could best be fostered by the use in teaching of the famous *Herbartian Steps*: each lesson was to follow the sequence of Preparation, Presentation, Comparison and Abstraction, Generalization, Application. In many ways this model was a reflection of the atomistic notions in the science of that time. It has been described as 'Mental Chemistry'. It was widely used in the teaching of geography, especially in

America, where it prevailed until it was submerged by the more pragmatic methods associated with John Dewey's philosophy.

Besides these older models of learning, there exist as a result of modern experimental work newer psychologies of learning. They fall into two groups: (1) Recent developments of Associationism, known as Stimulus-Response or S-R Associationisms; one of these forms the theoretical basis of programmed learning. (2) Gestalt-field theories or models. In so far as they give theoretical support to teaching processes each of these theories has its relative strengths and weaknesses.

## Stimulus-response (S-R) associationism

Early this century Herbartianism gave way gradually to a new form of associationism based not on mental but on physiological phenomena. The earlier associations were concerned with the linking or association of ideas in minds; modern physiological psychologists asserted that psychology could become a science only if it studied observable bodily processes in an experimental, verifiable manner; and ceased to base its findings on introspection. Much of their work was done on animals. Some of these 'behaviourists' studied only those aspects of animal life that were amenable to exact experimental inquiry. For them a living organism was defined as a 'self-maintaining mechanism'. S-R associationists of today may be classed as 'neo-behaviourists'. They are more interested in the observation and analysis of behaviour than in the neural mechanisms underlying it.

## Gestalt-field psychology

The central idea of this school is expressed in the German word *gestalt*, which may be taken to mean an organized pattern, or an organized whole, in distinction from an aggregate of parts. Max Wertheimer, in 1912, first outlined the principles of gestalt psychology. But the notion that the universe could be explained in terms of its laws of arrangement rather than by the study of its atomistic elements, dates back to the pre-Socratic Greeks. The expansion and development of gestalt psychology was largely the work of two of Wertheimer's students, Köhler and Koffka in the United States. Both criticized the learning theory and general ideas of behaviourism. Kurt Lewin further developed the gestalt theory as a cognitive-field theory or psychology of learning, using topological and vector terms. Essentially gestalt psychology stressed the primacy of the organized whole or 'form', in distinction from the atomistic views of the associationists, who viewed the world as consisting of minute indivisible elements, endowed with their own energies. This view triumphed in the natural sciences and in physiology. The organism was conceived as a combination of small elements or cells with the reflex as the basis element of movement. This mechanical concept which was taken over by the older psychology, was contested by the gestalt psychologists.

There are variations of view within these two schools, but much more critical differences between them. Each has its own philosophical assumptions, 'laws' and technical vocabulary. A brief analysis of these will reveal points of interest and importance to geographers.

Since any system of psychology rests on some beliefs about human nature, it is difficult to separate psychology from some philosophical standpoint. Among the leading contemporary schools, the Freudian regards man as an active creature of instincts, the S-R associationists see him as essentially passive and determined by environment, while for gestalt-field psychologists man acts with purpose in his psychological environment. S-R associationism, then, has close affinities with philosophical realism or positivism. It followed on the attempts to make psychology scientific. Gestalt-field theory is more closely related to a systematic *relativism*.

Realists consider the physical world, as experienced by us, is essentially as it appears to the senses, and that its existence is independent of our knowing it. It is assumed that natural laws operate in the physical world, inevitably in terms of cause and effect. Thus the universe is seen as governed by mechanical laws. This leaves the realist with little use for other kinds of explanation. His outlook and methods are critical and empirical and grounded in verifiable facts. His approach to teaching and education is that of a determinist and environmentalist. He assumes that the environment will largely control what the students do and learn. There is a tendency for the realist to favour teaching material that has the imprimatur of authority and is likely to be of use in the contemporary world. Predictable effects are thought to reside in particular subject matter apart from student activity in the solving of problems.

Relativists neither support nor deny absolute existence. They are concerned with psychological reality or what we make or form of what comes to us. The chief notion of relativism is that a 'thing derives its qualities from its relationship to other things' (Bigge, 1964, p. 68); that knowledge is a matter of human interpretation, and not a literal description of what exists 'out there', external to man. 'The relativistic test of truth is anticipatory accuracy, not correspondence to ultimate reality'.

Both realists and relativists agree on the importance of science as a method of inquiry, but their models of science differ. The former see psychological development as a matter of learning, or a conditioning in response to external stimuli; relativists regard development as springing from a relationship between man and his culture. From his interpretations of these interactions he will form a meaningful pattern or a reality on which his thoughts and actions are based.

Since these two schools of psychology differ in these respects, we shall expect them to differ in the use of a number of terms that are of special significance to the geographer: environment (here one refers to psychological environment), perception, experience, interaction and culture. In the following outline of the differences, it must be borne in mind that the statements are necessarily general, and in any specific context would require a critical evaluation. For the S-R associationists, perception is the reading and recording of the individual's physical and social environment. It is a two-stage process in which one gives meaning to what had been previously sensed. The person's physical and psychological environments in a sense correspond. One can only sense what is there. The interaction with the environment takes place through an alternation of stimulus and response. The term experience is little used by S-R associationist psychologists. It presupposes a world of consciousness which is but a 'mentalistic copy' of the physical world. Skinner, on whose operant conditioning psychology much of the theory of teaching machines is based, writes thus: 'the private event (that is, thought or consciousness) is at best no more than a link in a causal chain, and it is usually not even that. We may think before we act in the sense that we may behave covertly before we behave overtly, but our action is not an "expression" of the covert response (that is, thought) or the consequence of it. The two are attributable to the same variables'. In S-R terms motivation is an urge to act induced by a stimulus, whether from within or without. There is no need to make the student wish to learn; it is enough to engage him in appropriate activity and use reinforcement to produce learning.

The cognitive field theory of learning which derives from the gestalt psychology, does not distinguish between the sensation of an object and its meaning. It is thought of as a highly selective, simultaneous process of sensing and making meaning. A man's environment is psychological. It is what he makes of what is around him in an interaction that is simultaneous and mutual. Motivation comes from a dynamic psychological situation characterized by a person's desire to do something about a disequilibrium within his life space. The emphasis is on the situation and not on the historical antecedents of the situation. Experience is the outcome of insightful behaviour, of acting with a purpose in the expectation of probable results. In this there are active and passive elements. This model has the advantage of allowing for the notion of 'paying attention' to the environment.

## CONCLUSION

To the geography teacher, whether at school or university, the above paragraphs on psychological categories and models, may perhaps appear tedious and irrelevant to his customary activities. Such categories and models as the teacher uses in his professional activities, and these are not confined to learning models, come from varied sources; some from the conventions of his culture and upbringing, some from his specialist discipline and others from contemporary notions of sociology and psychology. Wherever derived, all have to pass the empirical test of the teaching situation. But in response to the needs of our technological society, the teaching situation, currently under strong criticism, is itself in process of change, as part of a complex of change, affecting in turn, the national structure of education, the role of the teacher and the content of the curriculum.

Quite apart from the influence of examinations and a traditional distrust of central control of the curriculum, teaching is a very conservative activity. Despite that, we have had at our command for some time research techniques as appropriate to curriculum study as to market research. But the publication of research results is only the first stage of curricular innovation. Loyalty to traditional ways has prevented the application of sufficient power and organization to the two succeeding stages in innovation; the institution of representative proving trials in schools and the diffusion of the new practices under ordinary classroom conditions. It was as recently as 1962 that a Curriculum Study Group was founded in the Department of Education and Science. Out of it grew the Schools Council, which directs, sponsors and co-ordinates educational research on curriculum and examinations from the kindergarten to university entry. The Nuffield Foundation is also engaged on curriculum research, at present mainly in mathematics and science. Their present commitments will probably prevent either body from promoting any direct research into the teaching of geography. The Foundation and the Council are, however, co-operating in a joint inquiry into courses in modern humanities or social studies, suitable for pupils of average and less than average ability, in the last two years of their schooling. In any such course on the problems of man, nature and society, relevant to the experience of the pupils, geography would have a place, but not as a body of subject matter taught in the usual academic fashion. Rather it would contribute its techniques to the co-operative treatment of selected case or sample studies. Some years ago the inclusion of geography in a comparable social studies grouping was severely criticized by geographers on the grounds that it threatened the identity of the subject, and that geography could not be effectively taught by non-geographers. Now in the light of subsequent experience of composite general courses in Sixth Forms, Further Education Colleges and Universities, interdisciplinary models of subject matter are much more acceptable. Not all, however, will go as far as Ackerman when he writes:

> We are no longer concerned about whether what we are doing is geography or not; we are concerned instead with what we contribute towards a larger goal, however infinite it may seem.

That larger goal is:

> Nothing less than an understanding of the vast, interacting system comprising all humanity and its natural environment on the surface of the earth. (Ackerman, 1963, p. 435.)

Referring more particularly to a research context, Ackerman goes on to state that systems analysis might have been ideally created as a technique for geographers. Adopted belatedly by geographers from the social and behavioural scientists, systems analysis, quite apart from its great possibilities as a technique, has the merit of re-enforcing the tendency to interdisciplinary studies.

Quantitative and analytical methods in geography are finding increasing favour at the university level. This cannot remain without influence on the teaching of geography in the grammar schools. Some of the first fruits of curriculum research in mathematics is beginning to appear in the schools. Here is a profitable field for collaboration in which geography may benefit from the application of some of these newer mathematical techniques to geographical problems.

## REFERENCES

ACKERMAN, E. A., [1963], Where is a Research Frontier?; *Annals of the Association of American Geographers*, 53, 429–440.

ARMSTRONG, H. E., [1903], *The Teaching of Scientific Method*, (London), 476 pp.

BIGGE, M. L., [1964], *Learning Theories for Teachers*, (New York), 366 pp.

BORING, E. G., [1957], *A History of Experimental Psychology*, (New York), 777 pp.

BOULDING, K. E., [1956], General Systems Theory – The Skeleton of Science; *Management Science*, 2 (3), 197–208.

BROOKS, H., [1965], Scientific Concepts and Cultural Change; *Daedalus*, 94 (1), 66–83.

COLLINGWOOD, R. G., [1939], *An Autobiography*, (London), 167 pp.

DEWEY, J., [1916], *Democracy and Education*, (New York), 434 pp.

HAYS, S. B., [1965], Social Analysis of American Political History (1880–1920); *Political Science Quarterly*, 80, 373–394.

KOFFKA, K., [1935], *Principles of Gestalt Psychology*, (New York), 720 pp.

KÖHLER, W., [1947], *Gestalt Psychology*, (New York), 369 pp.

KUHN, T. S., [1962], *The Structure of Scientific Revolutions*, (Chicago).

LEWIN, K., [1936], *Principles of Topological Psychology*, (New York), 231 pp.

LEWIS, C. S., [1964], *The Discarded Image*, (Cambridge), 232 pp.

PARSONS, T., [1965], Unity and diversity in the modern intellectual disciplines: The role of the social sciences; *Daedalus*, 94 (1), 39–65.

SPENCER, H., [1861], *Education: Intellectual, Moral and Physical*, (London), 190 pp.

# Index

D. R. STODDART

Since this book is primarily about ideas and people rather than places, no attempt has been made to index the many incidental references to place-names in the text. The intention of this index is, first, to locate references to people and their writings, and second, to trace the main ideas common to many of the papers.

Page-references to literature citations are given in italics, thus: *135*.